Blue Carbon Reservoir of the Blue Planet

Abhijit Mitra · Sufia Zaman

Blue Carbon Reservoir of the Blue Planet

Springer

Abhijit Mitra
Department of Marine Science
University of Calcutta
Kolkata, West Bengal
India

Sufia Zaman
Department of Oceanography
Techno India University
Kolkata, West Bengal
India

ISBN 978-81-322-2106-7 ISBN 978-81-322-2107-4 (eBook)
DOI 10.1007/978-81-322-2107-4
Springer New Delhi Heidelberg New York Dordrecht London

Library of Congress Control Number: 2014954343

Printed on acid-free paper

Springer is part of Springer Science+Business Media (www.springer.com)

Acknowledgments

The purpose of this book is to shed light on blue carbon, which is one of the *vital* sinks of carbon dioxide and represents 55 % of the biological carbon on the planet Earth. Our goal is to analyze the potential of different verticals of blue carbon (like mangroves, salt marsh grass, seagrass, etc.) in absorbing carbon dioxide from the ambient media. Also, the capacity of seaweeds, phytoplankton, and molluscan community in the process of carbon sequestration has been critically scanned on the basis of our data bank. Our aim is to bring the importance of blue carbon in the climate change domain and draw the attention of busy decision makers at all levels, whether political leaders establishing national priorities, funding officers allocating research projects, or young couples deciding whether to have a child. We hope this book will be a useful source of information for those working for a world that is ecologically sustainable.

This book is by no means a final work. It should be viewed as part of an ongoing process, of a continuing effort to understand a set of complex interrelated global issues. It is part of a continuous flow of climate change-related research papers and books. Indeed, some of the material in this book has been taken from our own earlier papers, where research scholars like Dr. Ananda Gupta, Dr. Amitava Aich, Dr. Kiran Lal Das, Dr. Aftab Alam, Dr. Harekrishna Jana, Dr. Rajrupa Ghosh, Mr. Prosenjit Pramanick, Mr. Shankhadeep Chakraborty, Mr. Atanu Roy, Ms. Bulti Nayak, Ms. Suhana Datta, Mr. Deepta Chakravartty, Ms. Kasturi Sengupta, Mr. Kunal Mondal, Mr. Subhasmita Sinha, Ms. Mahua Roychowdhury, Mr. Saumya Kanti Ray, Mr. Rahul Bose, and Mr. Saurav Sett contributed their findings through rigorous fieldwork.

Completion of a book of this scope leaves the author indebted to many people. At various points along the way Prof. Yusuf Ali Jamaddar and Dr. Atanu Kumar Raha assisted with the research and analysis.

Our debt to Dr. Kakoli Banerjee, Subhdra Devi Gadi, and Dr. Rajrupa Ghosh is uncommonly large. Their contribution went far beyond that of editor as they added texture and color to our prose. At times it was difficult to tell when editing ended and writing began.

Each of our colleagues Dr. Subhra Bikash Bhattacharya and Mr. Tanmay Ray Chaudhuri put great effort in updating the book. We are thankful to Dr. Pardis Fazli of Department of Biological and Agricultural Engineering, University Putra, Selangor, Malaysia for her effort in representing our data in graphical form.

We are also indebted to the Ministry of Environment and Forest, Ministry of Earth Science, Department of Science and Technology (DST), Government of India, and IUCN for funding our venture through various projects.

Several innovative programs that constitute the annexure sections of the present book may serve as roadmaps in the climate change mitigation and adaptation process. A few examples of such innovations are mangrove-based fruit products by Mr. Prosenjit Pramanick, carbon content in gastropods by Ms. Bulti Nayak, etc., which are valuable assets of the present knowledge reservoir.

Finally, Dr. Abhijit Mitra expresses his gratefulness to his wife Shampa, daughter Ankita, and mother Manjulika whose inspirations and encouragements acted as boosters to complete the manuscript. Dr. Sufia Zaman expresses her deepest gratitude to her mother Mrs. Ayesha Zaman for her unconditional love and practical day-to-day support, father, Mr. Salim-uz-Zaman who gave her immense moral support. Dr. Zaman also acknowledges the support of her beloved husband Dr. Sahid Imam Mallick. Dr. Zaman wishes to accord her deep sense of gratitude to her family members including her uncle (Mr. Pradip Kumar Mitra) and aunt (Late Mrs. Kanika Mitra), younger sister (Ms. Sharmilee Zaman), her in-laws and beloved grandmother (Mrs. Shibani Dhar) for their encouragement and inspiration throughout the strenuous period of manuscript preparation.

Contents

About the Authors

Dr. Abhijit Mitra Associate Professor and former Head, Department of Marine Science, University of Calcutta (India) has been active in the sphere of Oceanography since 1985. He obtained his Ph.D. as NET qualified scholar in 1994. He joined Calcutta Port Trust and World Wide Fund for Nature-India (WWF) in various capacities to carry out research programs on environmental science, biodiversity conservation, climate change, and carbon sequestration. Presently, Dr. Mitra is serving as the advisor of Oceanography Division of Techno India University, Kolkata. He has to his credit about 274 scientific publications in various National and International journals, and 26 books of postgraduate standards. Dr. Mitra is presently the member of several committees like PACON International, IUCN, SIOS, Mangrove Society of India, etc., and has successfully completed about 16 projects on biodiversity loss in the fishery sector, coastal pollution, aquaculture, alternative livelihood, climate change and carbon sequestration. Dr. Mitra also visited as faculty member and invited speaker several universities in Singapore, Kenya, Oman, and USA. In 2008, Dr. Mitra was invited as visiting fellow at University of Massachusetts at Dartmouth, USA to deliver a series of lectures on Climate Change. Dr. Mitra also successfully guided 22 Ph.D. students. Presently, his research areas include environmental science, mangrove ecology, sustainable aquaculture, alternative livelihood, climate change, and carbon sequestration.

Dr. Sufia Zaman presently serving as Adjunct Assistant Professor in the Department of Oceanography in Techno India University (Kolkata) started her career in the field of Marine Science in 2001. She worked in the rigorous region of the Indian Sundarbans and has wide range of experience in exploring the floral and faunal diversity of Sundarbans. She has published a book on carbon sequestration, 45 scientific papers, and contributed chapters in several books on biodiversity, environmental science, aquaculture, and livelihood development. Dr. Zaman is presently a member of Fisheries Society of India. She is also running projects on carbon sequestration by mangroves of Indian Sundarbans. She is the recipient of DST Women Scientist and Jawaharlal Memorial Doctoral fellowship awards. Her areas of research include aquaculture, fish nutrition, phytoplankton diversity, climate change, and mangrove ecology. Dr. Zaman is also the first researcher in the maritime state of West Bengal (India) to initiate trial experiments on iron fertilization and subsequent enhancement of primary (phytoplankton) and secondary (fish) productions in the brackish water ponds of Indian Sundarbans with financial assistance of Department of Science and Technology, Government of India.

1 The Blue Zone of the Planet Earth

...The sculptures on the ocean basin
Can never be created by human machine....
The Authors

1.1 Sculptures of the Ocean Basin and Estuary

The ocean basin has unique designs and sculptures similar to the land features above it. Mountain ranges, canyons, valleys, and vast plains are all the important components of the underwater landscape. These physical features of the ocean bottom are called bathygraphic features, and unlike their counterpart topographic features on land, they change at relatively slow pace. Erosion is slow in the relatively calm recesses of the ocean, and changes are mainly attributed to sedimentation, uplifting, and subsidence.

The structures and features of ocean basin are represented in Figs. 1.1 and 1.2.

Brief descriptions of these features are given here:

Continental Margin

The continental margin includes the continental shelf and slope. At the edge of a continent is the continental shelf (Fig. 1.2). Continental shelves are generally flat areas, averaging 68 km (40 miles) in width and 130 m (430 ft) in depth, that slope gently towards the bottom of the ocean basin. The width of a continental shelf is frequently related to the slope of the land it borders. Mountainous coasts, like the West Coast of the USA, usually have a narrow continental shelf, whereas low-lying land, like the East Coast of the USA, usually has a wide one. Continental shelves are actually part of the continents to which they are attached. The criterion for defining the seaward edge of the continental shelf is a marked change in slope. The continental shelf has a gradient of about 1:1000. Seaward from the continental shelf is the **continental slope**, which extends to a depth of 1.6–3.2 km. The continental slope has a gradient between 1:2 and 1:40. The extent of the sloping can vary from a gradual drop to a steep decline into an ocean trench, as illustrated by the slope that occurs off the western coast of South America. Because of the steepness of the angle, the continental slope usually has less sediment.

Some continental slopes have submarine canyons that are similar to canyons found on land. Many of these submarine canyons are aligned with river systems on land and were probably formed by these rivers during periods of low sea level. The Hudson River canyon on the East Coast of the USA is an example of this. Other submarine canyons have ripple marks on the floor, and at the ends of the canyons, sediments fan out, suggesting that they were formed by moving sediments and water. Oceanographers believe that these canyons were formed by turbidity currents. Turbidity currents are swift avalanches of sediment and water that erode a slope as they sweep down and pick up speed. At the end of the slope, the current slows and the sediments fan out. Turbidity currents can be caused by earthquakes or the accumulation of large amounts of sediments on steep slopes that overload the slope's capacity to hold them.

Continental Rise

At the base of a steep continental slope, there may be a gentle slope called a continental rise. A continental rise is produced by processes such as landslides that carry sediments to the bottom of the continental slope. Most continental rises are located in the Atlantic and Indian Oceans and

A. Mitra and S. Zaman, *Blue Carbon Reservoir of the Blue Planet*,
DOI 10.1007/978-81-322-2107-4_1, © Springer India 2015

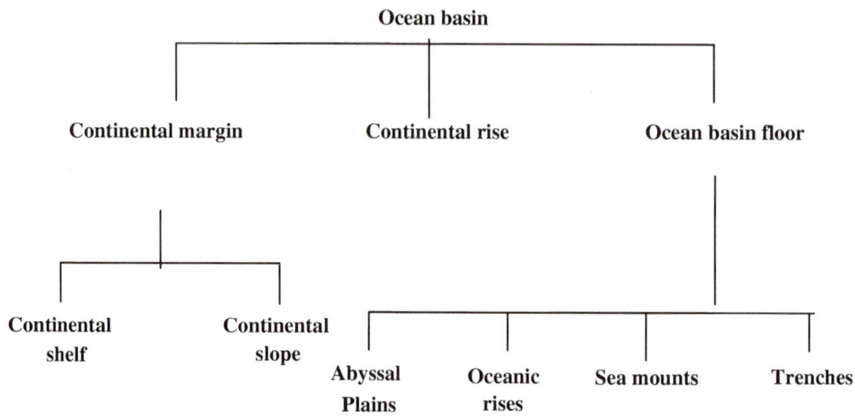

Fig. 1.1 Components of the ocean basin (Mitra 2000a)

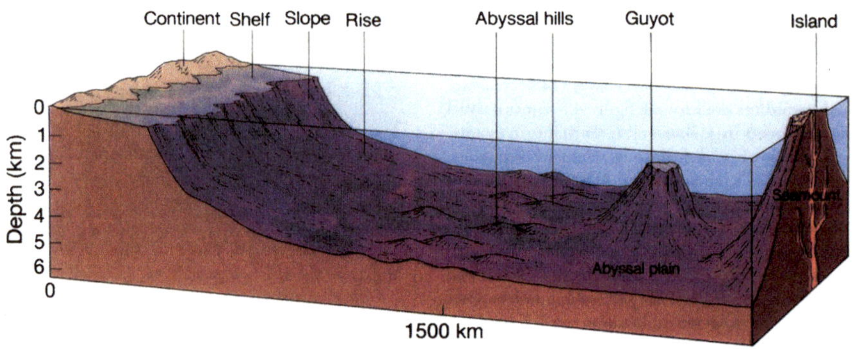

Fig. 1.2 Diagrammatic sectional view of the ocean basin

around the continent of Antarctica. In the Pacific Ocean, it is more common to find trenches located at the bottom of the continental slopes.

Oceanographers have studied that sediments from the continental shelf are carried into the deep ocean through the canyons present in the continental slope. At the foot of the continental slope, the suspended materials and sediments settle out that slope gently towards the ocean floor. This region is basically the continental rise and has a gradient between 1:50 and 1:800.

Ocean Basin Floor

The ocean basin floor extends seaward from the continental rise or the continental margins and includes the abyssal plains, oceanic rises, seamounts and trenches. **Abyssal plains** are found at the base of the continental rise and are relatively flat plains having a gradient less than 1:1000. They are formed due to even deposition of sediments from the continental rise carried down by the turbidity currents. **Oceanic rises** refer to rather isolated areas, which are elevated above the abyssal floor and are distributed sporadically on the ocean basin floor. They may vary from low hills to mountains as high as 1,525 m. The example of **Bermuda rise** is very prominent in this context on which the Bermuda Islands are formed. **Seamounts** are isolated peaks that rise several thousand metres above the sea floor. They are volcanic in origin and are found principally in the vicinity of the fault zones. Some seamounts show evidence of coral reefs and surface erosion, suggesting that at one time they were above the surface. Movements of the ocean floor, the natural process of compaction that volcanic material undergoes, subsidence due to cooling of the ocean floor, erosion, and the increased weight of sediments at the top may be the reasons for the sinking of these structures.

Trenches are long narrow depressions in the ocean floor that are over 6,100 m deep. The deepest known trench in the ocean compartment is the ***Mariana trench*** of the western North Pacific that is about 11,000 m deep. The Peru–Chile Trench extends for over 6,120 km (3,600 miles) along the coast of South America and is the longest of the ocean trenches. The Java Trench extends for a distance of almost 4,760 km (2,956 miles) along the coast of the islands of Indonesia. By comparison, the Atlantic has only two, relatively short trenches, the South Sandwich Trench and the Puerto Rico–Cayman Trench. Trenches are invariably associated with the systems of active volcanoes and are believed to be caused by down wrapping of the oceanic crust beneath the continental crust.

Overview of Estuarine Environment

An estuary may be defined as a transition zone between the freshwater and marine water with unique physico-chemical and biological attributes. According to Odum (1971), estuaries belong to different class of 'fluctuating water-level ecosystems'. Estuaries generally occupy those areas of the coasts, which are least subjected to marine action and thus are major sites for development of harbours, recreational activities and appropriate aquacultural activities. In estuaries, freshwater collected over vast regions of the land pours into an ocean, which sends salt water upstream far beyond the river mouth.

Each estuary has its own physical features that influence its ecology. These include the amount of river discharge, depth and general topography, specific circulation pattern, climatic regime and vertical tide range. Though the physico-chemical condition in estuaries is stressful, yet, the food availability is so favourable that the ecosystem is packed with life. Organisms living in this habitat therefore exhibit wide range of tolerances (i.e., they are mainly euryhaline). Estuaries rank first among the most productive regions of marine ecosystems as they typically contain high biomass of benthic algae, seagrasses and phytoplankton that support large number of fishes and birds. This is mainly because estuaries are enriched by nutrients that are contributed through land drainage. Moreover, the nutrient retention capacity within the estuary is also unique due to which the flora and fauna thriving within such system never face the scarcity of nutrients.

Several definitions have been forwarded to depict the features of estuarine compartment, some of which are highlighted here.

According to Cameron and Pritchard (1963), 'an estuary is a semi-enclosed coastal body of water which has a free connection with the open sea and within which seawater is measurably diluted with fresh water derived from land drainage'.

According to Pritchard (1967), 'an estuary is a semi-enclosed coastal body of water which has a free connection with the open sea; it is thus strongly affected by tidal action and within it seawater is mixed with fresh water from land drainage'.

Perillo (1995) defined estuary as 'a semi-enclosed coastal body of water that extends to the effective limit of tidal influence, within which seawater entering from one or more free connections with the open sea or any other saline coastal bodies of water is significantly diluted with fresh water derived from land drainage and can sustain euryhaline biological species, from either part or whole of their life cycle'.

Estuaries described by these definitions do not include the narrowness of the system, neither these definitions address the situation where evaporation exceeds the freshwater supply from rivers and from local rain. Considering all these gaps, in recent times, estuary has been defined as 'a narrow semi-enclosed coastal body of water which has a free connection with the open sea at least intermittently and within which salinity of the water is measurably different from the salinity in the open ocean' (www.es.flinders.edu.au/~mattom/ShelfCoast/chapter11.html).

Classification of Estuary

Estuaries may be of different types (Fig. 1.3).

Pritchard (1967) classified estuaries on the basis of the following four characteristics. They are (a) salinity (b) geomorphology (c) water circulation and stratification and (d) systems energetics.

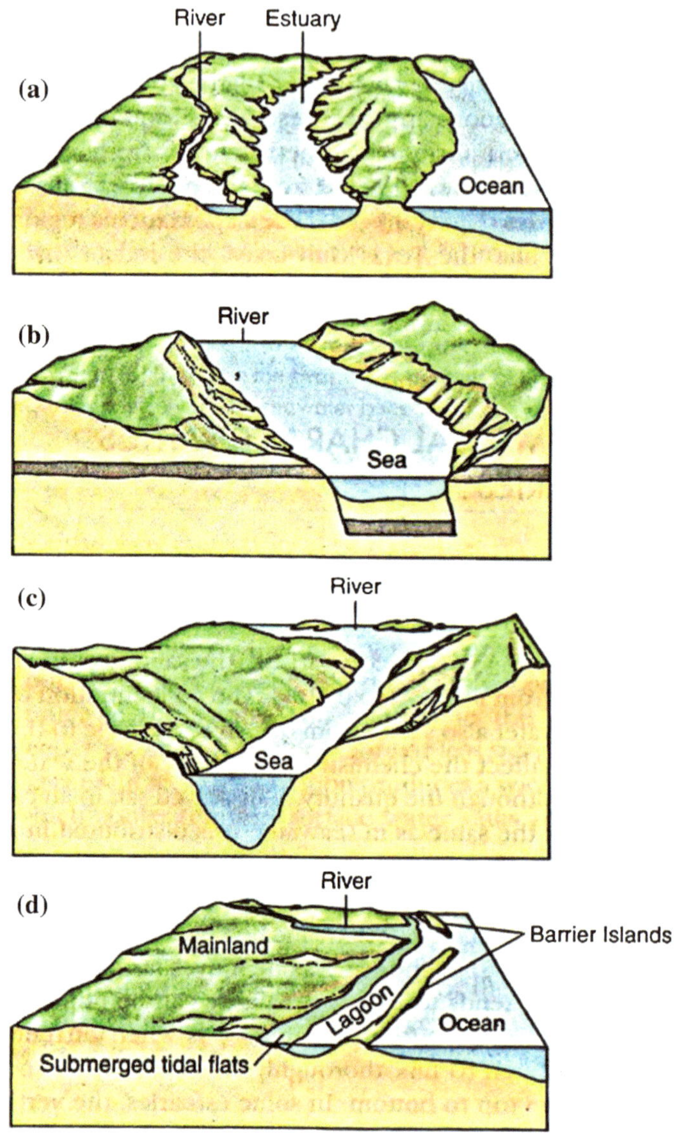

Fig. 1.3 Types of estuaries. **a** Coastal plain, **b** tectonic, **c** fjord, **d** bar-built

1. **On the basis of salinity, estuaries can be grouped into the following types**
   (a) **Oligohaline**: In this type of estuary, the freshwater mixes with the saline water in such proportion that the water ultimately becomes uniformly saline. Such estuaries are common where small rivers meet the seas, e.g., Haldi River of West Bengal (India).
   (b) **Mesohaline**: It is that type of estuary where the estuarine water has medium salinity due to proportional mixing of the freshwater and saline water, e.g., Mahanadi River of Odisha.
   (c) **Polyhaline**: In this type of estuary, there are areas, where there is distinct variation of salinity owing to the tidal water intrusion into the river mouth, e.g., estuaries adjacent to Bay of Bengal and the Indian Sundarbans. The estuarine complex of this mangrove-dominated deltaic is broadly divided into three salinity regimes (Table 1.1).

Table 1.1 Divisions of Indian Sundarbans estuarine system on the basis of salinity

Zone	Areas (blocks) included
High saline (average 24 psu)	Pathar Pratima, Namkhana, Sagar, Kultali, Gosaba and Basanti
Medium saline (average 12 psu)	Canning-I and II, Kakdwip, Mathurapur-II, Jainagar-II
Low saline (average 3 psu)	Kulpi, Mathurapur-I

According to a study conducted by Sengupta et al. (2013), the entire mangrove-dominated Indian Sundarbans can be divided into hyposaline western and eastern sectors and hypersaline central sector. This spatial variation of salinity is the outcome of a two decade study in 18 stations selected in the three sectors of the deltaic complex (Fig. 1.4). The western sector of the deltaic lobe receives the snowmelt water of mighty Himalayan glaciers after being regulated through several barrages on the way. The central sector on the other hand is fully deprived from such supply due to heavy siltation and clogging of the Bidyadhari channel since the fifteenth century (Chaudhuri and Choudhury 1994). The eastern sector of Indian Sundarbans is adjacent to the Bangladesh Sundarbans (which comprises 62 % of the total Sundarbans) and receives the freshwater from the River Raimangal and also from the Padma–Meghna–Brahmaputra river system of Bangladesh Sundarbans through several creeks and inlets. The study revealed that in the western and eastern sectors, the salinity decreased by 21.91 and 16.35 %, respectively, over a period of 24 years, whereas, in the central sector, there has been a steady increase in the salinity by 9.32 % during the same period (Fig. 1.5).

In recent times, we have analysed the salinity value of past 30 years in three sectors of Indian Sundarbans and attempted to forecast the trend of salinity in the next 30 years. Our analysis is based on 18 stations (6 in each sector of Indian Sundarbans) as marked in Fig. 1.6 and Table 1.2.

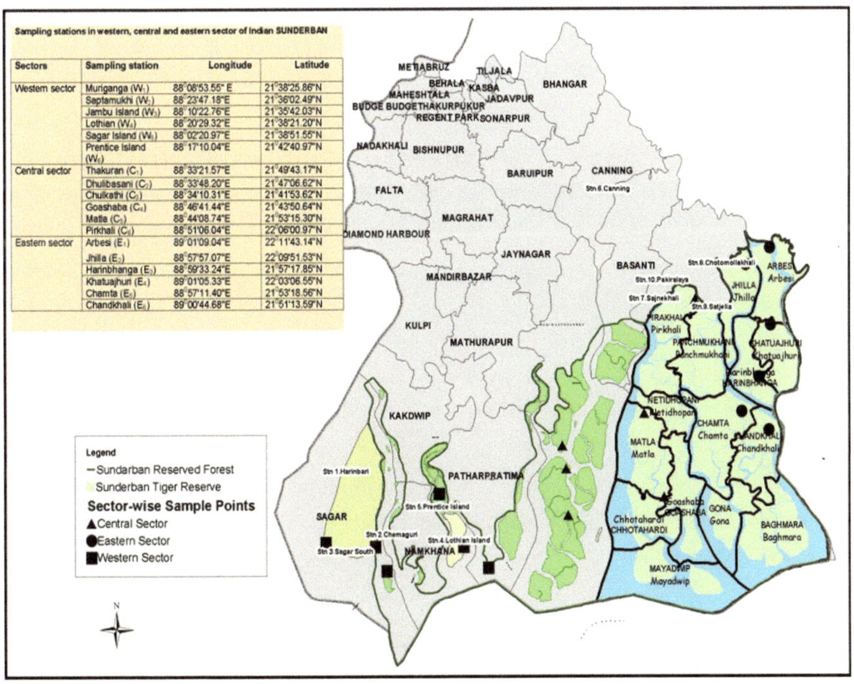

Fig. 1.4 Map of Indian Sundarbans showing the locations of sampling stations in the western, central and eastern sectors. *Source* Sengupta et al. (2013)

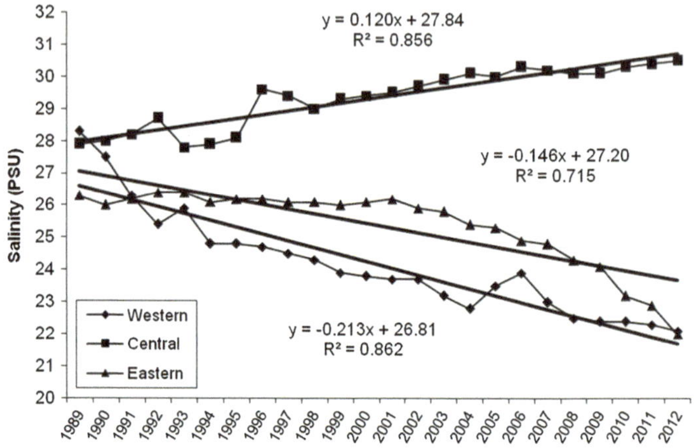

Fig. 1.5 Spatio-temporal variations of surface water salinity (in psu) in Indian Sundarbans. *Source* Sengupta et al. (2013)

It is interesting to note the significant spatio-temporal variation of surface water salinity in the study region. In the western sector, the salinity decrease ranged from 0.58 psu/year (at Jambu Island) to 1.46 psu/year (at Harinbari). Although station 2 (Saptamukhi) is situated in the western sector, but the salinity has increased by 0.51 psu/year (Fig. 1.7). Considering all the six stations in the western sector, the decrease of salinity is 0.63 psu/year, which represents a decrease of 7.50 psu per decade. The salinity has decreased from 17.30 % (in Jambu Island) to 43.76 % (in Harinbari) over a period of 30 years (Fig. 1.7).

The exponential smoothing method that produces maximum-likelihood estimate of the variable predicts a salinity value of 13.05 psu in 2043 (Fig. 1.8), which is a decrease of 38.4 % since 1984 (over a span of 60 years).

The central sector presents a completely reverse picture in terms of aquatic salinity. Irrespective of stations, salinity has increased (Fig. 1.9) between the range 1.05 psu/year (in Chulkathi) and 1.12 psu/year (in Matla and Pirkhali). Considering the salinity values of selected six stations, the increase is 1.09 psu/year, which is equivalent to 13.05 psu/decade. The percentage of

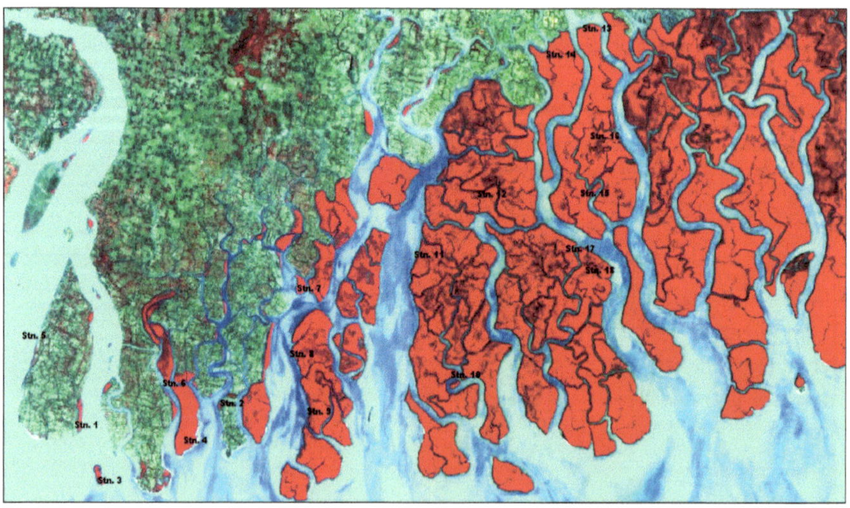

Fig. 1.6 Location of sector-wise sampling stations in Indian Sundarbans; the *red colour* indicates the mangrove vegetation

Table 1.2 Sampling stations in the western, central and eastern sectors of Indian Sundarbans in the lower Gangetic delta region

Sectors		Sampling stations	Latitude	Longitude
Western sector	Stn. 1	Chemaguri (W_1)	21° 38′ 25.86″N	88° 08′ 53.55″E
	Stn. 2	Saptamukhi (W_2)	21° 40′ 02.33″N	88° 23′ 27.18″E
	Stn. 3	Jambu Island (W_3)	21° 35′ 42.03″N	88° 10′ 22.76″E
	Stn. 4	Lothian (W_4)	21° 38′ 21.20″N	88° 20′ 29.32″E
	Stn. 5	Harinbari (W_5)	21° 44′ 22.55″N	88° 04′ 32.97″E
	Stn. 6	Prentice Island (W_6)	21° 42′ 47.88″N	88° 17′ 55.05″E
Central sector	Stn. 7	Thakuran Char (C_1)	21° 49′ 53.17″N	88° 31′ 25.57″E
	Stn. 8	Dhulibasani (C_2)	21° 47′ 06.62″N	88° 33′ 48.20″E
	Stn. 9	Chulkathi (C_3)	21° 41′ 53.62″N	88° 34′ 10.31″E
	Stn. 10	Goashaba (C_4)	21° 43′ 50.64″N	88° 46′ 41.44″E
	Stn. 11	Matla (C_5)	21° 53′ 15.30″N	88° 44′ 08.74″E
	Stn. 12	Pirkhali (C_6)	22° 06′ 00.97″N	88° 51′ 06.04″E
Eastern sector	Stn. 13	Arbesi (E_1)	22° 11′ 43.14″N	89° 01′ 09.04″E
	Stn. 14	Jhilla (E_2)	22° 09′ 51.53″N	88° 57′ 57.07″E
	Stn. 15	Harinbhanga (E_3)	21° 57′ 17.85″N	88° 59′ 33.24″E
	Stn. 16	Khatuajhuri (E_4)	22° 03′ 06.55″N	89° 01′ 05.33″E
	Stn. 17	Chamta (E_5)	21° 53′ 18.56″N	88° 57′ 11.40″E
	Stn. 18	Chandkhali (E_6)	21° 51′ 13.59″N	89° 00′ 44.68″E

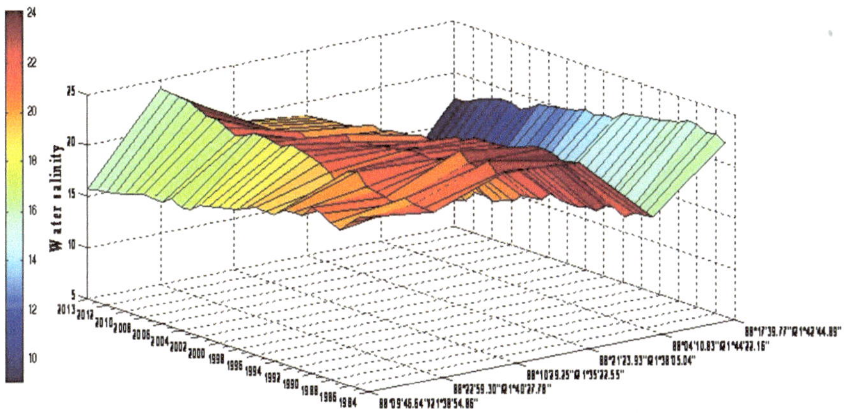

Fig. 1.7 Spatio-temporal variation of salinity in western Indian Sundarbans

salinity increase in this sector ranges from 31.49 psu (in Chulkathi) to 33.64 psu (in Matla) with an average of increase 32.62 % over a period of 30 years (Fig. 1.9).

Considering the observed data set of 30 years (1984–2013), we predict that salinity will be around 36 psu after a period of 30 years in the central sector of Indian Sundarbans (Fig. 1.10), which is an indication of alarming hypersaline condition (a rise by 67.1 %) in 2043 in this sector.

In the eastern sector, salinity has decreased (Fig. 1.11), which ranges from 0.54 psu/year (in Chamta) to 0.98 psu/year (in Jhilla). Considering all the six stations in eastern Indian Sundarbans, the average decrease of salinity is 0.86 psu/year, equivalent to a decadal decrease of 10.30 psu.

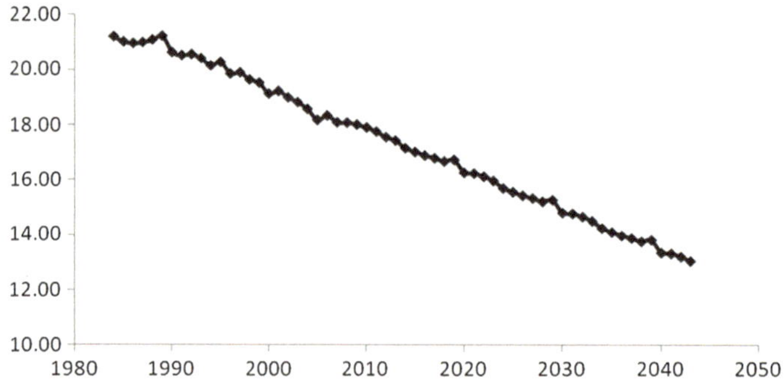

Fig. 1.8 Future trend of surface water salinity in western Indian Sundarbans

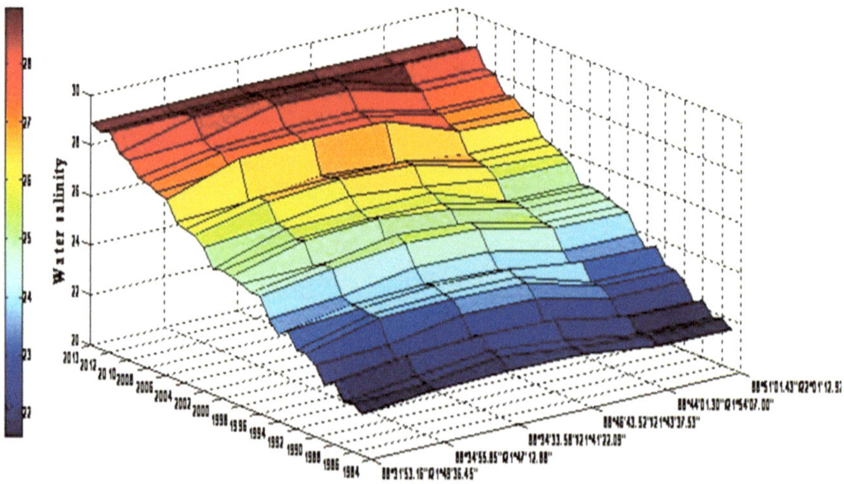

Fig. 1.9 Spatio-temporal variation of salinity in central Indian Sundarbans

Fig. 1.10 Future trend of surface water salinity in central Indian Sundarbans

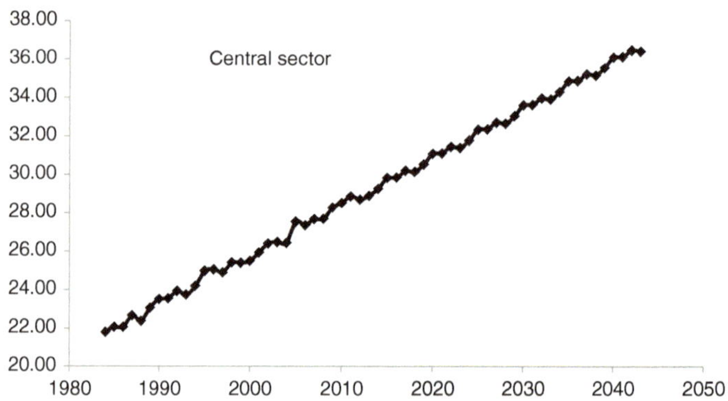

Over a period of 30 years, the average percentage decrease of salinity is 25.66 psu (Fig. 1.11).

On the basis of observed data, the prediction of salinity in 2043 is around 7.54 psu (Fig. 1.12), which is decrease of 52.4 % considering a time span of 60 years.

2. **On the basis of geomorphology, estuaries are of the following types**
   (a) **Drowned river valleys**: The coastlines with relatively low and extensively wide coastal plains, e.g., Chesapeake Bay on the mid-Atlantic coast of the USA.

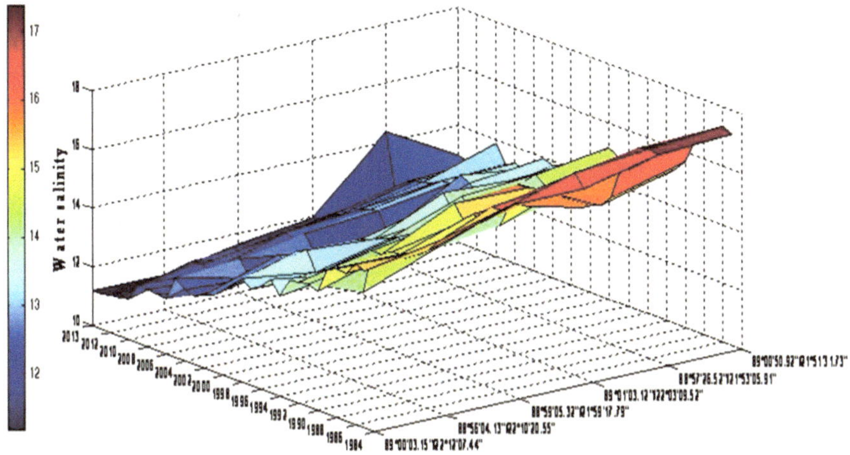

Fig. 1.11 Spatio-temporal variation of salinity in eastern Indian Sundarbans

Fig. 1.12 Future trend of surface water salinity in eastern Indian Sundarbans

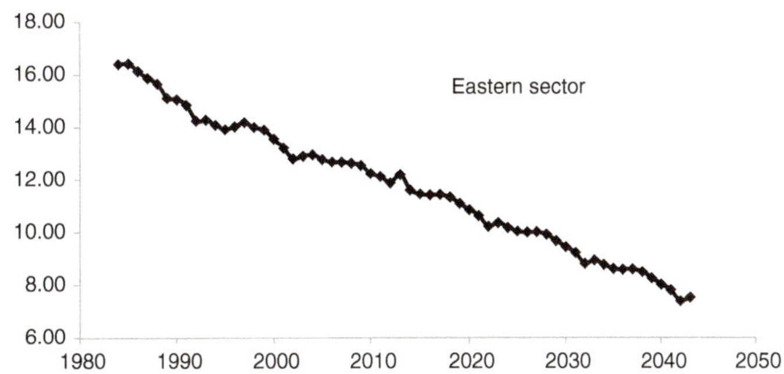

(b) **Fjord-type estuaries**: The deep U-shaped indentures due glacial erosion with a shallow sill at their mouth formed by terminal glacial deposits, e.g., Norway, British Columbia and Alaska.

(c) **Bar-built estuaries**: These are shallow basins often partly exposed at low tide enclosed by a chain of offshore bars or barrier islands, broken at intervals by inlets. These bars are sometimes deposited offshore or are remnants of former coastal dunes, e.g., North Carolina and Georgia.

(d) **Estuaries produced by tectonic processes**: Geological faulting or local subsidence results in the formation of coastal indented structures, e.g., San Francisco Bay.

3. **On the basis of water circulation and stratification**:

(a) **Highly stratified or 'salt-wedge' estuary**: When the freshwater discharge from the rivers dominates over the tidal action, the freshwater tends to overflow the heavier salt water which forms a 'wedge' extending along the bottom for a considerable distance upstream. This flow of freshwater is again governed by the Coriolis force, which forces the freshwater to move strongly along the right shore (if the observer faces the sea in the Northern Hemisphere and vice versa in the Southern Hemisphere). Such 'stratified' or 'bilayered' estuary will exhibit a salinity profile with a 'halocline' or zone of sharp change in salinity from top to bottom, e.g., Mississippi River.

(b) **Partially mixed or moderately stratified estuary**: When the freshwater and the tidal water mix in equal proportion due to turbulence caused by periodicity of waves, such estuary is formed. Due to this, the energy is dissipated in vertical mixing thus creating a complex pattern of layers and water masses, e.g., Chesapeake Bay.

(c) **Completely mixed or vertically homogenous estuary**: When the tidal action is more than the freshwater discharge, the water tends to mix well from top to bottom and the salinity is relatively high. When there is wide variation in salinity and temperature, then horizontal estuaries are formed, e.g., bar-built estuaries.

4. **On the basis of systems energetics**: Odum (1971) have described estuaries on the basis of systems energetics as stated here:

(a) **Physically stressed systems of wide latitudinal range**: These include mainly the high-energy breaking waves, strong tidal currents, severe temperature and salinity shocks, low dissolved oxygen level during night or high rate of sedimentation. Due to severe environmental stress in this system, few species (*opportunistic species*) are able to thrive in this fluctuating condition. It is because of this fact such systems exhibit extremely poor species diversity. The index of dominance is high in these systems owing to the ability of fewer numbers of species to cope up with adverse environmental conditions. However, in areas of intertidal zone, sharp zonation of species and seasonal replacement of communities is very distinct because adaptation is more efficiently accomplished by species replacement along a gradient than by adaptation within the species. Rocky sea fronts, intertidal rocks, sand beaches, high-velocity tidal channels, sedimentary deltas and hypersaline lagoons are included in this category.

(b) **Natural arctic ecosystems with ice stress**: Extreme cold conditions (in Arctic and Antarctic coasts) exert considerable stress on the intertidal zones of the areas in which light is limiting due to very short summer season. Productivity is extremely low in these systems due to poor diversity of phytoplankton. In most of the times of the year, the phytoplankton species remain in encysted condition to get rid of the chilled water.

(c) **Natural temperate coastal ecosystems with seasonal programming and reproduction**: Seasonal changes in primary productivity of the ocean and the behaviour of animals are the common characteristics in estuarine biology. The more subdued tides, waves and currents in the semi-enclosed basins provide energy subsidies rather than stresses in comparison to deeper sounds and offshore waters which often get charged with nutrients and organic matter from fertile shallow zones. Temperate estuaries, though very fertile, are vulnerable to damages caused by pollution, dredging, filling and diking. Some important features of the temperate estuaries include tide pools, salt marshes, eel grass (*Zostera* sp.) beds, seaweed bottoms, kelp beds, oyster reefs and mudflats, which harbour dense population of clams and sea worms.

(d) **Natural tropical coastal ecosystems of high diversity**: These systems are characterized by minimum environmental stress, and therefore, negligible energy is spent by the species for adaptive modifications. Due to congenial environment, the species diversity is very high in these systems. Examples of such systems are mangroves, seagrass meadows, salt marsh grass and seaweed community and coral reefs in shallow water zones. These biotic forms contribute nutrients due to which a wide spectrum of planktonic diversity exists in this system. The fishery potential is extremely high, and hence, the modern trend of monoculture (e.g., culture of *Penaeus monodon*) is perhaps a wrong choice in this system. These productive systems with varied ecological niches are suitable for polyculture practice.

(e) **Emerging new systems associated with man**: In modern societies, estuaries are treated as the ultimate sink of different categories of waste. Estuaries have varying capacities to handle 'degradable' material depending on the size of the system, flow patterns, hydrological parameters, microbial load, type of estuary and the climatic zone. Materials such as treated sewage and pulp mill wastes, seafood and other food processing wastes, petroleum wastes and dredging spoil can be decomposed and dispersed provided that (i) the system is not stressed by poison (insecticides, acids, etc.) and that (ii) the rate of input is controlled at low to moderate levels and not subjected to sudden shocks produced by periodic massive dumping. Of all the man made changes, impounding estuarine water has perhaps the greatest effect. It must be recognized that impounded waters are a completely different type of ecosystem that have nearly no natural capacity for waste treatment. In countries such as India and Bangladesh, huge numbers of impounded waterbodies have been created to promote shrimp or tiger prawn (*Penaeus monodon*) culture in the coastal areas, due to which the ecology and environment of the surrounding areas have been greatly damaged.

The estuaries adjacent to the Bay of Bengal are extremely important from the ecological and economic point of view. These are the live matrix of mangrove ecosystem of Sundarbans on which the unique spectrum of biological diversity is embedded. In Indian Sundarbans, approximately 2,069 km^2 area is occupied by tidal river system. The main estuaries, from west to east (Fig. 1.13), are highlighted in Table 1.3 along with their salient features.

Estuaries may also be classified as tide-dominated (Fig. 1.14) and wave-dominated (Fig. 1.15) estuary on the basis of several important (positive and negative) hydrological processes (Table 1.4).

Tide-dominated estuaries are very common in the tropical region. During high-tide condition, the coastal vegetation preferably the mangroves is partially drowned and the substratum becomes totally invisible (Fig. 1.16). However, during low-tide condition, the substratum gets exposed (Fig. 1.17), and several varieties of saline creepers and herbs are seen (Figs. 1.18 and 1.19).

The astronomical tide at the estuarine mouth determines the nature of flow in an estuary near its mouth. The tide along the coast of India is mixed, i.e., it consists of a mixture of oscillations with a period of about 12.5 h (semi-diurnal oscillation) and with a period of about a day (diurnal oscillation). The range is particularly high in gulfs such as the Gulf of Kutch or the Gulf of Khambhat or a channel like the Hooghly River, all of which get narrower away from their sea end.

1.2 Fluid Movements in the Oceans and Estuaries

1.2.1 Waves

Waves are generally produced by the action of wind blowing across the surface of the ocean and are formed as a result of the transfer of energy from the wind to the water. Winds can generate waves as small as a ripple or as high as 30 m.

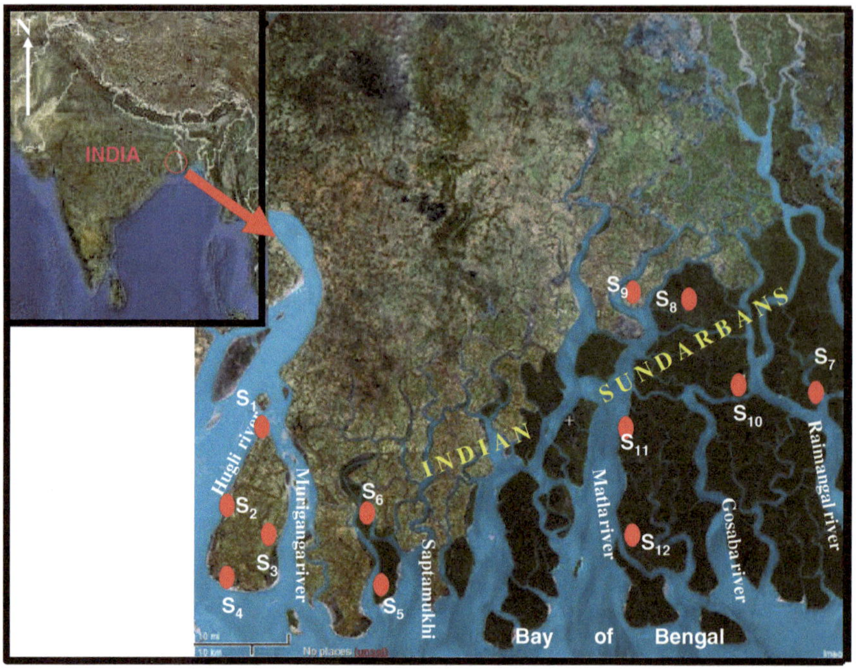

Fig. 1.13 Map showing the major estuaries of Indian Sundarbans; the *red circles* are the sampling stations. S_1–S_4 are in the western Indian Sundarbans, where the salinity is relatively low due to Farakka barrage discharge; S_5 and S_6 are in the central sector of Indian Sundarbans, where the rivers have lost their connections with freshwater and have become tide-fed in nature; S_7–S_{12} are in the eastern Indian Sundarbans (adjacent to Bangladesh Sundarbans), which receive certain volume of freshwater through creeks and inlets from the Bangladesh Sundarbans part

In a wave train, there is a regular succession of crests and troughs (Fig. 1.20). The horizontal distance between successive crests is the wavelength (L), while the wave height (H) is the vertical distance from a crest to a trough.

The period (T) is the time for passage of two successive wave crests passed through a fixed point and is usually expressed in seconds. Wave speed (V) can be calculated by V = L/T.

At the surface, water moves in circular vertical orbits with a diameter equal to the wave height. Away from the surface, orbital motion decreases, and orbits become smaller. At a depth of half a wavelength (L/2), orbital motion essentially vanishes. Thus, wave-generated water movements only occur near the surface. Where the water depth is less than half a wavelength, the bottom interferes with the water motion. Water particles near the bottom cannot move vertically. This causes elliptical-shaped orbits. Wave size and speed depend on wind speed, wind duration and distance of water over which the wind blows. Theoretically, a wave's height cannot exceed 0.14 of its wavelength. If this is exceeded, white caps and breaking waves are formed.

Ocean surface waves can be classified in several ways. The most common way is by the period of the waves (Table 1.5).

a. The shortest period waves under this classification are the **capillary waves**. When the wind starts to blow on the ocean surface, capillary waves are generated. They seem just like a fine structure of small ripples of nearly capillary dimension. They basically contribute the largest amount of energy from the wind to the ocean water. Their wavelength is up to 1.74 cm.
b. **Ultra-gravity wave** is another type of wave with periods between 0.1 and 1 s.

Table 1.3 Important rivers of Indian Sundarbans

Estuaries	Description
Hugli	It forms the western border of Indian Sundarbans
	It is the main river of West Bengal and along with the Ganges and Bhagirathi rivers comprise a national waterway
	Most of the coastal industries of West Bengal are concentrated along the western bank of this river
Muriganga	It is a branch of the Hugli River
	It flows along the east of Sagar Island
	Unique mangrove vegetation is found along the bank of this river
Saptamukhi	It has its origin at Sultanpur
	It is connected with the Muriganga (Bartala) branch of the Hugli River through Hatania–Duania canal
Thakuran	It begins near Jayanagar in South 24 Parganas and has a number of connections with the Saptamukhi
	It has perhaps earlier connected with the Calcutta canal through the Kultali and the Piyali rivers, which exist today in a dying state
Matla	This river originates at the confluence of Bidyadhari, Khuratya and the Rampur Khal close to the town of Canning in South 24 Parganas
	Matla is connected to Bidya and ultimately flows to the Bay of Bengal. The freshwater connection and discharge to this river have been lost in the recent times
	Salinity of the river water is relatively high (in comparison to Hugli or Muriganga) owing to freshwater cut-off from the upstream region
Bidyadhari	This was flourishing branch of the Bhagirathi during the fifteenth or sixteenth century, but now serves only as a sewage and excess rainwater outlet from the city of Kolkata
	The river bed is completely silted, and presently, it is almost in dying condition
Gosaba	The waters of Matla and Harinbhanga (Raimangal) through a large number of canals from it
	The river flows through the reserve forests
Harinbhanga	The river begins from Sahebkhali in North 24 Parganas and is connected with the Rampura khal by Barakalagachi River and with the Gosaba River through the River Terobhanki
	The Harinbhanga (also known as Ichamati and Raimangal) forms a natural demarcation between India and Bangladesh

c. The third division is ordinary **gravity waves** with periods between 1 and 30 s. They are composed of two types of waves, namely '**sea waves**' and '**swell waves**'.

d. The **infra-gravity waves** are generally long period waves having periods between 5 min and 12 h. 'Tsunamis' and 'storm surges' belong to this division.

e. The ordinary **tidal waves** are **astronomical waves** with fixed time period of about 12 h (semi-diurnal) and 24 h (diurnal).

f. The **trans-tidal** waves are those having periods greater than 24 h. Under this, the longer period components of astronomical tides are also included.

Waves can also be classified **on the basis of the relationship of the waveform to the depth of water**.

a. A '**deep-water waves**' is one in which the ratio depth (d)/wave/length (l) is more than 0.5. Capillary waves belong to this category.

b. A '**shallow water wave**' is one in which D/L is less than 1/20. 'Tsunamis' are shallow water waves even in the deepest portion of the ocean.

Another mode of wave classification is by the **nature of controlling force**.

a. The shortest of the waves, the capillary waves, have a restoring force due to the **surface tension** of water. When water has been displaced, the force brings it back to normal position due to surface tension.

b. In case of ultra-gravity waves and gravity waves, **gravity** acts as the major restoring force.

c. **Gravity** together with **Coriolis force** acts as the restoring force in case of long-period waves.

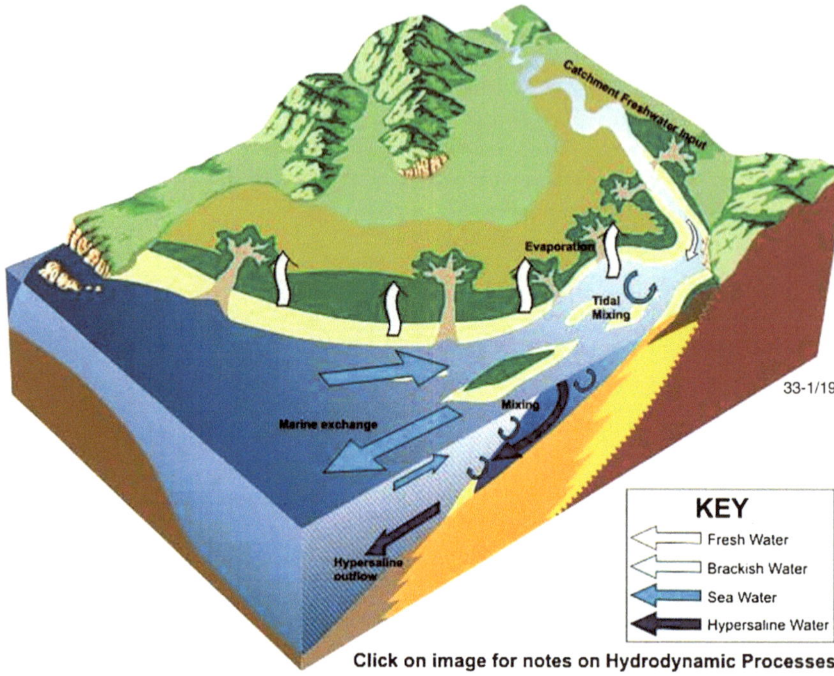

Fig. 1.14 Negative hydrology processes in tide-dominated estuaries

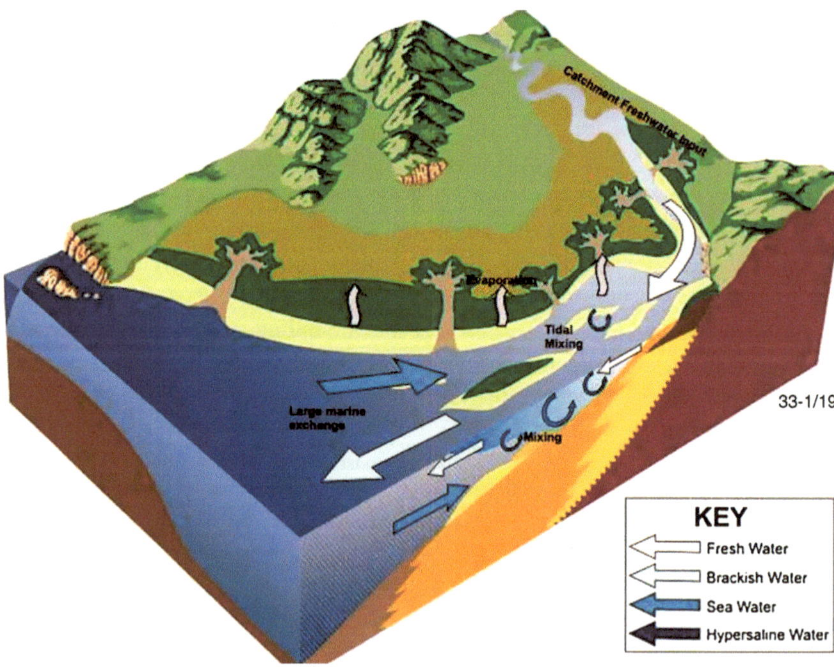

Fig. 1.15 Positive hydrology processes in wave-dominated estuaries

Table 1.4 Comparative study of tide- and wave-dominated estuaries

Point	Tide-dominated estuary	Wave-dominated estuary
Catchment's input	Freshwater entering from the catchments is relatively lower. In negative estuaries, the net inflow of marine water exceeds the outflow of catchment-derived (fresh) water. In such cases, the hypersaline water is usually exported to the ocean	Freshwater enters from the catchments. Although the volume of freshwater input varies spatially and temporally (depending on local catchments and climate conditions), it is often relatively high in positive estuaries
Freshwater input	The volume of freshwater entering the estuary is too low to cause significant level of stratification. High tidal ranges may tend to accelerate mixing of any freshwater inputs and marine water	Water circulation in wave-dominated estuaries generally ranges from well mixed to salinity stratified, depending on the degree of wave mixing, volume of freshwater input and climate. 'Positive' wave-dominated estuaries have lower salinity water towards their head. The volume of freshwater causes stratification (or layering) in the water column, which varies with seasonal flow. Buoyant low-salinity fresh water floats above the denser, high-salinity ocean water
Salinity	High rates of evaporation cause increases in salinity within the estuary. The resulting high-density hypersaline water sinks beneath the buoyant marine water which penetrates through the estuary mouth and flows out of the estuarine entrance into the coastal ocean through a process known as reverse stratification. A large degree of mixing occurs between the two layers	A 'salt-wedge' or intrusion of denser saline marine water penetrates through the entrance along the bed of the estuary. Some mixing occurs at the interface between the freshwater and marine water. The distance that the salt-wedge penetrates is dependant on tidal range and the amount of fluvial flow received by the estuary. During high fluvial flow events (which may be seasonal), fresh floodwater may push the salt water beyond the mouth. However, the large volume of central basins typical of wave-dominated estuaries tends to reduce this effect
Evaporation	Evaporation is the dominant process in negative tide-dominated estuaries due to arid climatic conditions and the extensive area of shallow intertidal environments. Aridity and the degree of evaporation may vary seasonally; however, by definition, evaporation in 'negative' estuaries is much larger than freshwater input. Consequently, negative estuaries tend to have longer residence times than positive estuaries	While significant evaporation can occur in wave-dominated estuaries characterized by positive circulation, evaporation (by definition) does not exceed the amount of freshwater input
Water exchange	Exchange of sea water and estuarine water occurs through the wide entrance of the estuary. Flood and ebb tides may follow different routes into and out of the estuary, and the tidal prism tends to be large. In negative estuaries, the net inflow of marine water exceeds the outflow of freshwater derived from the catchments. In such cases, the hypersaline water is usually exported to the ocean	Exchange of ocean water and estuarine water occurs through the entrance of the estuary, although the magnitude of exchange depends on the size and length of the entrance channel. In positive wave-dominated estuaries, the outflow of freshwater exceeds the inflow of marine water. During dry conditions, the entrance of the estuary may be intermittently closed

Fig. 1.16 A tide-dominated estuary with wide mouth (entrance) during high-tide condition

Fig. 1.17 A tide-dominated estuary with intertidal mudflat exposed during low-tide condition

Waves can also be classified on the basis of their **generating force**

a. **Winds** are responsible for the generation of capillary waves, ultra-gravity waves and infra-gravity waves.
b. **Storms** and **earthquakes or movement of the crustal plates** is responsible for long-period waves such as *Tsunamis.*
c. **Gravitational attraction of sun and moon** acts as generating force for ordinary tidal waves.
d. **Storms, sun, and moon** are effective causes for trans-tidal waves.

On the basis of motion of waveform, wave can be classified into two categories.

a. **Progressive waves** are those which propagate laterally.
b. **Standing waves** are those which do not propagate laterally but move in a vertical direction. A standing wave may form when two progressive waves move in opposite direction in such a way that interference results. 'Seiches' are standing waves.

Seismic sea waves (or tsunami) are generated by displacement of the ocean floor. Volcanic eruptions and catastrophic submarine mass

Fig. 1.18 *Ipomoea pescaprae*: a common sand binder (creeper) on the intertidal mudflat exposed during low-tide condition

Fig. 1.19 Estuarine herb on the intertidal mudflat exposed during low-tide condition

movements can produce them, but by far, the most significant cause is earthquake. The displacement of a large mass of ocean water, often at a great depth, generates waves of small amplitude (usually less than 1 m), considerable length (up to 200 km or more) and high velocity. As with translatory waves, the wave velocity is related to ocean depth and waves can travel at a velocity of about 600 km/h over water 3,000 m deep. Seismic waves can therefore cross the entire Pacific Ocean within a span of few hours.

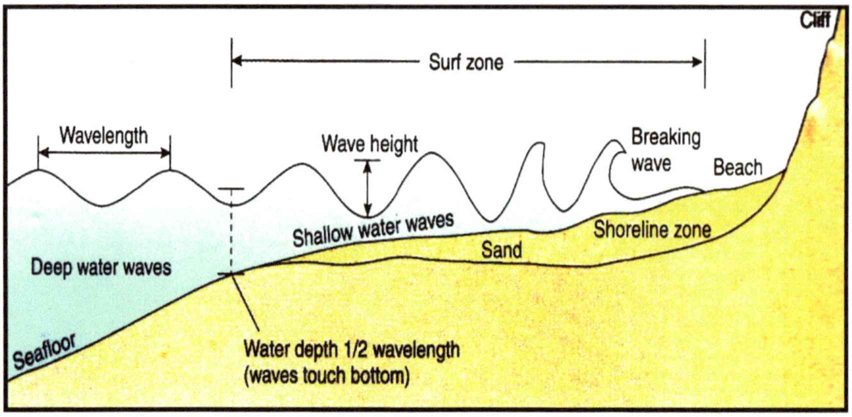

Fig. 1.20 Waves of the sea showing relevant terminologies

Table 1.5 Classification of ocean surface waves

Classification	Period range	Depth classification	Restoring force	Generating force
Capillary waves	<0.1 s	Deep-water wave	Surface tension	Winds
Ultra-gravity waves	0.1–1 s	Deep-water wave	Surface tension and gravity	Winds
Gravity waves	1–30 s	Deep-water wave to shallow water wave	Gravity	Winds
Infra-gravity waves	30 s to 5 min	Deep-water wave to shallow water wave	Gravity and coriolis force	Winds and gravity
Long-period waves	5 min to 12 h	Transitional to shallow water wave	Gravity and coriolis force	Storms and earthquake
Ordinary tidal waves	(semi-diurnal and diurnal) 12–24 h	Shallow water wave	Gravity and coriolis force	Gravitational attraction of sun and moon
Trans-tidal waves	>24 h	Shallow water wave	Gravity and coriolis force	Storms, sun and moon

In addition to the above categories of waves, several researchers have stated about **coastal trapped waves** in the Sydney region of Australia. These are not waves in true sense, since they have little observable effects at the sea surface. These are characterized by timescales of 7–20 days and scales of long-shore variability of many hundreds of kilometres (Church et al. 1986). They may be thought of as a current oscillation which moves equatorward from its generation source in Bass Strait affecting Sydney waters some two days after being generated by strong winds in the Bass Strait (Griffin and Middleton 1991). They are, therefore, predictable to some extent and can have the effect of producing strong currents in the near-shore region which may have either northward flowing or southward flowing.

Very little information is available on the classification of waves in the Bay of Bengal region adjacent to the Indian part of Sundarbans. Surface waves in the Bay of Bengal are caused due to wind system whose direction and velocity are mainly controlled by northeast and southwest monsoons. The wind from north and northeast starts blowing from the beginning of October to the end of March. The months of January and February are relatively calm with an average speed of around 3.5 km/h. It again commences to blow violently from southwest around the middle of March and continues till September. During this period, several low-pressure systems are formed in this region, among which few take the shape of depressions and cyclonic storms of varying intensity (Table 1.6). The wind speed

Table 1.6 Cyclonic storms which crossed the Sundarban coast (1907–1997)

Year	Duration of storm	Maximum wind speed (km/h)
1907	17–19 June	95
1907	24 and 25 June	95
1909	17 and 18 October	126
1912	5 and 6 September	81
1913	23–25 July	126
1916	21 and 22 September	68
1916	9 and 10 November	81
1917	3 May	126
1917	30 and 31 October	45
1919	24 and 25 September	126
1920	21–23 July	81
1921	26 and 27 July	68
1925	27 and 28 June	81
1927	28 and 29 July	126
1928	18–20 July	126
1929	15–17 July	56
1932	23 and 24 May	126
1933	20 and 21 September	56
1935	8 and 9 July	126
1936	26–28 May	113
1937	28 and 29 September	126
1940	30 June and 1 July	–
1940	7 and 8 July	–
1940	2 and 3 August	–
1940	21 and 22 October	–
1941	9 and 10 July	81
1941	8 and 9 August	–
1942	9 and 10 July	81
1945	1 and 2 July	81
1946	4 and 5 October	81
1950	10 and 11 June	95
		87
1952	4 July	–
1956	30 May to 2 June	–
1960	27–29 May	74
		76
1962	20 and 21 September	93
		80
1965	10 and 11 May	74
1997	20 and 21 August	120

(–) means not recorded; *Source* Mitra (2013)

during this time rises above 100 km/h and is usually accompanied by gigantic tidal waves causing much loss of lives, damage to properties, forests and habitats of animals and plants thriving in this area. Sea waves in this region rarely become destructive except during cyclonic storms. When the cyclonic incidences coincide with the spring tides, wave height sometimes exceeds over 5 m above the mean tide level. Ripple waves appear in the months of October–December, when wind-generated wave height varies approximately from 0.20 to 0.35 m. In the months of April–August, when the normal wind speed ranges between 15 and 40 km/h, large wavelets are formed in the shelf region, which starts breaking as they approach towards the coastal margin. In this condition, the wave height rises up to 2 m that causes maximum scouring action to adjoining of landmasses.)

1.2.2 Tides

Tides may be defined as the periodic rise and fall of the sea level and occur as a result of the gravitational pull of the sun and the moon. The magnitude of this attraction varies directly as the product of the mass of attracting bodies and inversely as the square of the distance between them. Though the sun's mass is 26×10^6 times bigger than that of moon, the ratio of the square of the distance from the sun to the earth and that from the moon to the earth is roughly 58×10^6. It is because of this, the moon's pull is about 2.2 times stronger than that of the sun. Depending on the position of the heavenly bodies, the tides may be of two types (Fig. 1.21).

Spring tides: These tides are witnessed during the new moon condition, when the sun, the moon and the earth are in a line (linear alignment) leading to maximum tidal range.

Neap tides: These tides occur when the sun and the moon are at right angles to each other during the first or third quarter of the lunar cycle. In Indian coasts, the tides are *semi-diurnal* in

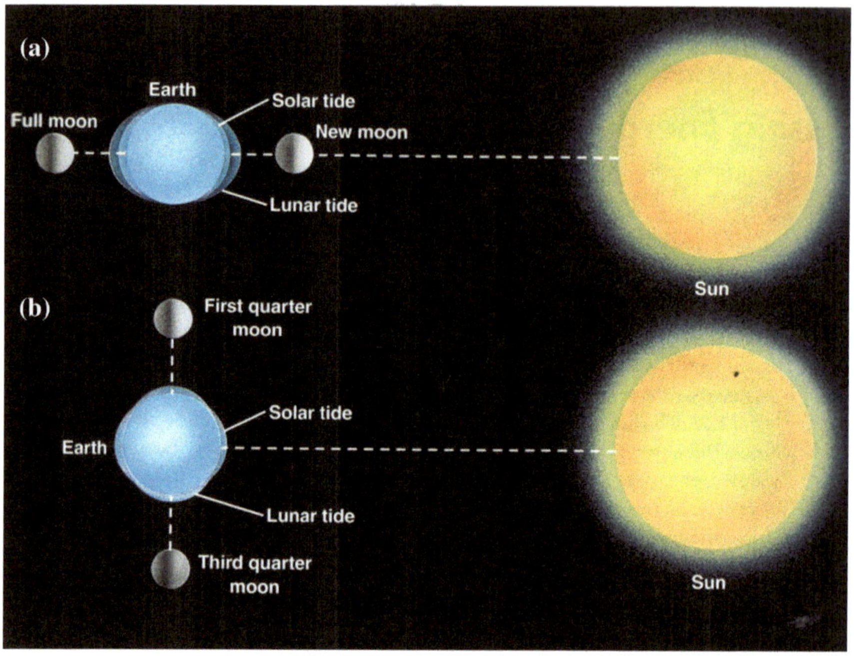

Fig. 1.21 Spring and neap tides. **a** Spring tide, **b** neap tide

nature, i.e., two high tides and two low tides (each of 6 h 25 min) on each lunar day.

It may so happen that, while the sun, the moon and the earth are in conjunction, the moon is at perigee, i.e., shortest distance in its orbit from the earth. The tidal forces are mightiest under such circumstances, and the tides are higher than the usual. If in addition, the sun is on the equator, and the perigee spring tides are the strongest. This happens in the **equinoxes** in March and September.

The tidal currents, produced by the ebb and flow of tides, have little significance in open sea, but in the near-shore zones (depending on the size, configuration and topography of the marine basins), they control the coastal morphology to a great extent. The coasts can be categorized into macrotidal (<2 m), mesotidal (2–4 m) and microtidal (>4 m) depending on the tidal range.

Tides are normally semi-diurnal in nature. However, diurnal and mixed tides are also observed in some places in the ocean and bays. Hence, tides are classified as one of three types, semi-diurnal, diurnal or mixed according to the characteristics of the tidal pattern.

In **semi-diurnal tide**, two high waters and two low waters occur in each day with relatively small differences in the respective highs and lows. Examples of such tides are the tides on the Atlantic coast of USA. In **diurnal tides**, only a single high and a single low water occur each tidal day. Tides of diurnal type occur along the northern shore of the Gulf of Mexico, in the Java Sea and the Gulf of Tonkin. **Mixed tides** are characterized by large diurnal inequalities in heights and/or times of successive high and/or low waters. In general, this type of tide intermediates between predominantly semi-diurnal and a predominantly diurnal. Such tides are prevalent along the Pacific coast of the USA and in many other parts of the world. At Los Angles, it is typical that the inequalities in the high and low waters are about the same. At Seattle, the greater inequalities are typically in the low water, while in Honolulu, it is the high waters that have the greater inequalities. Table 1.7 exhibits the tidal range at different corners of the globe.

In general, tides have two basic components.

- Magnitude.
- Time.

Table 1.7 Tidal ranges at different coast

Region	Tidal range (m)
Hugli Estuary	5
Sagar Island	4
Diamond Harbour	4.5
Gulf of Kutch	7.25
Gulf of California	9
End of Manas basin	15.6
Chignecto bay	14
Bay of Foundy	15.6

Magnitude components are described in terms of *tidal range and tidal prism,* and the time component is described by *tidal period.*

Majority of the coastal processes operate within the tidal range, and so if the values of **HTL** and **LTL** are known, then on the basis of these limits, one can demarcate the zones with different magnitude of tidal influence such as subtidal, lower intertidal and upper intertidal. These zones are important from geomorphologic and ecological point of views, because each zone is associated with its own set of physical and biological characteristics.

Tidal prism is usually computed to find out the volume of sediment load that is carried into the estuary during the flood and the volume carried out during ebb tide. If one multiply the suspended sediment concentration value by the tidal prism, the total volume can be calculated which can give an idea about positive or negative balance of sediment in the estuary or creek.

The semi-diurnal tide in the sea has a period averaging about 12 h 25 min, and the mean time of water rise roughly equals to that of the waterfall. However, inside bays or estuaries, these two time limits do not follow symmetric pattern, which goes a long way in interpreting the complexities of flow and sediment movement.

The understanding of tides is presently based on two important theories as discussed here.

Equilibrium Theory

According to equilibrium theory, the tidal bulge is caused by gravitational attraction and moon is the major component regulating the ocean tide. The tide here is exaggerated. The actual bulge is in the order of 1 m. Equilibrium tide is theoretical tide.

The effect of sun and moon's gravity and the rotation of the earth on tides are most easily explained by studying equilibrium tides on a smooth water-covered sphere. Let us consider the earth and the moon as a single unit, the earth–moon system that orbits the sun. The moon orbits the earth, held by the earth's gravitational force acting on the moon (B) which must be balanced by a force called centrifugal force (B́) acting to pull the moon away from the earth and send it spinning out into space, and this balance maintains the moon in its orbit. Similarly, the moon's gravitational force acting on the earth (C) must be balanced by the centrifugal force (Ć). These centrifugal forces are caused by the earth–moon system rotating about an axis at the centre of the system's mass, which is 4,640 km away from the earth's centre. Again, the earth–moon system is held by the sun's gravitational attraction (A). A centrifugal force again acts to pull the earth–moon system away from the sun (Á). To remain in this orbit, the earth–moon system requires that the gravitational forces equal the centrifugal forces.

The moon's gravitational force is stronger on particles on the side of the earth closest to the moon, while the centrifugal force is stronger on particles on the side of the earth furthest from the centre of mass of the earth–moon system. The distribution of forces tends to pull surface particles away from the centre of the earth and creates the tide-generating force field on the earth.

The stronger gravitational force of the moon acting on a unit mass of water at the earth's surface closest to the moon is proportional to $M \cdot r/R^3$, where 'M' is the mass of the moon, 'r' is the radius of the earth and 'R' is the distance between the centres of the earth and the moon, because the water covering the earth is liquid and deformable, and the moon's gravity moves water towards a point under the moon, producing a bulge in the water covering. At the same time, the centrifugal force acting on the surface opposite the moon creates a similar bulge of water. The earth model develops two bulges with two depressions in between the bulges or two crests and two trough or two high tide levels and two low tide levels. The wavelength is equal to half the circumference of the earth.

Discrepancies in Equilibrium Theory

Any physical theory is always provisional, in the sense that it is only a hypothesis when it cannot be proved directly. When a new experiment or observation agrees with the predictions, the theory gets the life, but if it disagrees, the theory has to be abandoned or modified. A comparison of tidal observations at different parts of the earth with equilibrium theory indicates a lot of discrepancies. They are

(i) Tidal high water occurs at a time not in conformity with the prediction as per equilibrium theory.
(ii) Tidal ranges are different in most cases than predicted by equilibrium theory.
(iii) Diurnal inequality often bears little resemblance to the theory.
(iv) The bulges as per equilibrium theory would be separated by half the earth's circumference (20,000 km) and would be moving at a velocity of more than 1,600 km/h. Tidal wave is a shallow water wave, and a depth of ocean of 22 km is necessary for the wave to attain this velocity.

The equilibrium theory, however, predicts the periodicity or nearly so and therefore cannot be abandoned but must be modified to account for several complicating factors. These complicating factors are as follows:

(i) Irregular shape of the ocean.
(ii) Varying depth of the ocean.
(iii) World is not entirely covered with water
(iv) Rotation of the earth.
(v) Inertia (in equilibrium theory, inertia was not considered, and therefore, water was supposed to respond immediately).

Dynamic Theory

The equilibrium theory of tides allows us to understand the basic concepts associated with the oceans' tide, but the actual tidal pattern that occurs in the discontinuous ocean basin requires a different approach for understanding. Tide waves become discontinuous due to the separation of the ocean by the continents except in the southern ocean around Antarctica. The tide wave has a long wavelength as compared to the ocean's depth; therefore, tides behave as shallow water wave. Laplace developed the dynamic theory of tide. In this theory, the tidal waves are considered instead of a bulge. The tidal waves are produced by tide-generating forces as considered in 'equilibrium theory' of tide. In contrast to 'equilibrium theory', the 'dynamic theory' of tides recognizes that water covers only three-quarters of our planet and is confined to seas and ocean basins that are fixed on a rotating earth and experiences forces of constantly changing magnitude and direction (the earth's crust is also affected by these forces, creating small but detectable 'earth tide').

The horizontal force fields that act on the water masses rotating with the earth are not at all random like winds blowing across the surface of a lake. Instead, they repeat definite patterns with a kind of 'push me, pull me' rhythm. Ocean waters respond with a wave-like motion characterized by definite period and tidal periods, identical to the ones derived from equilibrium theory. However, water cannot remain confined to basin engaged in wave motion at all like a 'tidal bulge' that supposedly sweep around the globe as depicted in equilibrium theory, without encountering a solid boundary that results in reflection of the wave from that then travel back the way it came. When a reflected wave meets a new waveform that does not go anywhere, but it merely oscillates up and down like a sea-saw.

The shallow water tide wave moves in theory at about 200 m/s, but at equator, the earth moves eastward under the tide wave at more than twice the speed at which the tide can travel freely. This eastward displacement due to friction continues until the portion of the moon's attractive force that pulls the tide wave eastward balances the friction force. This balance between these two forces holds the tide crest in a position to the east of the moon rather than directly under it.

The tide wave is also reflected from the edges of landmasses and mid-ocean ridges, refracted by the changes in water depth and diffracted as it passes through gas between continents. Together, all these factors produce the dynamic tides. In addition, different kinds of tides interact with each other along their boundaries, and the results are exceedingly complex.

1.2.3 Current

The geomorphic importance of currents lies in their role in transporting sediment and less significantly in erosion. Currents associated with tides can transport and erode sediment where flow velocities are high. This is usually confined to estuaries or other enclosed sections of coast which experience semi-diurnal tides with a high range. In mesotidal and microtidal regimes, fast tidal currents are confined to zones of constriction such as around coastal promontories and within narrow channels between islands.

Basically, the movements of water from one part of the ocean to another give rise to currents. The most important currents are the surface currents, which are partly caused by the great wind system of the globe. In addition to this, there are bottom currents as well as currents downward and currents upwards due to the uprise of warmer current. There are also local currents due to the tides or differences in evaporation and salinity. The water in the equatorial regions becomes hot and tends to flow polewards as warm surface currents, while the colder water from the poles creeps along as bottom currents to take its place. The evaporation in the hot regions causes bottom currents of cold waters to flow in below and replace the loss. The major ocean currents of the world oceans are shown in Fig. 1.22.

Cold Currents and Warm Currents

When current is flowing from a warm to a cold region, it is naturally a warm current, but when it flows from a cold to a warm region, it is a cold current. Notice that the terms 'warm' and 'cold' are relative to the temperature of the neighbouring sea water and that a warm current may swing round and becomes a cold current.

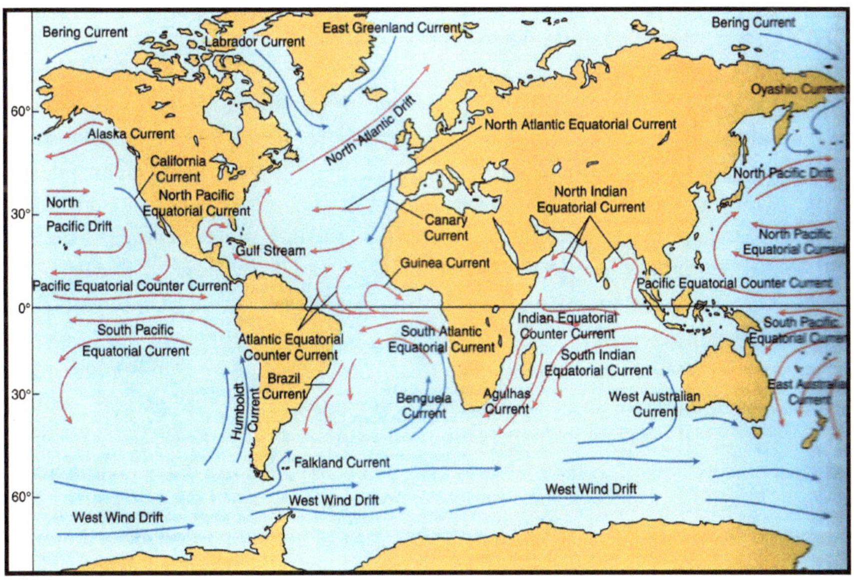

Fig. 1.22 Major currents of the world ocean

Origin and Evolution of Ocean Surface Current System

The water mass of the marine compartment remains constantly under the influence of some types of 'surface forces' as well as some form of 'body forces', which cause acceleration of that water mass.

These forces are basically of the following types:

- Differential (latitude-wise) solar heating (or cooling)—which in turn develops density variation both horizontally as well as vertically.
- Wind flow system (which also varies with latitude), with its frictional coupling imparts motion to the oceanic water mass, in a varying degree spatially.
- Atmospheric pressure gradient (inverted barometric effect) influencing the water level at the oceanic surface, which in turn brings in associated oceanic water circulation.
- Earth's spin and gravitational attraction also influencing the movement of ocean waters.

Besides these forces, other parameters that exercise control on ocean water circulation pattern are as follows:

1. Seasonal changes in the atmospheric wind setup and differing degrees of evaporation/precipitation/ice (or snow) melting/freshwater inflow, etc.
2. Position and shape of continental boundaries.
3. Topography of the ocean floor.

Atmosphere–ocean–earth system is a coupled 'heat engine' with appropriate 'sources' and 'sinks'. In this heat engine, differential heating takes place. Sun remains closer to the zenith in the Tropical region all the year round and consequently receives more heat, which progressively decreases polewards with increasing obliquity of sun's rays. A part of the heat under reradiation from earthly environment gets lost in space, at the same time. The balance sheet of the available solar energy shows positive (credit side) figures between equator and 40° latitude (N and S). Beyond 40° latitude towards polar region, the same shows negative (debit side) figures. Under such differential heating process as a natural course, heat redistribution from areas of credit side (i.e., from lower latitudes) to that of debit side (i.e., towards higher latitudes) establishes both wind flow system in the atmospheric and water flow system in the oceanic environment.

The different thermal capacities of the continental mass and that of its oceanic counterpart greatly influence the actual global pattern of above-stated interacting flow system *vis à vis* the idealized situation. Again the distributions of land areas *vis à vis* that of the oceans as at present vary greatly between the two hemispheres and hence exercise differing complex control.

Moreover, huge quantity of heat either is left locked up or is released (known as latent heat), whenever any change of state takes place in the hydrosphere (by way of evaporation from ocean surface, condensation during cloud/rain formations, freezing by some form of ice formation and their melting), they behave as hidden 'source and sinks', in this heat redistribution system.

With the establishment of the wind flow system, the wind flow pattern over ocean region transfers a part of the energy to the surface layers of the ocean. A part of this energy is utilized to generate oceanic currents. The wind flow drags the ocean waters to flow along with it, the rate of energy transfer on an average being about 3 % of the wind speed. In ideal situation under the influence of earth's spin, the oceanic surface current in the Northern Hemisphere gets deflected at an angle of about 40–45 to the right-hand side of the wind flow direction (and in the Southern Hemisphere by about the same amount to the left-hand side of the wind flow direction). Each layer of moving water sets the layer below it in motion. Under the influence of Coriolis force, deeper layers exhibit gradually shifting direction of flow with slower speeds (Ekman transport). The net flow on an average exhibits a deflection by 90° to the right (or to the left) with respect to the wind flow direction in the Northern (or Southern) Hemisphere (Fig. 1.23).

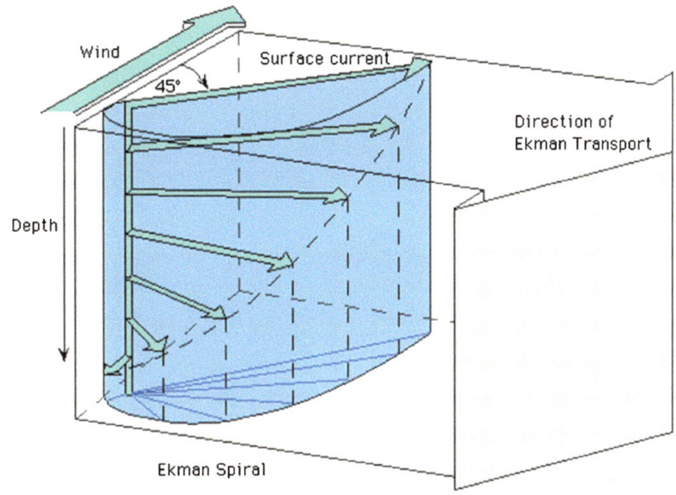

Fig. 1.23 Ekman spiral and Ekman transport

Layer Formation and Vertical Mixing of Fluid

Density is an important property of blue fluid in the ocean. Pure water has a density of 1 g/cm^3. Since sea water contains dissolved salts, it has a higher density than pure water (1.0270 g/cm^3). Variations in the surface temperature and salinity of the oceans control the water density. Density increases with an increase in salinity and a decreases in temperature (water reaches its maximum density at a temperature of 4 °C). Density decreases when there is decrease in salinity or increase in temperature. Salinity increases when evaporation occurs or when some of the sea water comes out of solution in freezing to form ice. Precipitation, influx of river water, melting ice or a combination of factors can all contribute to a decrease in salinity. Pressure can also affect the density of sea water, but at the surface, these effects are minor. The combination of different surface temperatures or salinities produces regions of oceans with low densities. Denser water sinks until it joins water of same density, whereas less dense water rises. This situation produces the situation of effect of layered water in the ocean.

Sea water that is not very dense, like the warm, low-salinity water around the equator (1.0230 g/cm^3), remains at the surface. Surface water at 30°N and 30°S is also warm, but it has a higher salinity and is denser (1.0267 g/cm^3) than the equatorial waters. This difference in density causes the water at 30°N to extend from the surface at this latitude to below the less dense surface water at the equator and then turn back again to the surface again at 30°S. The combination of colder temperatures and higher salinities produces even denser surface water at 60°N and 60°S (1.0276 g/cm^3). This water extends from the surface of one hemisphere below the other surface waters to the surface of other hemisphere. During winter at the poles, the lower water temperature and increased salinity that results from the formation of sea ice result in very dense water that sinks towards the floor of the ocean. The variations in the density of surface water, then, produce a layered effect on the ocean. The zone of sharp density variation is pycnocline.

The characteristics of temperature, salinity and density change with depth. The surface layer of the ocean extends down to about 100 m, is warmed by solar heating, and is well mixed by a variety of processes. From 100 to 1,000 m temperature decrease, creating layers of waters of increasing density. This zone of rapid temperature change is called as thermocline. Similarly, below the surface waters in the temperate zone, the salinity increases with depth to about 1,000 m. This zone is called a halocline. The changes in temperature and salinity in the region from 100 to 1,000 m produce a pycnocline, a

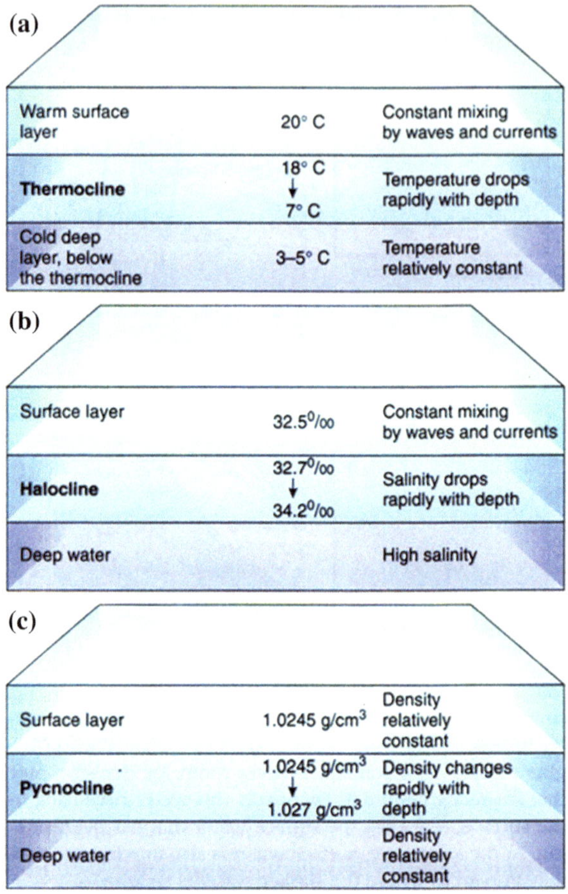

Fig. 1.24 Thermocline (**a**), halocline (**b**) and pycnocline (**c**) in ocean compartment

zone where density increases rapidly with depth. Below the thermocline, temperatures are relatively stable, with only small decreases in temperature towards the ocean bottom. Likewise, below the halocline, salinities are constant down to the ocean floor. As a result, water that is deeper than 1,000 m has relatively constant density (Fig. 1.24).

Vertical Mixing

When the density of water increases with depth, the water column from the surface down is said to be stable. If the top water in a water column is denser than the water column below it, then the water column is unstable. Unstable water column do not persist, because the denser water at the top sinks and the less dense water below rises to the surface. This changeover in a water column produces a vertical overturn. If the water column has the same density from top to bottom, it is isopycnal. Such neutral stability means there is no tendency for water in the column to either sink or rise. A water column that is neutrally stable can be easily mixed in the vertical direction by such forces as wind, wave action or currents.

Any process that increases the density of surface water will cause vertical movement of water or vertical mixing. Vertical mixing is an important process for the exchange of water from top to bottom throughout the world's oceans. Since bottom water usually contains abundant nutrients from the settling of organic matter from above and the process of decomposition, vertical mixing provides a means of exchanging this nutrient-rich bottom water with oxygen-rich surface water. As density is controlled by temperature and salinity, this type of circulation is also known as thermohaline circulation.

The open oceans in temperate latitudes exhibit a seasonal effect. During the summer months, the surface water is warm and the water column is relatively stable. In the fall, as the surface water cools, its density increases, the water column becomes unstable, and the surface water sinks. The mixing process continues with the aid of winter storms and the continued cooling of surface waters. The thermocline developed during the previous summer is ultimately eliminated and the upper regions become isopycnal to the deeper water. With spring come warmer temperatures, and the thermocline begins to re-establish itself, stabilizing the water column. This process of stabilization then continues through the summer.

In the open ocean, surface temperature is more important than salinity in determining the water density. For instance, the surface water in tropics has highest salinity in the open ocean, but the water is so warm that it remains less dense than the water below it and it does not sink. On the other hand, water in the North Atlantic has a lower salinity but is much colder, and as a result, the surface water is denser and it sinks.

Salinity becomes a more important factor in determining the density of water close to the shore. This is especially evident in semi-enclosed bays that receive a large amount of freshwater run-off. Wave action also contributes to the density and mixing of ocean water close to the shore, where the water is relatively shallow.

Upwelling and Downwelling

Since the quantity of water in the oceans is essentially fixed, any movement of water from one place to another would cause a similar but opposite movement to replace the water that leaves an area. For instance, when dense surface water sinks to a depth at which it no longer denser than water beneath it, it stops sinking. At this point, the water begins to move horizontally, making room for denser water that is sinking behind it. Eventually, this water rises back to the surface, replacing the surface water that has been sinking. At the same time, surface water also moving horizontally into areas where the surface water is sinking.

Downwelling zones represent areas where surface water is sinking, whereas upwelling zones are areas where bottom water rises to the surface. These processes are extremely important for the organisms that live at various depths. Because downwelling carries oxygen-rich surface areas to deeper areas, many organisms can live in deep-water where downwellings occur. This process is essentially important for the organisms that live below the photic zone (deeper than sunlight can penetrate). Upwelling on the other hand brings water rich in nutrients from sedimentation and decay processes from great depths to the surface, where it supplies the needs for photosynthetic organisms and the zooplankton that are vital parts of oceanic food chains.

Upwelling and downwelling can also be produced by wind-driven surface currents. When current drives two masses of water together, water is forced downward. In other regions of the ocean, surface currents may push two water masses in opposite directions, drawing deeper water to the surface. During the slow movement produced by these currents, adjacent layers of ocean water are mixed with nutrients and chemicals.

1.3 Chemical Characteristics of Ocean and Estuarine Water

Current theories hold that our solar system along with our blue planet Earth was formed about 4.6 billion years ago. It is generally believed that for the first billion years of its existence, the earth was mainly composed of silicon compounds, iron, magnesium oxide, and some other elements in small amounts. Geologists and earth scientists believe that originally, this blue planet was composed of cold matter. Gradually with the passage of time, several factors, such as energy from space and the decay of radioactive elements, contributed to raising its temperature. The process of heating continued for several hundred million years until the temperature at the centre of the earth was high enough to melt iron and nickel. As these elements melted, they moved to

Table 1.8 Comparative account of the major ocean

Ocean	Surface area (km^2)	Volume (km^3)	Average depth (m)
Atlantic	82,400,000	323,600,000	3,926
Pacific	165,200,000	707,600,000	4,282
Indian	73,400,000	291,000,000	3,962

the earth's core, displacing lighter, less dense elements and eventually raising the core temperature to approximately 2,000 °C. Molten material from the earth's core moved to the surface and spread out, creating some of the features of the early earth's landscape. It is probable that the process of melting and solidifying happened repeatedly, ultimately separating the lighter elements of the earth's crust from the deeper, denser elements. This separation of elements eventually produced the various layers of the blue planet. In these early times, any water present on the planet was probably locked up in the earth's minerals. As the cycles of heating and cooling took place, water, in the form of water vapour, was carried to the surface, where it cooled, condensed and formed the oceans.

Today, 1.37 billion cubic kilometres (approximately 362×10^{18} gallons) of water covers 71 % of the earth's surface and forms its oceans. There are three principle oceans of the world: the Atlantic, Pacific and Indian Oceans. Of the three, the Pacific Ocean is the largest, the Indian Ocean is the smallest, and the Atlantic Ocean is the shallowest. Each ocean has its own characteristic surface area, volume and depth (Table 1.8).

The chemical composition of the ocean water is, however, uniform (Table 1.9) except some regional or local effects such as anthropogenic or industrial discharges, natural disasters, etc.

According to Pytkowicz and Kester (1971), the dominant components of sea water are sodium and chloride ions. The entire spectrum of sea water components may be divided into major (Table 1.10) and minor (Table 1.11) components.

The salinity of sea water (mean value 35 psu) is made up by all the dissolved salts shown in Table 1.8. Interestingly, their proportions are always the same, which can be understood if salinity differences are caused by either evaporating freshwater or adding freshwater from rivers.

Salinity affects the marine organisms through the process of osmosis. A fish with a cellular salinity of 1.8 % will swell in freshwater and dehydrate in salt water. So, marine fish species drink water copiously while excreting excess salts through their gills. Freshwater fishes act in opposite manner by not drinking, but excreting copious amounts of urine while losing little of their body salts.

Marine plants (seaweeds) and many lower organisms (like plankton) have no mechanism to control osmosis, which makes them very sensitive to the salinity of the water in which they live.

The main nutrients for plant growth are nitrogen (N as in nitrate NO_3^-, nitrite NO_2^-, ammonia NH_4^+), phosphorous (P as phosphate PO_4^{3-}) and potassium (K) followed by sulphur (S), magnesium (Mg) and calcium (Ca). Iron (Fe) is an essential component of enzymes and is copiously available in soil, but not in sea water (0.0034 ppm). This makes iron an essential nutrient for phytoplankton growth. Plankton organisms (like diatoms) that make shells of silicon compounds furthermore need dissolved silicon salts (SiO_2) which at 3 ppm can be rather limiting.

The main salt ions that contribute 99.9 % of sea water salinity are highlighted in Table 1.12.

It is to be noted by the readers that the figures stated in the tables may differ slightly in differing publications. Also, landlocked seas such as the Black Sea and Baltic Sea have varying concentrations.

The world map in Fig. 1.25 shows how the salinity of the oceans changes slightly from around 32 psu (3.2 %) to 40 psu (4.0 %). Low salinity is found in cold seas, particularly during summer season when ice melts. High salinity is found in ocean deserts in a band coinciding with the continental deserts. Due to cool dry air

Table 1.9 Detailed composition of sea water at 3.5 % salinity

Element	Atomic weight	ppm	Element	Atomic weight	ppm
Hydrogen H_2O	1.00797	110,000	Molybdenum Mo	0.09594	0.01
Oxygen H_2O	15.9994	883,000	Ruthenium Ru	101.07	0.0000007
Sodium NaCl	22.9898	10,800	Rhodium Rh	102.905	–
Chlorine NaCl	35.453	19,400	Palladium Pd	106.4	–
Magnesium Mg	24.312	1,290	Argentum (Silver) Ag	107.870	0.00028
Sulphur S	32.064	904	Cadmium Cd	112.4	0.00011
Potassium K	39.102	392	Indium In	114.82	–
Calcium Ca	40.08	411	Stannum (tin) Sn	118.69	0.00081
Bromine Br	79.909	67.3	Antimony Sb	121.75	0.00033
Helium He	4.0026	0.0000072	Tellurium Te	127.6	–
Lithium Li	6.939	0.170	Iodine I	166.904	0.064
Beryllium Be	9.0133	0.0000006	Xenon Xe	131.30	0.000047
Boron B	10.811	4.450	Caesium Cs	132.905	0.0003
Carbon C	12.011	28.0	Barium Ba	137.34	0.021
Nitrogen ion	14.007	15.5	Lanthanum La	138.91	0.0000029
Fluorine F	18.998	13	Cerium Ce	140.12	0.0000012
Neon Ne	20.183	0.00012	Praseodymium Pr	140.907	0.00000064
Aluminium Al	26.982	0.001	Neodymium Nd	144.24	0.0000028
Silicon Si	28.086	2.9	Samarium Sm	150.35	0.00000045
Phosphorus P	30.974	0.088	Europium Eu	151.96	0.0000013
Argon Ar	39.948	0.450	Gadolinium Gd	157.25	0.0000007
Scandium Sc	44.956	<0.000004	Terbium Tb	158.924	0.00000014
Titanium Ti	47.90	0.001	Dysprosium Dy	162.50	0.00000091
Vanadium V	50.942	0.0019	Holmium Ho	164.930	0.00000022
Chromium Cr	51.996	0.0002	Erbium Er	167.26	0.00000087
Manganese Mn	54.938	0.0004	Thulium Tm	168.934	0.00000017
Ferrum (Iron) Fe	55.847	0.0034	Ytterbium Yb	173.04	0.00000082
Cobalt Co	58.933	0.00039	Lutetium Lu	174.97	0.00000015
Nickel Ni	58.71	0.0066	Hafnium Hf	178.49	<0.000008
Copper Cu	63.54	0.0009	Tantalum Ta	180.948	<0.0000025
Zinc Zn	65.37	0.005	Tungsten W	183.85	<0.000001
Gallium Ga	69.72	0.00003	Rhenium Re	186.2	0.0000084
Germanium Ge	72.59	0.00006	Osmium Os	190.2	–
Arsenic As	74.922	0.0026	Iridium Ir	192.2	–
Selenium Se	78.96	0.0009	Platinum Pt	195.09	–
Krypton Kr	83.80	0.00021	Aurum (gold) Au	196.967	0.000011
Rubidium Rb	85.47	0.120	Mercury Hg	200.59	0.00015
Strontium Sr	87.62	8.1	Thallium Tl	204.37	–
Yttrium Y	88.905	0.000013	Lead Pb	207.19	0.00003
Zirconium Zr	91.22	0.000026	Bismuth Bi	208.980	0.00002
Niobium Nb	92.906	0.000015	Thorium Th	232.04	0.0000004
Plutonium Pu	244.00	–	Uranium U	238.03	0.0033

Remark ppm = parts per million = mg/l = 0.001 g/kg. *Source* Karl (1968)

Table 1.10 Major components of oceanic water

Component	Concentration (mg/kg of sea water at 35 ‰ salinity)
Calcium	412
Magnesium	1,294
Nitrogen	15
Potassium	399
Silicon	2.9
Sodium	10,760
Strontium	7.9
Bicarbonate	145
Boron	4.6
Bromide	67
Chloride	19,350
Fluoride	1.3
Sulphate	2,712

Source Pytkowicz and Kester (1971)

Table 1.11 Minor components of oceanic water

Component	Concentration (μg/kg of sea water at 35 psu salinity)
Aluminium	2
Argon	4
Arsenic	3
Barium	20
Iron	2
Iodide	60
Lithium	180
Molybdenum	10
Phosphorus	60
Rubidium	120
Titanium	1
Uranium	3
Vanadium	2
Zinc	3
Cadmium, chromium, caesium, copper, krypton, manganese, neon, nickel, antimony	2.4*
Bismuth, cobalt, gallium, germanium, mercury, niobium, lead, selenium, tin, thallium, xenon, zirconium	0.4*
Silver, gold, beryllium, cerium, helium, hafnium, lanthanum, tantalum, thorium, tungsten, yttrium, radium	0.05*
Dysprosium, erbium, europium, gadolinium, holmium, lutetium, praseodymium, samarium, terbium, thulium, ytterbium	0.005*
Indium, protactinium, scandium	0.0007*

*Concentrations are sum totals for all elements
Source Holland (1978)

descending and warming up, these desert zones have very little rainfall and high evaporation. The Red Sea located in the desert region but almost completely closed shows the highest salinity of all (40 psu), but the Mediterranean Sea follows as a close second (38 psu), and lowest salinity is found in the upper reaches of the Baltic Sea (0.5 %). The Dead Sea is 24 % saline, containing mainly magnesium chloride ($MgCl_2$). Shallow coastal areas are 2.6–3.0 % saline and estuaries 0–3 %.

The components of sea water at the surface, coastal waters and estuarine waters often suffer characteristic seasonal changes due to factors such as evaporation, precipitation and run-off. In the north-western Bay of Bengal, the pre-monsoon period (summer) is usually characterized with high saline surface water and relatively high pH. The picture in monsoon period is totally opposite. Due to increased precipitation and run-off from the adjacent land masses, the salinity of the surface water lowered along with pH, but the concentrations of the nutrients increase, which may be attributed to increased run-off from the adjacent agricultural fields, municipality, urban sectors, etc. (Mitra 2000b). The composition of coastal waters also gets affected by wastewater discharges from industrial and anthropogenic factors. Table 1.13 lists the population around and volume of wastewater discharged in major estuaries of India.

A case study highlighting the seasonal variations of hydrological parameters in the Hooghly estuary (a continuation of Ganga–Bhagirathi River system) is depicted in Tables 1.14 and 1.15.

A comparison between Tables 1.14 and 1.15 exhibits significant tidal variations in relation to the hydrological parameters and therefore suggests the consideration of tidal effects while monitoring the water quality of an estuary.

Table 1.12 Major salt ions contributing to sea water salinity with their respective valency and weight

Chemical ion	Valency	Concentration (ppm or mg/kg)	Part of salinity (%)	Molecular/atomic weight	μmol/kg
Chloride ion Cl	−1	19,345	55.03	35.453	546
Sodium Na	+1	10,752	30.59	22.990	468
Sulphite SO_4	−2	2,701	7.68	96.062	28.1
Magnesium Mg	+2	1,295	3.68	24.305	53.3
Calcium Ca	+2	416	1.18	40.078	10.4
Potassium K	+1	390	1.11	39.098	9.97
Bicarbonate HCO_3	−1	145	0.41	61.016	2.34
Bromide Br	−1	66	0.19	79.904	0.83
Borate BO_3	−3	27	0.08	58.808	0.46
Strontium Sr	+2	13	0.04	87.620	0.091
Fluoride F	−1	1	0.003	18.998	0.068

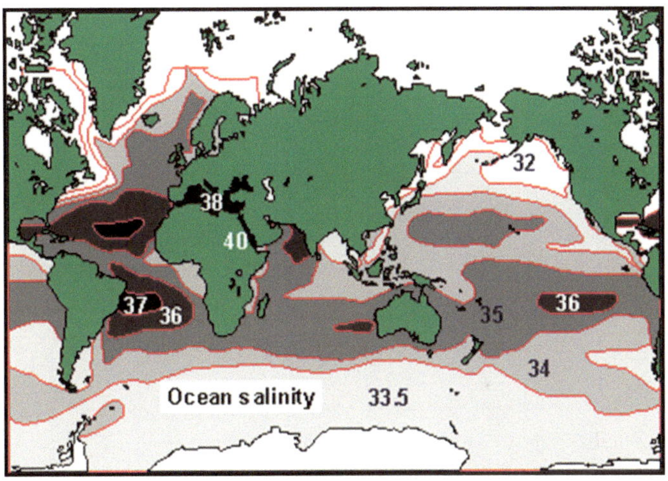

Fig. 1.25 Salinity profile of important oceanic regions

Table 1.13 Major Indian estuaries with the major cities along their banks, city population and wastewater discharge (wherever known)

Estuary/river	Major city	Population (10^6)	Wastewater discharge (10^6 m^3/day)
Sabarmati	Ahmadabad, Gandhinagar	4.7	14
Narmada	Jabalpur, Bharuch	1.3	Not available
Tapi	Sural	2.8	Not available
Mahim, Ulhas, Thane	Mumbai	16.5	2.8
Cochin Backwaters	Kochi	1.4	0.1
Cauvery	Thiruchirapalli, Erode	1.3	0.1
Ennore, Adyar	Chennai	6.7	0.01
Krishna	Hyderabad, Vijayawada	6.7	Not available
Hooghly	Kolkata	13.2	15.3

Source Satish (2011)

Table 1.14 Physico-chemical characteristics of surface water at high tide condition during May 2012

Station no.	Water temperature (°C)	Salinity (‰)	pH	Alkalinity (mg/l)	DO (mg/l)	BOD (mg/l)	COD (mg/l)	NO_3 (μg at/l)	PO_4 (μg at/l)	SiO_3 (μg at/l)	Extinction coefficient (m^{-1})	SO_4 (mg/l)	Na (mg/l)	K (mg/l)	Cl (mg/l)	Total N (μg at/l)
1	35.5	2.13	7.85	148	5.71	5.6	69	23.78	4.34	99.87	5.50	490	706	26	1,219	28
2	35.5	3.97	7.68	159	5.43	6.4	118	26.55	3.98	103.65	5.61	502	1,189	44	2,175	31
3	35.6	5.65	7.70	162	7.02	5.2	120	24.50	3.20	111.90	4.98	584	1,714	62	3,099	42
4	35.6	12.14	8.00	190	6.55	4.3	79	21.09	3.09	115.20	4.79	713	3,685	134	6,599	37
5	35.5	11.24	8.10	160	4.80	6.7	143	30.02	4.81	99.40	5.30	665	3,394	124	6,054	59
6	35.5	12.80	8.00	169	4.91	5.2	81	19.78	4.18	98.34	4.80	681	3,870	142	6,991	49
7	35.6	14.41	8.10	200	4.80	4.1	107	17.44	2.98	98.70	4.99	742	4,397	159	7,861	23
8	35.6	14.70	8.10	185	4.75	3.1	93	15.40	2.41	110.05	6.12	858	4,231	163	8,005	19
9	35.6	16.02	8.10	210	4.53	6.5	138	25.99	4.72	79.78	5.52	962	4,872	177	8,575	40
10	35.6	18.34	8.15	250	4.68	3.4	63	13.88	1.49	71.22	5.00	1,041	5,605	203	9,200	28
11	35.6	19.55	8.20	260	5.05	3.9	98	15.34	2.20	66.10	5.13	1,123	5,877	217	10,079	31
12	35.6	23.05	8.22	278	5.02	4.6	120	17.84	2.79	61.00	5.05	1,291	6,703	249	11,733	44

Table 1.15 Physico-chemical characteristics of surface water at low tide condition during May 2012

Station no.	Water temperature (°C)	Saliniy (‰)	pH	Alkalinity (mg/l)	DO (mg/l)	BOD (mg/l)	COD (mg/l)	NO_3 (μg at/l)	PO_4 (μg at/l)	SiO_3 (μg at/l)	Extinction coefficient (m^{-1})	SO_4 (mg/l)	Na (mg/l)	K (mg/l)	Cl (mg/l)	Total N (μg at/l)
1	35.4	1.28	7.60	130	6.01	6.0	73	26.85	4.59	85.60	6.09	333	429	20	795	33
2	35.4	2.66	7.55	146	5.72	7.0	125	29.01	4.28	92.00	6.12	418	802	39	1,460	46
3	35.4	3.98	7.55	145	6.79	5.9	128	27.02	3.67	92.87	5.63	496	1,200	59	2,150	50
4	35.4	10.96	7.85	170	6.98	4.6	82	23.00	3.28	94.98	5.27	645	3,259	129	5,943	45
5	35.4	10.05	7.90	144	4.91	7.4	152	34.00	5.31	79.20	6.20	578	2,998	119	5,507	69
6	35.5	11.50	7.80	154	5.06	5.6	82	21.56	4.38	80.87	5.50	590	3,469	133	6,150	52
7	35.5	12.85	8.00	183	5.19	4.5	109	19.85	3.22	78.66	5.73	600	3,856	152	6,991	31
8	35.5	13.02	8.00	169	5.02	2.9	94	17.45	2.61	97.33	6.91	759	3,870	158	7,139	25
9	35.5	14.50	8.00	193	4.97	7.3	148	29.75	5.04	69.54	6.31	830	4,355	170	7,875	51
10	35.5	15.98	8.00	231	4.46	3.8	69	14.66	1.63	61.22	5.45	922	4,894	197	8,445	39
11	35.5	17.49	8.06	247	5.15	4.3	102	16.83	2.44	60.64	5.63	916	5,245	212	9,263	43
12	35.5	21.11	8.17	260	5.19	5.0	123	19.05	3.05	54.14	5.66	1,115	6,175	241	10,389	57

Table 1.16 Dissolved gases in sea water

Gas molecule	% in atmosphere	% in surface sea water	ml/l of sea water	mg/kg (ppm) in sea water	Molecular weight	M mol/kg
Nitrogen N_2	78	47.5	10	12.5	28.014	0.446
Oxygen O_2	21	36.0	5	7	31.998	0.219
Carbon dioxide CO_2	0.03	15.1	40	90	42.009	2.142
Argon	1	1.4	–	0.4	39.948	0.01

Dissolved Gases in Sea Water

The gases dissolved in sea water are in constant equilibrium with the atmosphere, but their relative concentrations depend on each gas solubility, which depends also on salinity and temperature. As salinity increases, the amount of gas dissolved decreases because more water molecules are immobilized by the salt ion. As water temperature increases, the increased mobility of gas molecules makes them escape from the water, thereby reducing the amount of gas dissolved.

Inert gases such as nitrogen and argon do not take part in the process of life and are thus not affected by plant and animal life. However, non-conservative gases such as oxygen and carbon dioxide are influenced by sea life. Plants reduce the concentration of carbon dioxide in the presence of sunlight, whereas animals do the opposite in either light or darkness.

Among the gases stated in Table 1.16, the conservative gases nitrogen and argon do not contribute to life processes, even though nitrogen gas can be converted by some bacteria into fertilizing nitrogen compounds (NO_3, NH_4). Surprisingly, the world under water is very much different from the above in the availability of the most important gases for life: oxygen and carbon dioxide. Whereas in air about one in five molecules is oxygen, in sea water, this is only about 4 in every thousand million water molecules. Whereas air contains about one carbon dioxide molecule in 3,000 air molecules, in sea water, this ratio becomes 4 in every 100 million water molecules, which make carbon dioxide much more common (available) in sea water than oxygen. Note that even though their concentrations in solution differ due to differences in solubility (ability to dissolve), their partial pressures remain as in air, according to Henry's Law, except where life changes this. Plants increase oxygen content while decreasing carbon dioxide and animals do the reverse. Bacteria are even capable of using up all oxygen.

All gases are less soluble as temperature increases, particularly nitrogen, oxygen and carbon dioxide which become about 40–50 % less soluble with an increase of 25 °C. When water is warmed, it becomes more saturated, eventually resulting in bubbles leaving the liquid. Fish like sunbathing or resting near the warm surface or in warm water outfalls because oxygen levels there are higher. The elevated temperature also enhances their metabolism, resulting in faster growth and perhaps a sense of well-being.

Likewise, if the whole ocean was to warm up, the equilibrium with the atmosphere would change towards more carbon dioxide (and oxygen) being released to the atmosphere, thereby exacerbating global warming.

Sediments of Oceans and Estuaries

A large portion of the sea bottom is covered with sediments, which contains rock fragments, animal debris, sand, silt, clay, etc. Sediments may be categorized as sand, if the diameter of the individual particle exceeds 62 μm, or mud, if the diameter of the individual particle is less than 62 μm. Geologists use a phi or Ø scale, based on the **negative of the power of 2** to differentiate between the sizes of the sediment particles (Table 1.17).

Apart from describing the marine sediments as mud or sands, the particles may be categorized by origin such as biogenous, hydrogenous and lithogenous. The biogenous sediment is derived from the marine biota, particularly the debris of the shell. If the sediment contains more than

Table 1.17 Classification of sediments on the basis of particle size

Division	Subdivision	Diameter (mm)	Phi (Ø)-size	Sinking rate (cm/s)
Gravel	Boulder	>256	<−8	>4.29 × 10^6
	Cobble	64–256	−6 to −8	2.68 × 10^5 to 4.29 × 10^6
	Pebble	4–64	−2 to −6	1.05 × 10^3 to 2.68 × 10^5
	Granule	2–4	−1 to −2	2.62 × 10^2 to 1.05 × 10^3
Sand	Very coarse	1–2	0 to −1	65.6–262
	Coarse	0.5–1	+1 to 0	16.4–65.5
	Medium	0.25–0.5	+2 to +1	4.09–16.4
	Fine	0.125–0.25	+3 to +2	1.02–4.09
	Very fine	0.0625–0.125	+4 to +3	0.256–1.02
Mud	Silt	0.0039–0.0625	+8 to +4	<9.96 × 10^{-4}
	Clay	<0.0039	>8	

Table 1.18 Types of sediment on the basis of origin

Type	Source	Favourable deposition site	Examples
Biogenous	Biota	Seabed under neritic and oceanic zone	Coral, siliceous and calcareous oozes
Cosmogenous	Outer space	Seabed under pelagic zone	Meteorites
Hydrogenous	Chemical precipitation from sea water	Seabed under neritic and oceanic zone	Phosphates, manganese nodules
Lithogenous	Eroded rock, volcanoes and airborne dust	Seabed under neritic and oceanic zone	Boulder–sand, sand–silt, silt–clay

30 % of the biogenous particle, then only the sediment may be referred to as biogenous type. Essentially, biogenous particles may be rich in calcium carbonate (calcareous), silicate (siliceous) or phosphate (phosphatic) depending upon whether they are derived from the shells of foraminifers or diatoms or from the bones/scales of marine vertebrates. The hydrogenous particles are derived from inorganic chemical reactions in water, which comprises the manganese nodules of the ocean floor in the Pacific Ocean. The lithogenous particles originate from the erosion of rocks, the particles of which are transported in the seas or rivers. A summary of the major types of sediment is given in Table 1.18. On the continental slope, the particles are sorted by size according to the action of waves and currents. The largest particles are deposited, whereas the smallest grains move seaward in deep and are eventually deposited in deep-water.

In coastal Bay of Bengal, the intertidal zone registers a gradual change in the textural characteristics from high water level to low water level, indicating a sediment change from sandy to silty nature. The sand flats of the mixed and open-sea intertidal zone consist of 90–95 % of fine to very fine sand. Sorting of sediments is very systematic due to natural agents in the present geographical locale. The surface of the sand flat is covered with crescentic ripples, asymmetrical ripples, backwash ripples, ladder back or cross ripples, swash marks, rhomboid

Fig. 1.26 Patterns created by sediment fauna

Fig. 1.27 Holes of crabs on the sea beach

marks, rill marks and current crescents. Bioturbation structures are special attractions to nature lovers that are witnessed in the sediment bed of muddy and sandy beaches. These structures result from the activities of near-shore benthic organisms and reflect a unique fabric work of the biotic community on the abiotic matrix. The main agents behind these marvellous fabric works and sculptures are the anemones, gastropods, crustaceans, decapods, bivalves, holothuroids and hemichordates. In some pockets of Indian Sundarbans, animals such as *Macoma birmanica*, *Telescopium telescopium*, *Uca* sp. and *Cerithedia* sp. prepare a muddy substratum and produce bioturbation in the form of mounts, trails and burrows (Figs. 1.26, 1.27 and 1.28).

Fig. 1.28 Ripple marks and bioturbation

References

Cameron WM, Pritchard DW (1963) Estuaries. In: Hill MN (ed) The sea, vol 2. Wiley, New York, pp 306–324

Chaudhuri AB, Choudhury A (1994) Mangroves of the Sundarbans, vol 1. IUCN—The World Conservation Union, India, p 165

Church JA, White NJ, Clarke AJ, Freeland HJ, Smith RL (1986) Coastal trapped waves on the east Australian continental shelf. Part II: model verification. J Phys Oceanogr 16:1945–1958

Griffin DA, Middleton JH (1991) Local and remote wind forcing of New South Wales inner shelf currents and sea level. J Phys Oceanogr 16:1652–1664

Holland HD (1978) The chemistry of the atmosphere and the oceans. Wiley, New York

Karl KT (1968) Oceans. Prentice-Hall, Englewood Cliffs

Mitra A (2000a) Foundation of environmental science. Narendra Publishing House, Delhi

Mitra A (2000b) The northeast coast of the Bay of Bengal and deltaic Sundarbans, Chap. 62. In: Sheppard C (ed) Seas at the millennium—an environmental evaluation. Elsevier Science, London, pp 143–157

Mitra A (2013) Sensitivity of mangrove ecosystem to changing climate, 2013 edn. Springer, New Delhi. ISBN-10: 8132215087; ISBN-13: 978-8132215080; Copyright Springer, India 2013; ISBN 978-81-322-1509-7

Odum HT (1971) Environment, power and society. Wiley, New York 331

Perillo GME (1995) Definitions and geomorphological classifications of estuaries. In: Perillo GME (ed) Geomorphology and sedimentology of estuaries. Developments in sedimentology, vol 53. Elsevier, Amsterdam, pp 17–47

Pritchard DW (1967) What is an estuary: physical stand point? In: Lauff GH (eds) Estuaries, vol 83. American Association for the Advancement of Science. Washington, DC, pp 3–5

Pytkowicz RM, Kester DR (1971) The physical chemistry of seawater. In: Barens H (ed) Annual review oceanography marine biology, vol 9. Harper and Row, New York, pp 11–60

Satish RS (2011) Indian estuaries: dynamics, ecosystems and threats. Natl Acad Sci Lett 34:7–8

Sengupta K, Roy CM, Bhattacharya SB, Raha A, Zaman S, Mitra A (2013) Spatial variation of stored carbon in *Avicennia alba* of Indian Sundarbans. Discov Nat 3 (8):19–24

Internet Reference

www.es.flinders.edu.au/~mattom/ShelfCoast/chapter11.ht

2 Biodiversity of the Blue Zone

...Come along, come with us
Take a dive into the deep blue mass
Amazing are the creatures
With their unique adaptive features....
The Authors

2.1 Producer Community

The producer community consists of organisms that are capable of synthesizing their own food through photo- or chemosynthesis. Most of the primary production in the marine and estuarine ecosystems is carried out by phytoplankton. Apart from phytoplankton that manufacture their organic food through photosynthesis (in the photic zone), there are several strains of bacteria that can prepare their food through the process of chemosynthesis. Chemosynthetic bacteria (also known as chemoautotrophs) can produce their own food from inorganic compounds without sunlight, a process called chemosynthesis. They use energy derived from chemical reactions that involve substances such as ammonia (NH_3), sulphides (S^{2-}), nitrates (NO_3^-) and sulphates (SO_4^{2-}). This energy is then used to manufacture organic food molecules. Chemosynthetic bacteria are found in areas where there is little or no sunlight and where the inorganic substances that they require are in abundance. The bacteria that live around deep-sea hydrothermal vents are examples of chemosynthetic bacteria. These bacteria use sulphide ions that spew from the vents. They oxidize the sulphide to sulphur and sulphates and use the energy released by this process to produce food. Like other autotrophs, they form the base of a productive food chain that consists of a diverse assembly of organisms, including worms, clams, and crabs, that constitute the hydrothermal vent community.

Several groups of bacteria, including the purple bacteria, the green bacteria and cyanobacteria, are able to photosynthesize. Purple bacteria and green bacteria are usually strict anaerobes (organisms that live in environments that lack oxygen) and are found in areas of the marine environment such as mud flats that provide light, anaerobic conditions and sulphur-containing compounds. These organisms cannot use water in their photosynthetic process, and therefore, they do not produce any oxygen (Fig. 2.1). Instead, they often use hydrogen sulphide (H_2S) and produce elemental sulphur or sulphate (Fig. 2.2). Purple bacteria and green bacteria are an important source of food for some zooplankton and filter feeders.

The seaweeds, seagrasses, salt marsh grasses and mangroves are also among the producer list of the marine and estuarine ecosystems (Figs. 2.3, 2.4 and 2.5)

2.1.1 Phytoplankton

Phytoplankton are free-floating tiny floral components that are widely distributed in the marine and estuarine environments. Like land plants, these tiny producers require sunlight, nutrients or fertilizers, carbon dioxide gas, and water for growth. The cells of these organisms contain the pigment chlorophyll that traps the solar energy for use in photosynthesis. The photosynthetic process uses the solar radiation to convert carbon dioxide and water into sugars or high-energy organic

A. Mitra and S. Zaman, *Blue Carbon Reservoir of the Blue Planet*,
DOI 10.1007/978-81-322-2107-4_2, © Springer India 2015

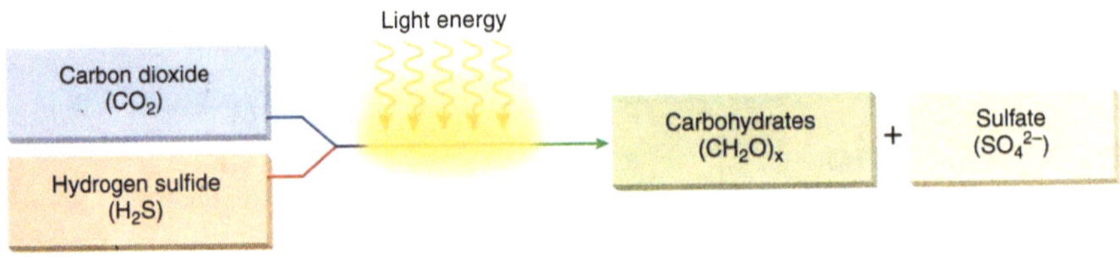

Fig. 2.1 Free oxygen is not generated during food preparation by purple and green bacteria

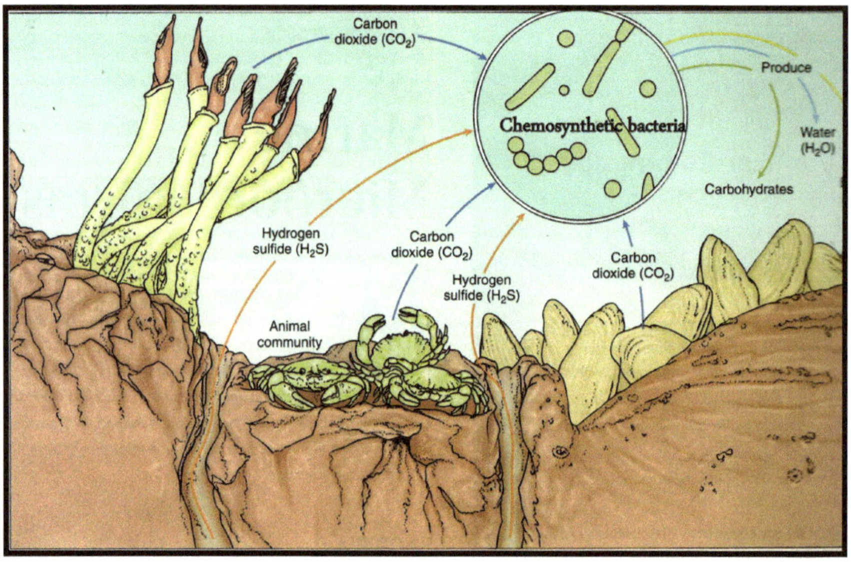

Fig. 2.2 Chemosynthetic bacteria survive in the deep-sea vent zones and are capable of synthesizing food from carbon dioxide and hydrogen sulphide using the energy derived from chemical reactions

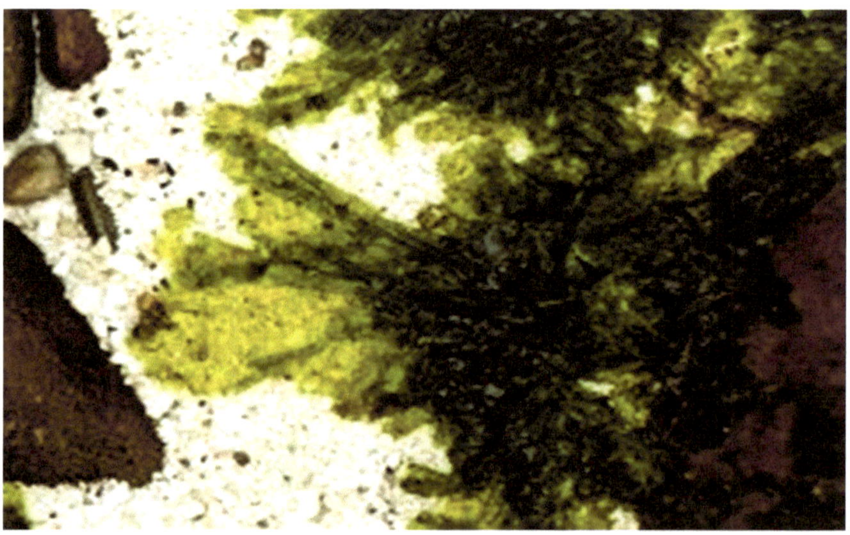

Fig. 2.3 Thallus structure of seaweeds in the intertidal zone

Fig. 2.4 Salt marsh grasses contribute to primary production in the coastal and estuarine system

Fig. 2.5 Mangrove floral community constitutes the primary producer tier in the coastal and estuarine system

compounds from which the cell forms new materials. The synthesis of organic material by photosynthesis is termed primary production. Since phytoplankton are the dominant producers in the ocean, their role in the marine food chain is of paramount importance. Approximately 4,000 species of marine phytoplankton have been described and new species are continually being added to this total (Lalli and Parsons 1997). Phytoplankton exhibit remarkable adaptations to remain in floating condition in the sea water. In fact, all marine phytoplankton tend to stay in the photic zone to utilize the solar radiation for performing the process of photosynthesis. In order to retard the process of sinking, this group of organisms adopts various mechanisms. These include their small size and general morphology, as the ratio of cell surface area to volume determines frictional drag in the water. Colony or chain formation also increases surface area and slows the rate of sinking. Most species carry out ionic regulation, in which the internal concentration of ions is reduced relative to their concentration in sea water. Diatoms also produce and store oil, and this metabolic by-product further reduces cell density. In experimental conditions, living cells tend to sink at rates ranging from 0 to 30 m day^{-1}, but dead cells may sink more than twice as fast. In nature, turbulence of surface waters is also an important factor in maintaining phytoplankton near the surface where they receive abundant sunlight. Excessive phytoplankton population imparts greenish colour to the brackish water bodies that are often witnessed in aquaculture farms (Fig. 2.6).

Phytoplankton present in the marine and estuarine environments use carbon dioxide for photosynthesis and hence play an important role in maintaining the carbon dioxide budget of the atmosphere. The larger the world's phytoplankton population, the more carbon dioxide gets pulled from the atmosphere. This lowers the average temperature of the atmosphere due to lower volumes of this greenhouse gas. Scientists have found that a given population of phytoplankton can double its numbers in the order of once per day. In other words, phytoplankton respond very rapidly to changes in their environment.

Phytoplankton sometimes may cause adverse impact on the marine and estuarine environment. During excessive bloom of phytoplankton, the light energy is intercepted, which could otherwise reach fixed plants such as eelgrass (*Zostera* sp.) and kelp. Furthermore, when the phytoplankton eventually die back and break down, an excessive amount of oxygen is required to fuel this process, and hence, areas may become deprived of oxygen. Excessive nutrients, and/or changes in their relative concentrations, may be one factor in a chain of events leading to changes in the species composition of the phytoplankton communities. Increased occurrence of toxic algal blooms may accelerate toxin production. Toxic phytoplankton, when consumed by shellfish or other species, can affect the marine food chain, including poisoning of seabirds, mammals and even humans.

It has been established that phytoplankton naturally contains DMS (dimethyl sulphide), which is released from dead phytoplankton into the atmosphere. This compound can transform into sulphuric acid, which eventually may contribute to acid rain (http://oceanlink.island.net/ask/pollution.html).

Nearly all marine plants, whether unicellular or multicellular, even those attached to substrata (sessile) or free floating, pass some part of their life cycle in floating condition as phytoplankton. However, those organisms which always remain

Fig. 2.6 Brackishwater with a greenish tinge due to phytoplankton bloom

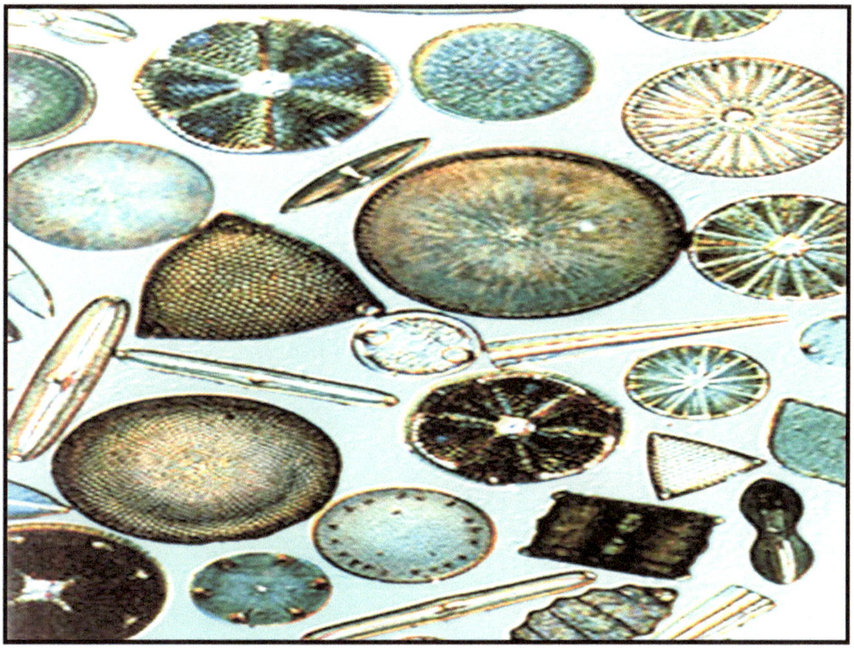

Fig. 2.7 Common brackishwater phytoplankton with unique shape and configuration

planktonic through out the life cycle are as follows: (1) diatoms, (2) dinoflagellates, (3) coccolithophores, (4) selective species of blue-green algae and (5) some species of green algae. The members of phytoplankton community have unique size, shape and morphological features (Fig. 2.7).

Diatoms

These floating plants are all microscopic in size and are characterized by the presence of shell or frustule. The shell or frustule is composed of translucent silica. The cell wall of diatom has two parts resembling a pillbox bottom and lid. The lid is called the **Epitheca** and the bottom is known as **hypotheca**. These shells have great importance from the geological point of view and constitute the diatomaceous crust. The diatoms exhibit remarkable varieties and forms and many species possess beautifully sculptured shells. Examples of diatoms are *Skeletonema costatum* and *Coscinodiscus eccentricus*.

Depending on the nature of valves and pattern of ornamentation in the valve surface, the diatoms are grouped into **centric** and **pennate** diatoms. The major differences between these two groups are given in Table 2.1.

Dinoflagellates

These are important producers of the marine environment and rank second in importance in the economy of the sea. Typically, these are unicellular; some are naked, while others are armoured with plates of cellulose. The dinoflagellates possess two flagella for locomotion. Several of them are luminescent and produce light. *Prorocentrum* sp. and *Ceratium trichoceros* are two common dinoflagellates of marine and estuarine waters.

Coccolithophores

These are among the smallest category of phytoplankton having a size range between 5 and 20 μm. Some coccolithophores have flagella, while others are devoid of them. Their soft bodies are shielded by tiny, calcified circular plates or shields of various designs. These are normally found in the open sea, but their profuse

Table 2.1 Differences between centric and pennate diatoms

Point	Centric diatom	Pennate diatom
Cell shape	Discoid, solenoid or cylindrical	Elongated and fusiform, oval, sigmoid or roughly circular
Ornamentation	Radial in nature, i.e., the arrangement of the markings is radiating from the centre	Bilateral in nature, i.e., the arrangement of the markings is on either side of the apical (main) axis

occurrence has been recorded in coastal waters. They form important diet components of filter-feeding animals. *Coccolithus* sp. and *Isochrysis galbana* are common coccolithophores of oceans, seas and estuaries.

Blue-Green Algae

These include both unicellular and multicellular organisms. The blue colour in them is due to the presence of a pigment known as phycocyanin. Of the various organisms belonging to this category, the most important is *Trichodesmium erythraeum* because in certain seasons of the year, its biomass increases greatly resulting in the formation of clumps.

Green Algae

Microscopic green algae present in the planktonic community largely occur in coastal waters. The green colour in them is due to the presence of chloroplasts. They are widely distributed in the warmer (tropical) seas, and only few species are found in the Arctic and Antarctic oceans. Some common species of microscopic green algae that are planktonic in nature are *Chlorella marina*, *Chlorella salina*, etc.

Classification of Phytoplankton

The phytoplankton community consists of a variety of organisms, namely diatoms, dinoflagellates, blue-green algae, silicoflagellates, and coccolithophores which ranges in terms of size from 0.001 to 0.2 mm.

Phytoplankton may be classified variously from different angles which are discussed here:

(a) On the basis of size, the phytoplankton may be grouped under five categories (Table 2.2).
(b) Phytoplankton can also be classified on the basis of the cell characteristics (Table 2.3).

2.1.2 Seaweeds

Seaweeds or benthic marine algae are the group of plants that live either in marine or brackish water environment. They are macroalgae and contain photosynthetic pigments. Like the terrestrial producer community, the seaweeds can prepare their own food with the help of sunlight and nutrient present in the sea water. Seaweeds are found in the coastal region between high tide to low tide and in the subtidal region up to a depth where 0.01 % photosynthetic light is available. They require hard substratum for their growth, which may be tree trunks (preferably mangroves) that get submerged during the high tide or even brick or boulders that are often laid to enter the island from the adjacent bays or estuaries (Fig. 2.8).

Plant pigments, light, exposure, depth, temperature, tides and the shore characteristic combine to create different environment that determines the distribution and variety among seaweeds. The important criteria used to distinguish the different algal groups based on the recent biochemical,

Table 2.2 Classification of phytoplankton on the basis of size

Plankton category	Maximum dimension (μm)
Ultraplankton	<2
Nanoplankton	2–20
Microplankton	20–200
Macroplankton	200–2,000
Megaplankton	>2,000

Table 2.3 Classification of phytoplankton on the basis of cell characteristics

Class	Common name	Area(s) of predominance	Common genera
Cyanophyceae (cyanobacteria)	Blue-green algae	Tropical	*Oscillatoria, Synechococcus*
Rhodophyceae	Red algae	Cold temperate	*Rhodella*
Cryptophyceae	Cyptomonads	Coastal	*Cryptomonas*
Chrysophyseae	Chrysomonads	Coastal	*Aureococcus*
	Silicoflagellates	Cold waters	*Dictyocha*
Bacillariophyceae (Diatomophyceae)	Diatoms	All waters, especially coastal	*Coscinodiscus* *Chaetoceros* *Rhizosolenia*
Raphidophyceae	Chloromonads	Brackish	*Heterosigma*
Xanthophyceae	Yellow-green algae	Brackish	Very rare
Eustigmatophyceae	Yellow-green algae	Estuarine	Very rare
Prymnesiophyceae	Coccolithophorids	Oceanic	*Emiliania*
	Prymnesiomonads	Coastal	*Isochrysis*
			Prymnesium
Euglenophyceae	Euglenoids	Coastal	*Eutreptiella*
Prasinophyceae	Prasinomonads	All waters	*Tetraselmis*
			Micromonas
Chlorophyceae	Green algae	Coastal	Rare
Pyrrophyceae (Dinophyceae)	Dinoflagellates	All waters, especially warm	*Ceratium*
			Gonyaulax
			Protoperidinium

Fig. 2.8 Seaweed (*Ulva lactuca*) on the brick path towards island village on the intertidal zone

physiological and electron microscopic studies are as follows: (a) photosynthetic pigments, (b) storage food products, (c) cell wall component, (d) fine structure of the cell and (e) flagella. Accordingly, algae are classified into three main groups, i.e. green (Chlorophyta), brown (Phaeophyta) and red (Rhodophyta).

Marine biologists have documented that the distribution of seaweeds is limited by the availability of sunlight at various depths. The evolution of a variety of accessory pigments that absorb those wavelengths of light that are able to reach the deeper zones of the ocean allows the algae to survive in these bottom habitats. Sea water selectively absorbs light with longer wavelengths, such as the reds and yellows, so the light that penetrates to the greatest depths is the short-wavelength blues and greens. Pigments such as fucoxanthin and phycoerythrin, which absorb blue and green light, allow algae to grow at greater depth than those algae that do not possess these or similar accessory pigments.

The distribution of macroalgae is also affected by temperature. The greatest diversity of algal species is in tropical waters. Farther north or south of the equator, the number of species decreases, and the species themselves are different. Many marine macroalgae found at the colder altitudes are perennials, which means they live longer than 2 years. During survival, a few cells, but most often a mass of stem-like structures appear. When the temperature warms up in the spring, this body part initiates new growth. Temperature is not usually a limiting factor for algae that live in tropical and subtropical seas, although for some species, the temperature in intertidal areas may be too warm.

Seaweeds are similar in the form with the higher vascular plants but the structure and function of the parts significantly differ from the higher plants. Seaweeds do not have true roots, stem or leaves, and whole body of the plant is called thallus that consists of the holdfast, stipe and blade. The holdfast resembles the root of the higher plants, but its function is for attachment and not for nutrient absorption. The holdfast may be discoidal, rhizoidal, and bulbous or branched depending on the substratum it attaches. The stipe resembles the stem of the higher plants, but its main function is for support of the blade for photosynthesis and for absorption of nutrients from surrounding sea water. The blade may resemble leaves of the higher plants and have variable forms (smooth, perforated, segmented, dented, etc.). The important functions of the blade are photosynthesis and absorption of nutrient. The blades may be flat, ruffled, feathery or even encrusted with calcium carbonate. The most significant difference of seaweeds from the higher plants is that their sex organs and sporangia are usually one celled or if multicellular, their gametes and spores are not enclosed within a wall formed by a layer of sterile or non-reproductive cells.

Carbohydrates produced by these floral communities provide nutrition to faunal members of the benthic habitats. The large biomass of seaweeds in the intertidal zone (Fig. 2.9) makes it an important primary producer of the marine and estuarine ecosystems.

The macroalgae are noted for their primary production. Efficiencies of solar energy trapped showed a maximum in *Enteromorpha intestinalis* (0.64 %) and *Ulva lactuca* (0.43 %) with an average of 0.35 % by this group. A research conducted on this aspect indicates that in the deltaic complex of Indian Sundarbans, *Enteromorpha intestinalis* and *Ulva lactuca* are the most productive species, followed by *Enteromorpha prolifera* and *Rhizoclonium grande* (Chaudhuri and Choudhury 1994). The gross and net primary productions and energetics of benthic macroalgae in this mangrove-dominated ecosystem are highlighted in Tables 2.4 and 2.5.

2.1.3 Seagrass and Salt Marsh Grass

Seagrass refers to the marine flowering plants (angiosperm) that grow luxuriantly in tidal and subtidal marine environment. They belong to the families Hydrocharitaceae and Potamogetonaceae. There are about 13 genera and 58 species of seagrass available all over the world. Of these, six genera (*Amphibolis*, *Heterozostera*, *Phyllospadix*, *Posidonia*, *Pseudalthenia* and *Zostera*) are mostly restricted to temperate seas and the remaining seven genera (*Cymodocea*, *Enhalus*,

Fig. 2.9 Seaweed biomass contributes to primary production

Table 2.4 Gross primary production (GPP) and energetics of benthic macroalgae

Species	GPP gC/m^2/day	Glucose g/m^2/day	Energy kcal/m^2/day	Efficiencies (%)
Chlorophyceae				
Enteromorpha intestinalis	3.84	9.60	35.91	0.64
E. prolifera	2.10	2.75	10.29	0.19
Ulva lacuta	2.54	6.35	23.75	0.43
Rhizoclonium grande	0.84	2.10	7.86	0.14
Rhodophyceae				
Bostrychea radicans	2.26	3.15	12.78	0.21
Bostrychea sp.	0.61	2.53	5.72	0.10
Catenella nipae	2.14	2.85	10.66	0.19
C. adnata	0.24	0.60	2.25	0.04
C. leprieurii	0.27	0.68	2.55	0.05
Gracilaria verrucosa	0.96	2.40	8.98	0.16
Total	12.80	32.01	119.72	2.14
Mean	2.28	3.20	12.97	0.22

Halodule, *Halophila*, *Syringodium*, *Thalassia* and *Thalassodendron*) are distributed in tropical seas. Seagrass species have world-wide distribution. They are found in tropical (hot), temperate (cool) and the edge of the Arctic (freezing) regions. They thrive luxuriantly in coastal water region (Fig. 2.10). Seagrasses are mostly found in patches, but these patches can expand to form huge seagrass beds, or meadows. The beds may sustain one species of seagrass, or multiple species.

Macroalgae or seaweeds also colonize the coastal zone, and they are often confused with seagrass species. The basic differences between seagrass and seaweeds are highlighted in Fig. 2.11 and Table 2.6.

Seagrasses anchor themselves to the seafloor with their root systems. A very strong root structure allows seagrasses to withstand strong currents and waves especially during storm events (Fig. 2.12). They accomplish their underwater reproduction by producing filamentous pollen grains that can be transported by water currents.

Salt marsh grasses are special type of halophytes that are adapted to continual and periodic flooding. These are found primarily throughout

Table 2.5 Net primary production (NPP) and energetics of benthic macroalgae

Species	NPP ($g/m^2/day$)	Glucose ($g/m^2/day$)	Energy ($kcal/m^2/day$)	Percentage of GPP	Net efficiencies
Chlorophyceae					
Enteromorpha intestinalis	3.26	8.150	30.48	85.00	0.55
E. prolifera	0.98	2.450	9.14	89.09	0.16
Ulva lacuta	2.25	5.630	22.04	88.58	0.38
Rhizoclonium grande	0.32	0.800	2.99	80.01	0.05
Rhodophyceae					
Bostrychea radicans	0.86	2.154	8.04	68.25	0.15
Bostrychea sp.	0.38	0.950	3.56	62.29	0.06
Catenella nipae	0.86	2.150	8.04	75.43	0.15
C. adnata	0.12	0.300	2.12	50.00	0.02
C. leprieurii	0.21	0.530	2.97	77.77	0.03
Gracilaria verrucosa	0.62	2.550	5.80	64.98	0.10
Total	9.86	24.660	92.18	76.99	2.65
Mean	0.97	2.470	9.22	77.18	0.17

Fig. 2.10 Seagrass patch in the coastal zone

the tropical, temperate and sub-Arctic regions. The tide is the dominating characteristic of a salt marsh. The salinity of the aquatic phase defines the plants and animals species that can survive in the marsh area. The vertical range of the tide determines flooding depths and thus the height of the vegetation, and the tidal cycle controls how often and how long vegetation is submerged. Two areas are delineated by the tide: the low marsh and the high marsh. The low marsh generally floods and drains twice daily with the rise and fall of the tide; the high marsh, which is at a slightly higher elevation, floods less frequently.

Salt marshes usually are developed on a sinking coastline, originating as mud flats in the shallow water of sheltered bays, lagoons and estuaries, or behind sandbars. They are formed where salinity is high, ranging from 20 to 30 psu. Proceeding up the estuary, there is a transitional zone where salinity ranges from 20 to less than 5 psu. In the upper estuary, where river input dominates, the water has only a trace of salt. This

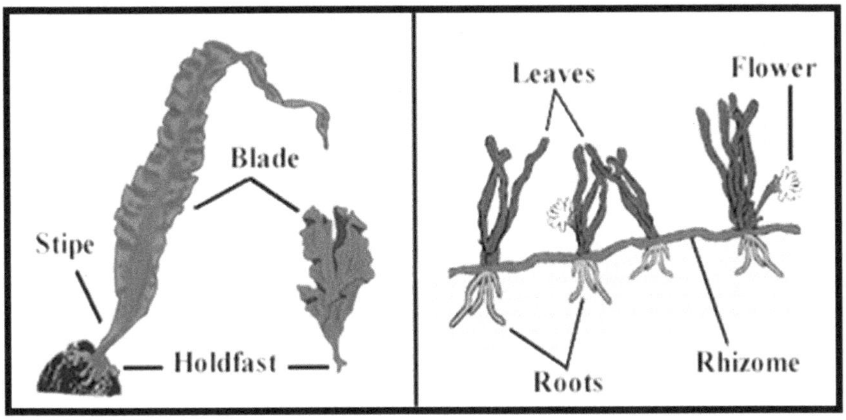

Fig. 2.11 Seaweed (*left*) and seagrass (*right*) structures

Table 2.6 Comparative study of seagrass and seaweeds

Point of difference	Seagrass	Macroalgae (seaweed)
Number of species world-wide	55	5,000–6,000
Plant category	Flowering plants (angiosperm)	Autotrophic organism (thallophytes)
Structure	Possesses roots, leaves and underground stems to anchor to substrate	Does not possess roots, leaves, but simple holdfast anchor them to hard substrates
Absorption of nutrients	Roots and rhizomes extract nutrients from sediment and leaves absorb the nutrients from water	Nutrients enter the tissue by diffusion mechanism
Transportation	The nutrients and dissolved gases are transported through a network of xylem and phloem distributed in the plant	Transportation does not occur
Reproduction	Reproduction occurs through flowers, fruits and seeds	Reproduction occurs through spores

Fig. 2.12 A typical seagrass showing aboveground and belowground structures

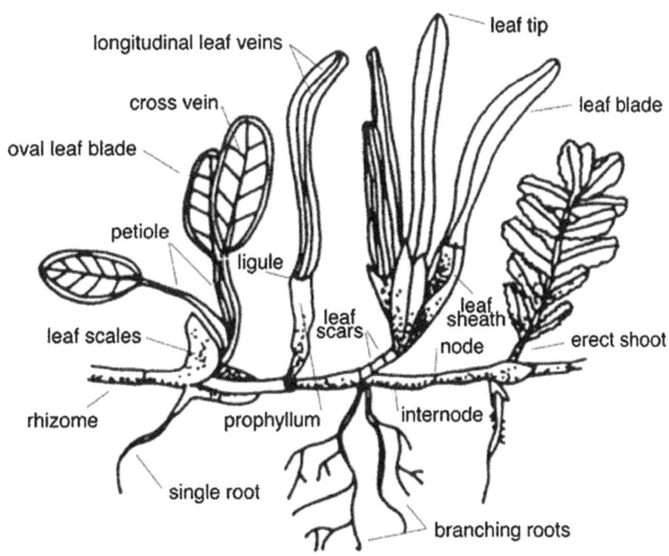

varying salinity produces changes in the marsh—in the kinds of species and also in their number. Typically, the fewest species are found in the hypersaline zone and the greatest numbers of species are found in the freshwater tidal marsh. In Indian Sundarbans region, salt marsh grass *Porteresia coarctata* grows in Nayachar Island (21° 45′ 24″N and 88° 15′ 24″E), where the salinity of water touches almost 2 psu during monsoon. Also the species is abundant in the intertidal mudflats of islands located in the high-saline zone such as Sagar Island (21° 39′ 04″N and 88° 01′ 47″E), Gosaba (22° 15′ 45″N and 88° 39′ 46″E) and Satjelia Island (22° 11′ 52″N and 88° 50′ 43″E), where the salinity varies between 15 and 28 psu (Fig. 2.13). In this mangrove-dominated ecosystem, the species is the neighbour of mangroves (Fig. 2.14) and plays an important role to maintain stability of the islands (Mitra and Banerjee 2005). It is the

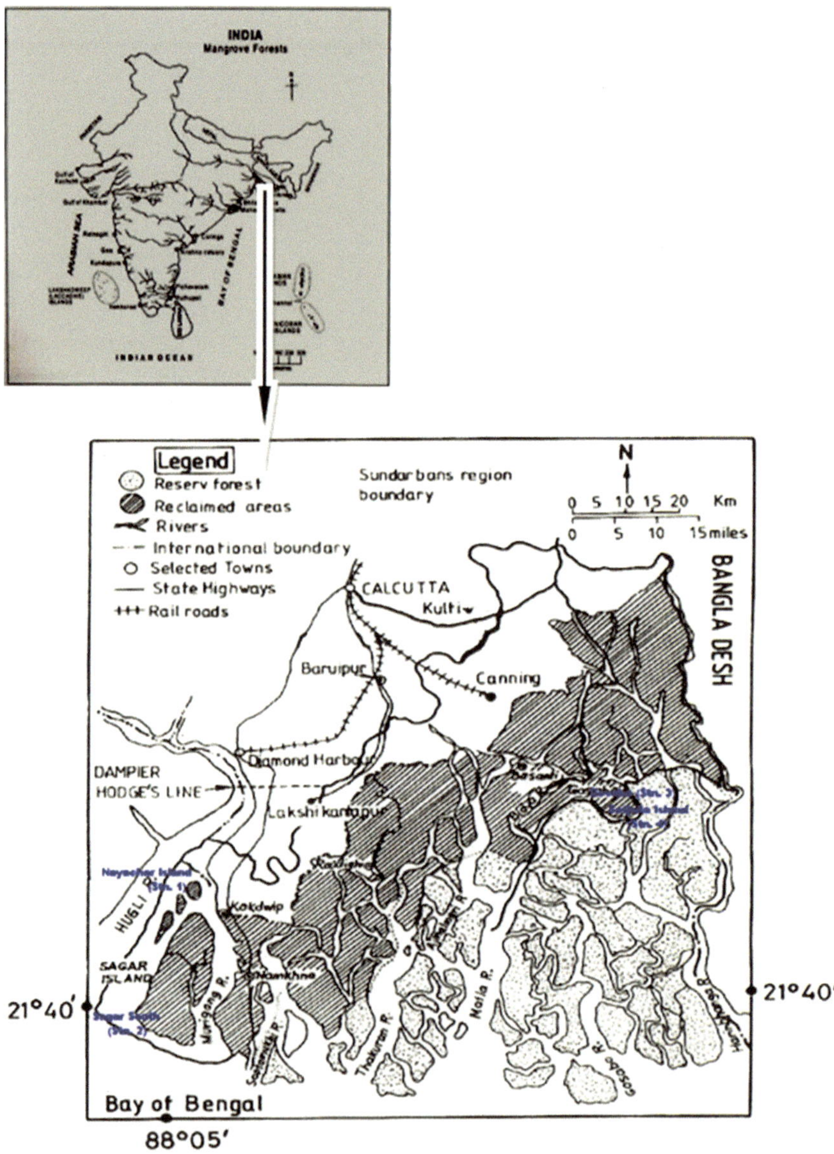

Fig. 2.13 Location of sampling stations (marked in *blue*) in the Indian Sundarbans

Fig. 2.14 View of *Porteresia coarctata* with few mangrove seedlings

pioneer species in the process of ecological succession. Studies carried out by researchers showed that salinity has a regulatory influence on the biomass and carbon content of *P. coarctata*. *P. coarctata* Tateoka (=*O. coarctata*) is a tetraploid ($2n = 4x = 48$) monotypic genus occurring as a halophyte, distributed along the coastal mangrove belt of India and Bangladesh. Taxonomically Tateoka (1965) classified *O. coarctata* as *P. coarctata* based on morphological and embryo anatomy. The plant contains useful traits including salt and submergence tolerance, perenniality, short internodal length conferring mechanical strength, all of which are little studied.

The salt marsh is one of the most productive ecosystems in nature and serves as a sediment sink, a nursery habitat for finfish and shellfish species, a feeding and nesting site for waterfowl and shorebirds, a habitat for numerous unique plants and animals, a nutrient source, a reservoir for storm water, an important erosion control component, and a site for aesthetic pleasures.

Seagrasses are monocotyledon that are not true grasses (true grasses belong to the family of Poaceae) (Fig. 2.14), but more closely related to lily family. Salt marsh grass such as *Porteresia coarctata* is under family Poaceae.

Scientists have identified four habitats where seagrasses and salt marsh grasses can grow along the coast. Different factors regulate their growth and survival in these highly delicate and vulnerable (to physical processes such as waves, tides, storms, high water) coastal habitats.

1. **River estuaries** are dominated by run-off (e.g. in Queensland, run-off is infrequent and on a massive scale) from the land that carries freshwater, sediment and nutrient into the ocean. Run-off can produce 'flood-plumes' that extend many kilometres from the coast and affect large areas. The examples of estuaries in the Ganges, Brahmaputra and Meghna basins are important in this context (Fig. 2.15), where salt marsh grasses are widely available. A comparative account of these three important basins is presented in Table 2.7. The estuaries of these basins are mostly connected with Himalayan glaciers due which many of them receive freshwater along with sediment load. Nutrient overloading is also an important issue in many of these estuaries, particularly the estuaries of the

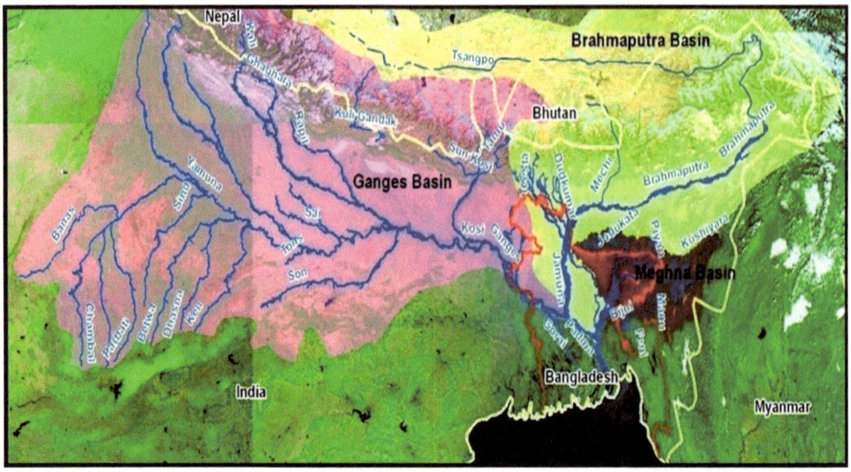

Fig. 2.15 Location of Ganges, Brahmaputra and Meghna basins in India, Bangladesh and Myanmar, respectively

Table 2.7 Comparative account of Ganges, Brahmaputra and Meghna basins

Features	Ganges basin	Brahmaputra/Jamuna basin	Meghna basin	Total
Catchments area (km^2)	1,000,000	573,000	77,000	1,650,000
Mean annual rainfall (mm)	1,200	1,900	4,900	8,000
Mean annual discharge (cumec)	11,000	20,000	4,600	35,600
Maximum discharge (cumec)	78,000	100,000	20,000	198,000
Sediment transport (m ton/year)	550	590	13	1,153

Source Delta plan 2100; http://www.partnersvoorwater.nl/wp-content/uploads/2012/07/MoUBangladeshDeltaPlan2100-22May20120001.pdf

Ganges basin. However, the abundance of salt marsh grass in all these estuaries indicates that unlike seagrasses (which are primarily dependent on transparency), salt marsh grass can thrive well in turbid water.

2. **Coastal** seagrass and salt marsh grass habitats are dominated by physical disturbance caused by periodic events such as cyclones, storms and floods. Major floods in the Hervey Bay area have caused the temporary loss of the extensive seagrass meadows and the subsequent death of hundreds of dugong. Also strong currents that cause erosion damage the seagrass and salt marsh grass patches (Fig. 2.16).
3. **Deep-water** seagrass meadows are limited in their growth due to minimum availability of light. Sunlight is filtered by the water column with less light reaching the bottom in deeper water. As a result, seagrass growth is limited by the clarity of the sea water above it. Seagrasses can grow even at a depth of 60 m as in the Great Barrier Reef lagoon where the transparency is very high.
4. **Reef** seagrasses are limited by substrate type and shelter from waves. Most coral reef waters and sediments also have very limited nutrients available for seagrass growth.

2.1.4 Mangroves

Along the edges of continents, around islands, wherever land meets sea, seashores are found. At nearly 440,000 km, the coastline across the Earth's runs over a staggering distance and can encircle the equator almost 13 times over. These miles and miles of shoreline have fascinating diversity and are not the same everywhere. On the coasts along the tropics on either side of the equator are the regions where mangroves flourish. These areas are warm, humid and swampy inter-tidal zones where tropical rivers flow into the sea.

Fig. 2.16 Salt marsh grass bed exposed to tidal surges

The word 'mangrove' finds its origin from the Portuguese and English Words 'Mangue' and 'Grove', respectively, and indicates mangroves plants as well as a group of trees. According to Mepham and Mepham (1984), the term has been inconsistent and confusing in the past. Mangroves are basically evergreen sclerophyllous, broad-leaved trees with aerial root such as pneumatophore or stilt root and viviparously germinated seedlings (UNESCO 1973). They grow along protected sedimentary shores especially in tidal lagoons, embayment and estuaries (Macnae 1968). They also can grow far inland, but never isolated from the sea. These emergent, evergreen canopies are found along the sedimentary shores of both tropical and subtropical regions in association with intertidal flora and fauna commonly known as mangrove ecosystem, and the community of these mangroves was coined by Macnae (1968) as **Mangal**. Lear and Turner (1977) expressed the word 'mangrove' of coastal ecosystem in a holistic manner, including its common habitat or inhabiting fauna. The term 'mangrove' also denotes both the ecological group of flowering halophytic shrubs and trees of up to 30 m high belonging to several unrelated families, and the complete community or association of plants which fringe sheltered tropical shores. About 60–75 % of tropical coastline is fringed with mangroves (Reimold and Queen 1974). Duke (1992) defined mangroves as '…A tree, shrub, palm and ground fern, generally exceeding one half metre in height and which normally grows above mean sea level in the intertidal zone of marine coastal environments or estuarine margins…'. This definition is acceptable except that ground ferns should be considered as mangrove associates rather than true mangroves. The term 'mangrove' often refers to both the plants and the forest community. To avoid confusion, Macnae (1968) stated that 'mangrove' should refer to halophytic plant species, while the entire forest community including micro- and macro-organisms should be considered as 'mangal'. The mangal is therefore a broad domain encompassing the entire biotic community comprising of individual plant species, associated microbes (such as bacteria and fungi) and animals. The mangal and its associated abiotic factors constitute an ideal mangrove ecosystem, which is a unique ecosystem of the planet Earth.

The number of mangrove species and associated plants vary across different parts of the world. They are most prolific around South-East Asia, and most live within the Tropic of Cancer and Tropic of Capricorn. Yet there are a few species to be found even in the cooler, temperature climates. But whether in the hot, humid marshes or far away from the tropical sun, mangroves across the globe have one similarity. They have a remarkable ability to adapt and survive in their suffocating, salt-laden environment.

Among the less obvious role of mangroves is that they have a far-reaching effect on mitigating the effects of global warming. They not only serve as carbon sinks in tropical and subtropical coastal regions, but they also protect communities in these parts from storms and surges associated with global warming. Coastal studies that have analysed the carbon budget of mangroves have found that mangroves are highly effective carbon sinks. These trees absorb carbon dioxide, thus taking carbon out of circulation and reducing the amount of greenhouse gas in the environment. Studies on their photosynthesis, sap flow and other processes in leaves have shown that mangroves have an exceptional ability to assimilate carbon. Mangrove habitats are responsible for carbon sequestration where much of the carbon ends up in forest sediments and remains therefore thousands of years. Nearly 38 % of the biomass of mangrove forests is below the ground, which represents a potentially vital carbon sink. When mangroves are removed for other purposes the region changes from being a carbon sink

Mangroves are salt-tolerant forest ecosystems found mainly in the tropical and subtropical intertidal regions of the world. They encompass swamps, forest land within, and the surrounding water bodies. It is a matter of great surprise that mangrove floral species can thrive luxuriantly in saline habitat (which is basically physiologically dry in nature) through orientation of their morphological, anatomical and physiological systems. Thus, this vegetation is the most efficiently adapted biotic community in response to climate change induced sea level rise.

Mangroves are circumtropical in distribution and this forest community occupies approximately 75 % of the total tropical coastline. Northern extension of this coastline occurs in Japan (31° 22′N) and Bermuda (32° 20′N), whereas southern extensions are in New Zealand (38° 03′S), Australia (38° 45′S) and on the east coast of South Africa (32° 59′S). Globally, mangroves are distributed in 112 countries and territories. It is interesting to note that mangrove plants are not native to the Hawaiian Islands—six species have been introduced there since the year 1900. The mangrove diversity is more in South-East Asian countries (Fig. 2.17). The region holds nearly 75 % of the world's mangrove species with

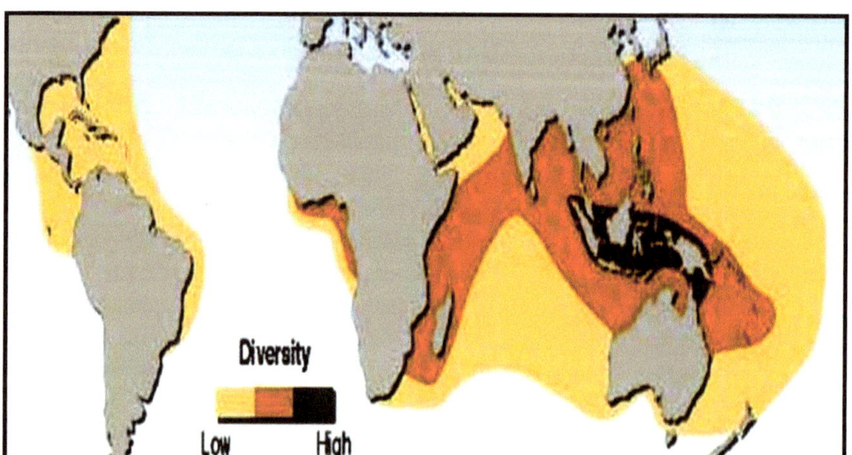

Fig. 2.17 Global distribution of mangrove diversity (after UNEP 2002)

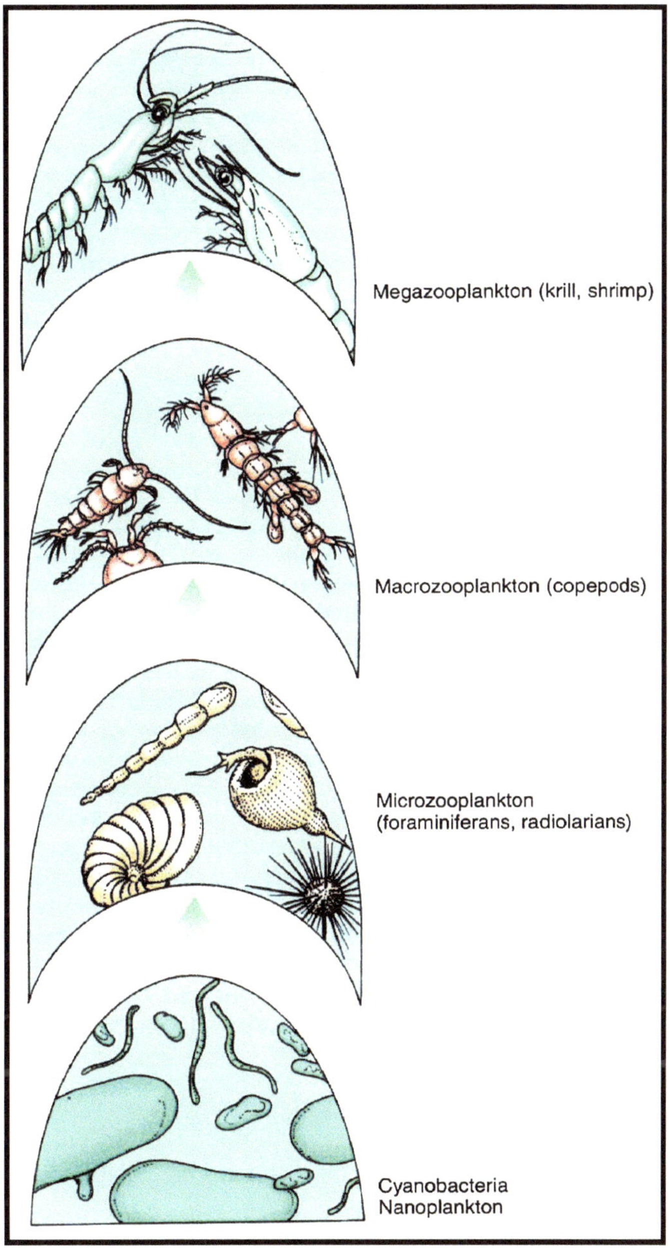

Fig. 2.18 Variation of zooplankton size

the highest species diversity found in Indonesia with 45 species, followed by Malaysia (36 species) and Thailand (35 species). India is no less in terms of the number of mangrove species (34 species of true mangroves) and hence is considered as one of the mega biodiversity countries in the world.

The total global coverage of mangroves has been variously estimated as 14–15 million hectares (Schwamborn and Saint-Paul 1996), 10 million hectares (Bunt 1992) and 24 million hectares (Twilley et al. 1992). Spalding et al. (1997) gave a recent estimation of global mangrove coverage around 18 million hectares with 42.4 % in South

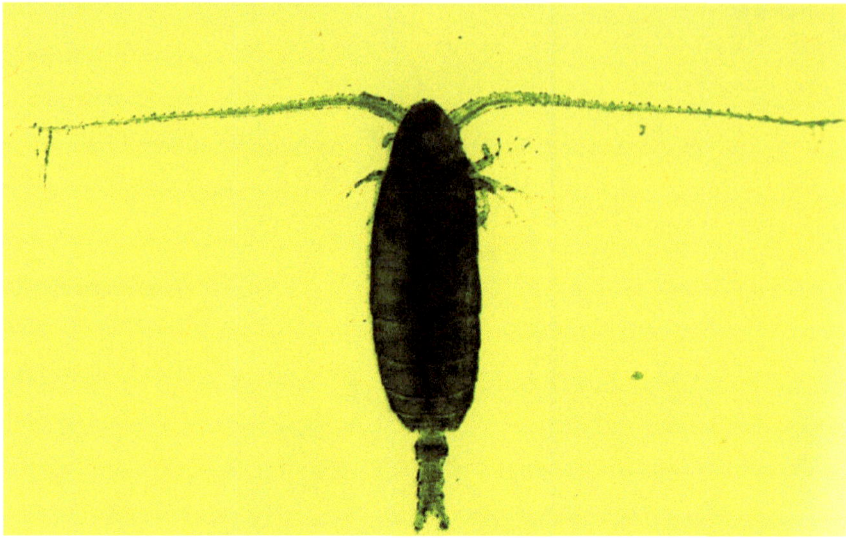

Fig. 2.19 Copepod: a common holoplankton

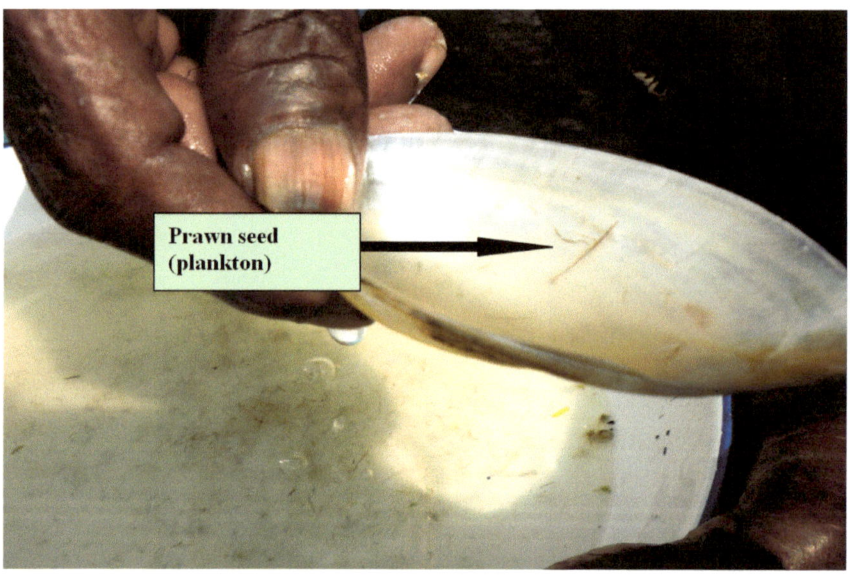

Fig. 2.20 Larval stage (plankton) of *Macrobrachium rosenbergii*

and South-East Asia and additional 23.5 % in Indonesia. The most recent estimates suggest that mangroves presently occupy about 14,653,000 ha of tropical and subtropical coastline (Table 2.8) (Wilkie and Fortuna 2003).

A list of mangrove-dominated countries in and around Indian Ocean is shown in Table 2.9.

In the Indian Ocean region, the mangroves are found in a variety of coastal settings, ranging from arid areas through estuaries, lagoons and deltas to coastal fringes. The functional types of mangroves in the Indian Ocean region are as follows:

1. Overwash mangrove forests—small mangrove islands, frequently overwashed by the tides.
2. Fringing mangrove forests—found along the waterways influenced by daily tides.

Fig. 2.21 Adult stage (prefers substrate) of *Macrobrachium rosenbergii*

Table 2.8 Estimates of mangrove areas

Reference	Reference year[a]	No. of countries included	Estimated total area
FAO/UNEP (1981)	1980	51	15,642,673
Saenger et al. (1983)	1983	65	16,221,000
FAO (1994)	1980–1985	56	16,500,000
Groombridge (1992)	1992	87	19,847,861
ITTO/ISME[b] (1993)	1993	54	12,429,115
Fisher and Spalding (1993)	1993	91	19,881,800
Spalding et al. (1997)	1997	112	18,100,077
Aizpuru et al. (2000)	2000	112[c]	17,075,600
FAO (2003)	2003	112	14,653,000

[a] Except for FAO/UNEP (1981) and Aizpuru et al. (2000), the reference year is the year of the publications in which the estimate is cited, not the weighted average of all the national area estimates

[b] Combined figure from three publications by Clough (1993), Diop (1993) and Lacerda et al. (1993)

[c] New estimates were provided for 21 countries, and for the remaining countries the study relied on Spalding et al. (1997)

3. Basin mangrove forests—stunted mangroves, located in the interior of swamps.
4. Hammock mangrove forests—similar to basin type, but existing in more elevated sites.
5. Scrub mangrove forests—dwarf stands of mangroves, existing on flat coastal fringes.

The sheltered coasts support a luxuriant growth of mangroves and a higher biodiversity and this is because of the favourable conditions such as muddy sediment, frequent water exchange, high rainfall and high humidity, prevailing in the areas. The best examples are mangroves of Sundarbans (India and Bangladesh), Malaysia and Indonesia. In contrast, the arid regions of Arabian Gulf countries, Pakistan and Gujarat (India), where the sediment is sandy, highly saline and poor in nutrients have only dwarf mangrove stands.

Table 2.9 Mangrove-rich countries in the Indian Ocean region

Country	Area (in km^2)
Indonesia	42,500
Myanmar	6,950
Malaysia	6,410
India	4,871
NW Australia	4,513
Bangladesh	4,500
Madagascar	4,200
Mozambique	4,000
Pakistan	2,600
Thailand	1,900

The mangrove forests are highly productive ecosystems with productivity about 20 times more than the average oceanic production (Gouda and Panigrahy 1996). Moreover, it is a '**detritus-based**' ecosystem unlike other coastal ecosystems, which are usually '**plankton-based**'. The detritus supplied by the ecosystem saturates the ambient water with nutrients, which triggers the growth and development of planktonic community in the water bodies on which the fishery resource is also dependent. The greatest concentration of mangrove species is observed usually at the mouth of tidal creeks and rivers where salt and fresh water mix in ideal proportion and floodwaters deposit plenty of material to build up the banks. This unique coastal ecosystem of the world sustains a rich spectrum of floral and faunal community in and around its vicinity. The mangroves enrich the coastal waters with nutrients, yield commercial forest products, protect coastlines and support coastal fisheries (Kathiresan and Bingham 2001). Generally, the mangrove vegetations are well adapted to extreme conditions of salinity, tides, winds and temperature, although they show a preference for fresh water. There are no floral groups in the plant kingdom, which possess such well-organized and highly developed morphological, biological and physiological as well as ecological adaptations to extreme environmental conditions.

Mangrove plants tolerate salinity of the soil and water through three basic processes as listed here:

(i) **Salt excretion**: Mangrove plants take saline water as such through roots. However, in the tissues of some species of mangroves, only water molecules and essential salts are retained. Excess salts are excreted through **salt glands** that are present in the leaves. The salt-excreting species of mangrove community such as *Avicennia alba*, *Aegiceros corniculatum*, *Acanthus ilicifolius*, *Aegialitis rotundifolia* regulate their internal salt levels through foliar glands. In salt-secreting (excreting) mangroves, the NaCl concentration of xylem sap is relatively high, about one-tenth of the concentration of salt in sea water. So, the salt-excreting species allow more salt into the xylem than do the non-excretors, but still exclude about 90 % of the salts (Scholander et al. 1962; Azocar et al. 1992). Salt is only partially excluded at the roots. The absorbed salt is primarily excreted metabolically via specialized salt glands in the leaves. The salt in solution can crystallize by evaporation and either can be blown away or washed off. Since, in salt-excreting mangroves, superfluous salts are excreted by guttation through special salt glands, all these salt-excreting halophytes are often referred to as **crinohalophytes**.

    It is interesting to note that salt excretion is an active process, as evidenced by ATPase activity in the plasmalemma of the excretory cells (Drennan and Pammenter 1982). The process is probably regulated by leaf hypodermal cells, which may store salt as well as water (Balsamo and Thomson 1995).

(ii) **Salt exclusion**: In some of the mangrove plants the roots possess an **ultra-filtration** mechanism called **reverse osmosis** by which water and salts in the sea water are separated in the root zone itself and only water is taken inside and the salts are rejected. Many mangrove species can exclude 90 % of salt in the ambient sea water or estuarine system (http://www.epa.qld.gov.au/nature_conservation/habitats/mangroves_andwetlands/mangroves). *Rhizophora mucronata*, *Ceriops decandra*, *Bruguiera gymnorrhiza*, *Kandelia candal*, etc., are few salt excluders of mangrove community. Scholander (1968) demonstrated experimentally that the salt separation process in mangroves occurs at or near the root surface. This is mediated by physical processes alone, since it is not inhibited by poisons or high temperature, which may cause an inhibitory effect on metabolic process. In the root area, the physical mechanism for salt separation involves ultra-filtration which occurs either at the root surface (Epiblema) or at the root endodermis. However, the latter region might be the most preferable site (Tomlinson 1986) because the ultimate absorbing roots in most of the mangroves lack root hairs (e.g. capillary rootlets of *Rhizophora* sp.). This indicates that the absorbing area of mangroves is reduced in comparison to non-mangrove plants.

(iii) **Salt accumulation**: In this type of mangrove plants, the species possess neither salt glands nor ultra-filtration system, but they have the capacity to accumulate a large amount of salts in their leaves. This imparts succulence to their leaves. *Sonneratia apetala*, *Lumnitzera racemosa*, *Excoecaria agallocha*, *Sesuvium portulacastrum*, *Sueada maritima* and *Sueada nudiflora* are included in this category. Leaf succulence in mangroves has a simple explanation in terms of salt balance. The osmotic potential of the leaf cells of mangroves is high (Scholander et al. 1964) which is essential if mangroves are to draw water from the sea with its high negative water potential. However, Scholander (1968) noted that the salt concentration of mangrove leaves remain constant and independent of age. Measurement of salt content in xylem sap demonstrates incomplete salt exclusion at the roots. But mangroves accumulate salt, and so this accumulation is partly compensated by salt glands, mainly in the less efficient salt excluders. Since salt concentration is constant and independent of leaf age, salt must accumulate by an increase in the volume of the leaf cells inducing succulence. The leaf succulence in mangroves may therefore be accepted as a part of their adaptation in an environment that provides ample water at the expense of some compensation for high aquatic salinity.

Studies on salt tolerance in *Aegiceros corniculatum* and *Sesuvium partulacastrum* generated few interesting findings as follows: (i) NaCl salinity has considerable effect on the degree of succulence. With the increase of NaCl salinity in the ambient media, the mass and volume of the leaves increases due to increment in the water content. (ii) Effect of NaCl and Na_2SO_4 is more pronounced in *Sesuvium* sp. than *Aegiceros* sp. (iii) In *Sesuvium* sp., effect of chloride salinity is more prominent than that of sulphate salinity. (iv) NaCl is most effective salt in promoting succulence (Van Eijk 1939). Succulence is due to expansion of the cell wall leading to increase in the size of cells. (v) Accumulation of NaCl is more in *Sesuvium* than *Aegiceros* due to their difference in the mode of salt regulation. (vi) Chlorophyll content decreases sharply at high concentration of NaCl in both the plant species. (vii) High concentration of Na_2SO_4 stimulates the synthesis of chlorophyll in *Aegiceros corniculatum*, but inhibits the same in *Sesuvium* sp.

The mangroves not only stabilize the shoreline and act as a bulwark against the encroachment by the sea, but they also act as the abode of several species of fin fish, nursery of a wide range of finfish and shellfish juveniles and biopurifying matrix of wastes generated as a result of industrialization and urbanization. In mangrove ecosystem, different kinds of unrelated fauna and flora get themselves adapted to thrive under the influence of tidal inundation and brackish water. This ecosystem is thus a zone of adaptive convergence, which is a critical issue in the sphere evolutionary biology.

2.2 Consumer Community

2.2.1 Zooplankton

Zooplankton are found in almost all the layers of the photic zone of the ocean. They are potentially limited by two factors in the coastal and estuarine zones: firstly by turbidity which can limit phytoplankton production and thus restrict the ration supply for the zooplankton community, and secondly by currents which, particularly in small estuaries are dominated by high river flow that usually carries the zooplankton out to the sea. The zooplankton biomass can increase the fishery productivity because they chiefly consume the primary producers (phytoplankton) and form the major food source for members of higher trophic levels in which several species of osteichthyes and chondrichthyes exist. Zooplankton are classified according to their habitat, depth distribution, size and duration of planktonic life (Tables 2.10, 2.11, 2.12 and 2.13).

2.2.2 Osteichthyes

Bony fishes are found at all depths and in all the oceans, but their distribution is determined directly or indirectly by the abundance and biomass of primary producers. This is the basis of evaluating potential fishing zone (PFZ), which is detected from the satellites. It is very important to note in this context that satellites do not observe fish stocks directly, but measurements, such as sea-surface temperature (SST), sea-surface height (SSH), ocean colour, ocean winds and sea ice, characterize critical habitat that influences marine resources including fish stocks. Most of the spatial features that are important to marine ecosystems, such as ocean fronts, eddies, convergence zones, river plumes and coastal regions, cannot be adequately resolved without satellite data. Chlorophyll, present within the phytoplankton, is the only biological component of the marine ecosystem accessible to remote sensing (via ocean colour) and as such provides a key metric for evaluating the health and productivity of marine ecosystems on a global scale. Long-term ocean colour satellite monitoring provides an important tool for better understanding of the marine processes, ecology, fish

Table 2.10 Classification of zooplankton on the basis of habitat

Type	Description
Oceanic plankton	These are marine zooplankton that inhabit beyond the continental shelf
Neritic plankton	These zooplankton inhabit waters overlying continental shelves. These waters are often very productive as they receive the runoff from the adjacent landmasses that triggers the phytoplankton growth in these regions
Brackish water plankton	These zooplankton inhabit estuarine regions, where there is a continuous mixing of fresh water and sea water. The zooplankton species of this category have wide range of tolerance to different dilution factors. Such zooplankton are very common in the shrimp culture farms and form important diet of the prawns

Table 2.11 Classification of zooplankton on the basis of depth distribution

Type	Description
Neuston	The zooplankton of this category are restricted at the top few millimetres (usually 10 mm) of the surface microlayer
Pleuston	These are widely distributed at the surface of the sea (with parts of the body sometime projecting above the water)
Epipelagic	These are distributed between 0 and 300 m water column, e.g. siphonophores, arrow worms
Mesopelagic	The zooplankton of this category are restricted within the depth 300–1,000 m, e.g. euphausiids, chaetognath
Bathypelagic	These are restricted within the depth 1,000 and 3,000 m, e.g. foraminifera, euphausiids
Abyssopelagic	The waters overlying the vast abyssal plains of the ocean are inhabited by a variety of zooplankton species, which is often referred to as abyssopelagic zooplankton. These zooplankton are thus restricted between 3,000 and 4,000 m

Source Santhanam and Srinivasan (1998)

Table 2.12 Classification of zooplankton on the basis of size (see Fig. 2.18)

Type	Size range
Nannozooplankton	<20 μm
Microzooplankton	20–200 μm
Mesozooplankton	200 μm to 2 mm
Macrozooplankton	2–20 mm
Megazooplankton	>20 mm

Source Santhanam and Srinivasan (1998)

stock and the coastal environmental changes (Tang and Kawamura 2001). Modern oceanographic vessels are therefore linked to satellites via computers allowing scientists to use immediate data to plan their sampling programmes while at sea.

Fishes are mostly concentrated in upwelling areas, shallow coastal areas and estuaries. The surface waters support much greater populations of fish per unit volume of water than the deeper zones, where food resources are very less in terms of quality (diversity) and quantity. The presence of mangroves and other associate floral species also regulate the distribution of fishes in the marine and estuarine compartments.

The marine and estuarine ecosystem is the dwelling place of a wide spectrum of **ornamental fishes**, which have unique economic linkage. Marine aquarium trade is rapidly expanding and there is a growing demand for tropical marine aquarium fishes in the international market. Globally, the aquarium industry is valued at $4 to $15 billion. In USA, 89 million freshwater fishes are being maintained in 12.1 million tanks, while 5.6 million tropical marine fishes in 2.1 million tanks. In India, after the scientific and technological advancements in the sector of ornamental fishery, there has been an increased demand for tropical marine aquarium fishes in recent years. This has opened up new avenues for alternative livelihood and a lucrative money spinning trade for marine ornamental fishes.

Coral reefs (Fig. 2.22) also provide diverse ecological niches for shelter as well as food and thus sustain a wide variety of marine ornamental fish species (Fig. 2.23). About 400 species of ornamental fishes belonging to 175 genera and 50 families are reported from Indian waters.

2.2.3 Chondrichthyes

Elasmobranchs (**chondrichthyes**) constitute a vital segment of marine and estuarine nekton and are of great commercial importance all over the globe, apart from being a major component in marine food web. About 350 species of sharks and 320 species of rays are known to exist. Nearly all are marine, although a few species inhabit estuaries, and a very few are permanent inhabitants of fresh water. It has been observed that sharks usually prefer swimming in open waters, whereas rays tend to be found on or near the bottom.

Sharks are thus dominant species of elasmobranchs and play a vital role in both ecology and economics. The various direct and indirect products obtained from sharks are today used in food, tourism and pharmaceutical industries. Because of such multiple uses, the community is presently under threat due to overexploitation. Deterioration of water quality due to anthropogenic activities has increased the magnitude of threat. Few species of sharks are so sensitive in nature that they cannot withstand the alteration of

Table 2.13 Classification of zooplankton on the basis of duration of planktonic life

Type	Description
Holoplankton	This group includes organisms which are planktonic throughout their life cycle, e.g. tintinnids, cladocerans, copepods, chaetognaths, etc. Figure 2.19 shows a very common holoplankton found in the marine and estuarine waters of tropics
Meroplankton	This group encompasses those organisms which remain planktonic only for a portion of their life cycle, e.g. larvae of benthic invertebrates and fish larvae (ichthyoplankton). In the estuaries of Indian Sundarbans, larvae of prawn (*Macrobrachium rosenbergii*) are found in plenty (Fig. 2.20), which finally become a semi-benthic species in the adult stage (Fig. 2.21)

Source Santhanam and Srinivasan (1998)

Fig. 2.22 A view of coral reef: an amazing housing complex for marine flora and fauna

Fig. 2.23 Coral reefs are favourable habitats for ornamental fishes

water quality caused by rapid industrialization and urbanization.

In the Indian coastal waters, the common shark species are highlighted in Table 2.14.

2.2.4 Reptiles

Reptiles are predominant on both land and in the sea. The morphological physiological and anatomical modifications, which were responsible for the ancestors of reptiles to conquer land, were also useful in allowing their descendants to return to the saline water of the marine and estuarine systems. This was possible after the evolution of an **amniotic egg** some 340 million years ago, which is covered by protective shell and contains liquid-filled sac called **amnion** filled with yolk. The egg also contains a disposal sac (where the wastes accumulate), called the **allantois** and a membrane called **chorion** (which provides surface for gas exchange). This

Table 2.14 Common shark in Indian coastal waters

Sl. No.	Species	Distribution	Diet	Length	Reproduction
1	*Carcharhinus limbatus* (blacktip shark) (IUCN status: lower risk)	Cosmopolitan in distribution near the inshore regions of tropical waters. It is capable of tolerating reduced salinity, but never penetrates into freshwater	Fish such as sardine, mackerel, croaker, sole along with cephalopods and crustaceans	Matured ones are 2.5 m in length. Male matures at 140–150 cm and female matures at 150–160 cm	Produces an average of 6 embryos per litter. Size at births is 55–60 cm
2	*Carcharhinus sorrah* (spot-tail shark)	Commonly found in coral reefs region.	Bony fish such as mackerels, sardines and shellfish such as squids and prawns	Short in length, about 2.5 m. Male matures at 115 cm and female matures at 120 cm	Litter size is 2–6 usually delivered during March to May. The size at birth is around 40 cm
3	*Carcharhinus dussumieri* (Whitecheek shark) (IUCN status: lower risk)	Common species in inshore waters	Small fishes, squids and crustaceans	Matured ones are 1 m. Male matures at 65 cm and female matures at 75 cm	Produces an average of 2 embryos per litter. Size at birth is around 35 cm
4	*Carcharhinus melanopterus* (blacktip reef shark) (IUCN status: lower risk)	Indo-Pacific tropical shark migrates into estuaries and brackish waters for delivering its pups	Fish such as mullets, silver bellies, anchovies, hilsa, skate, prawns and squilla	Matured ones are about 2.5 m	Size at birth ranges between 45 and 50 cm
5	*Carcharhinus macloti* (hardnose shark)	Widely distributed in the inshore region	Small fishes, squids and crustaceans	Matured ones are 1 m in length. Male matures at 60 cm and female matures at 70 cm	Produces an average of 2 embryos per litter. Size at birth is 35 cm approximately
6	*Galeocerdo cuvier* (tiger shark) (IUCN status: lower risk)	Widely distributed tropical shark. It is capable of cruising in mid ocean and shows nocturnal movement into bays and estuaries	Eels, cat fish, parrot fish, flat fish, flying fish, skates, rays, marine turtle, sea snake, sea birds, sea lion, dolphin, etc.	Matured ones are about 7.4 m in length	Development is oviparous and the litter size is between 10 and 82. The size at birth is 50–75 cm
7	*Scoliodon laticaudus* (Indian dog shark) (IUCN status: lower risk)	West and south coasts of India, dominant in east coast and in Indian Sundarbans	Small fishes, crustaceans and squids	Majority of males is 50–55 cm and females are about 65 cm. Males and females mature at 30 cm and 35 cm, respectively	Produces up to 20 embryos per litter. The size at birth is 14.5 cm

(continued)

Table 2.14 (continued)

Sl. No.	Species	Distribution	Diet	Length	Reproduction
8	*Rhizoprionodon acutus* (Brazilian sharpnose shark)	Abundant in the shore waters, particularly in the west coast of India during September to February and east coast during summer months	Small fishes, squids, cuttle fish, crabs and shrimps	Matured ones are about 1 m in length	Produces an average of 2–6 embryos per litter, which are 26–27 cm long
9	*Sphyrna lewini* (hammerhead shark) (IUCN status: vulnerable)	Abundantly found in the Indian seas, notable for its unique migratory behaviour	Sardine, anchovies, mackerel, eel, milkfish, sole fish. It also feeds on sharks and rays	Matured ones are up to 4.2 m in length	Development is viviparous and produces an average of 15–30 embryos per litter. The size at birth is 45–55 cm
10	*Rhincodon typus* (whale shark) (IUCN status: vulnerable)	Appears occasionally at Indian coastal waters	Filter-feeder and is believed to sieve plankton as small as 1 mm diameter through the fine mesh of their gill rakers	Largest length up to 1,200 cm but mean length is around 700 cm	Ovoviviparous, embryo length varies from 48–58 cm
11	*Atelomycterus marmoratus* (coral cat shark)	Appears occasionally at Indian coastal waters	Squids, crabs, etc.		
12	*Glyphis gangeticus* (river shark) (IUCN status: extremely endangered; Wildlife Protection Act: Schedule I)	Hooghly–Ganges river system of West Bengal coast	Fish	Maximum length is around 200 cm. Matured males are about 178 cm	Viviparous. Newly born individuals are 56–61 cm long
13	*Glyphis glyphis* (sharptooth shark or speartooth shark)	Confined to turbid waters of rivers, estuaries and inshore waters of coastal West Bengal	Fish	Minimum length is around 100 cm	Live-bearing probably with yolk-sac placenta
14	*Glyphis siamensis* (Irrawaddy river shark)	Confined to turbid waters of rivers, estuaries and inshore waters of coastal West Bengal	Fish	Minimum length is around 63 cm	Live-bearing probably with yolk-sac placenta

Source Mitra and Banerjee (2005)

development of amniotic egg favoured the juveniles of reptiles to develop within the egg and prevent them from predation. It also allowed the eggs to be laid in dry place, out of reach from the aquatic predators.

The evolution of copulatory organs by reptiles allowed for increased efficiency of internal fertilization prior to the ovum being encased by a shell and being laid by female. There are further many other features of reptiles, which enable them to get adapted to both land and ocean. The circulatory system of reptiles is more advanced than fishes. The circulation through lungs and circulation through rest of the body is completely separated, which allows for efficient method of supplying oxygen to the animal tissues to support their active lifestyles. Their kidneys are very efficient in elimination of wastes and conserve water during dry regions and salty environment of the ocean. Reptiles usually have scales on their bodies with no glands, which allow them to resist losing water in the marine environment.

The ancestors of modern reptiles first began to appear around 100 million years ago. Modern-day reptiles include crocodiles, turtles, lizards and snakes, all of which are present in the marine environment.

Marine Crocodiles

Several species of crocodiles such as American crocodile (*Crocodylus acutus*), Nile crocodile (*Crocodylus niloticus*) and Asian saltwater crocodile (*Crocodylus porosus*) are best suited to the marine environment. The saltwater crocodile inhabits estuaries from India and South-East Asia to northern Australia covering more than several hundred kilometres. They are known for their long-distant travel. It has been documented that some individuals are able to travel more than 1,100 km (683 miles) from Malaysia to Cocos Island in the Indian Ocean.

Saltwater crocodiles are very large in size growing up to a length of 6 m (20 ft). They feed mainly on fishes and are sometimes known to attack and kill sharks close to their own size. They are very aggressive and often attack and kill human beings within their range. It drinks saltwater eliminating the excess salt through salt glands on their tongues. This animal lives in burrows along the shore, where it makes its nest and lay eggs.

Crocodylus porosus is one of the largest saline water reptiles, which is common in the estuaries adjacent to Bay of Bengal. They are also sighted in estuarine creeks (Fig. 2.24), where the water salinity drops down to 10 ‰ during the

Fig. 2.24 Crocodiles on the mudflat of Indian Sundarbans (photograph taken by Mr. Saumya Kanti Ray, Environmentalist under the supervision of Dr. Pradeep Vyas, IFS)

monsoon season (July to October). The species was indiscriminately killed for the purpose of making luxury goods from its skin in late 1960s and early 1970s. The level of poaching became so severe that the population subsequently declined, making the species endangered. The initiation of crocodile breeding programme by the Govt. of India in 1976 at Indian Sundarbans reversed the situation.

Sea Turtles

Sea turtles are cold-blooded and lung-breathing nekton belonging to class reptilian (Fig. 2.25). They appeared on the planet Earth millions of years ago and got adapted to an aquatic life.

Sea turtles are widely used by humans as food, ornaments and leather. They are now endangered and protected. Of the World's 12 living families and approximately 250 species of turtles, only 2 families are marine dwellers. They are Dermochelyidae (common leatherback comprising of **single** species) and Cheloniidae (comprising of **seven** species). These eight species have several common characteristics, including relatively non-retractile extremities, extensively roofed skull (Fig. 2.26), limbs converted to paddle-like flippers with one or two claws and little independent movements of the digits. All are large turtles; adults weighing about 35–500 kg and exhibit various adaptations to marine environment, such as presence of large salt gland to excrete the excess salt ingested with sea water and food.

Fig. 2.25 Views of sea turtle species

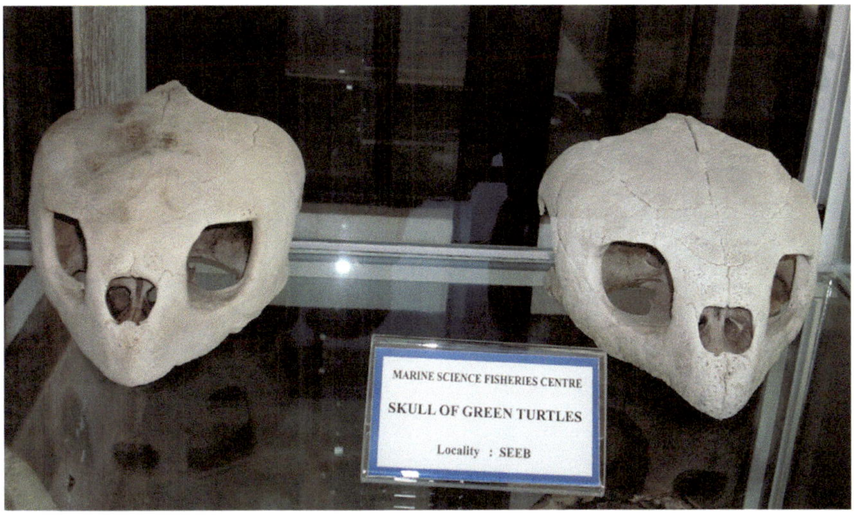

Fig. 2.26 Structural design of green turtle skull [photographs taken by Ms. Ankita Mitra (environmentalist) on 10 November 2013 with the kind permission of Dr. Lubna Hamoud Al-Kharusi, Director General of Fisheries Research under Ministry of Agriculture and Fisheries Wealth, Muscat, Sultanate of Oman]

Sea turtles have specialized feeding habits, for example, green turtle is herbivore and the hawksbill subsists largely on sponges. Reproductive behaviour patterns are similar among the species, but some interesting variations are known. Sea turtles are famous for their remarkable feats of navigation. They return at two, three, four-year intervals to lay eggs on the beaches at which they are hatched. Homing behaviour can be a great advantage to any animal. If the parent survived its earliest childhood at this location, it will probably be a suitable place for hatching the next generation. The navigation of green sea turtles to tiny Ascension Island, an emergent point of the Mid-Atlantic Ridge between Brazil and Africa, has been extensively studied. Researchers have found that the turtles use solar angle (to derive latitude), wave direction, smell and visual cues first to find the island, then to discover the spot on the beach where they hatched perhaps 20 years before. The migration of Olive Ridley (*Lepidochelys olivacea*) to the mangrove-dominated beaches of Odisha (in the east coast of India) for laying eggs is also an interesting event, which has become a subject of interest for many researchers. A comparative account of the five species of marine turtles, widely distributed in and around Indian coastal waters, is highlighted in Table 2.15.

Marine Lizard

The marine iguana (*Amblyrhynchus subcristatus*), the only marine lizard, is native to the Galapagos Islands. These large lizards are descendents of green, vegetarian iguanas that still inhabit the tropical forests of the mainland. It is believed that in the distant past, chunks of riverbank from Central America may have broken loose and carried into the sea along with different other flora and fauna into the Galapagos islands, which might be the probable cause of this animal to have settled here. The iguanas travelled and wandered in the islands in search of food and vegetation that they used to feed in the forest. But though the Galapagos Island was dramatically different from the forests, it had adequate resources for its survival. The conditions of the island thus favoured the species to survive and propagate. The unusual lifestyle of the marine iguana was due to its adaptability in precarious condition, when it had

Table 2.15 Comparative account of the different species of turtles distributed in Indian coastal waters

Species	Description	Distribution	Food habits	Nesting	Status
Leatherback sea turtle (*Dermochelys coriacea*)	It is the largest of all living sea turtles, attaining a length of 150–170 cm in straight carapace (upper shell) length (SCL) and weights around 500 kg (rarely 900 kg)	This species is mostly found along the coast of Tamil Nadu, Andhra Pradesh, Kerala, Goa, Gujarat, Andaman and Nicobar Islands, Lakshadweep	They are water column feeders and feed especially on coelenterates and jelly fishes	Nesting season mainly from May to August which takes place at night	It is included in the Red Data Book of IUCN, Appendix 1 of CITES and in the Schedule 1 of Indian Wildlife Protection Act, 1972, which includes species threatened by extinction
Hawksbill (*Eretmochelys imbricata*)	It is the smallest of five species, with SCL less than 95 cm. Adults are easily recognized by their thick carapace scutes, often with radiating streaks of brown and black on a amber background and a strongly serrated posterior margin of the carapace	It typically forages near rock or reef habitats in clear shallow tropical waters and found mainly along the coast of Tamil Nadu, Andhra Pradesh, Odisha, Kerala, Gujarat, Andaman and Nicobar Islands and Lakshadweep	Hatchlings are carnivorous and the adult omnivorous subsisting mainly on sponges, molluscs, jelly fishes, seagrass and algae	Major nesting period is around monsoon months and it takes place at night hours	It is included in IUCN Red Data Book, Appendix 1 of CITES and also in the Schedule 1 of Indian Wildlife Protection Act, 1972
Olive Ridley (*Lepidochelys olivacea*)	A large head with upper jaw hooked, more than five pairs of pleural and prominent pores on the sides of the plastron. Presence of three distinct keels on carapace of young and average length of adult is 27.5 in.	This species is widely found along the coasts of Tamil Nadu, Andhra Pradesh, Odisha, West Bengal, Kerala, Goa, Gujarat, Maharashtra, Andaman and Nicobar Islands and Lakshadweep	They feed on molluscs, crustaceans, fishes and jelly fishes	The female turtle comes ashore above the high water mark at night to nest and lays about 100–150 eggs and the length of the incubation period varies from 50–55 days	It is included in Red Data Book of IUCN, Appendix 1 of CITES and Schedule 1 of the Indian Wildlife Protection Act, 1972
Green turtle (*Chelonia mydas*)	It is the largest hard-shelled sea turtle. The SCL is 102.5 cm and it weighs around 136 kg	Mostly found along the coasts of Tamil Nadu, Andhra Pradesh, Kerala, Maharashtra, Gujarat, Andaman and Nicobar Islands and Lakshadweep	Hatchling is carnivorous and after about one year of age becomes herbivorous in nature mainly subsisting on marine algae and seagrass	The nesting season is throughout the year, with peak from May–September and it takes place during the night hours	It is included in Red Data Book of IUCN, Appendix 1 of CITES and also in the Schedule 1 of Indian Wildlife Protection Act, 1972
Loggerhead sea turtle (*Caretta caretta*)	The SCL is about 92 cm and the mean body weight is about 113 kg. The brown hatchling weighs about 20 g, 45 mm long	It is rare or absent from mainland shores. More number has been recorded in Tamil Nadu coast especially in Mandapam area	They mainly feed on sponges, molluscs, crustaceans, jelly fishes and fishes	It lays 4–5 clutches a season, at an intervals of 12–15 days and the length of incubation period is around 60 days	It is included in Red Data Book of IUCN, Appendix 1 of CITES and in the Schedule 1 of Indian Wildlife Protection Act, 1972

to feed on the marine algae, which got exposed only during low tide. Although it was a foreign food, but its subsequent generations resisted and persisted this condition. This caused the present-day iguana to be completely different from its relatives on the mainland.

The marine iguana is 3-feet-long lizard and is entirely black; some are mottled red and black showing some hint of green during the breeding season. This dark colouration allows them to absorb more heat energy to raise their body temperature so that they can swim and feed in the cold Pacific waters. They are very good swimmers, using lateral undulations of their body and tail to propel them through waters. They avoid heavy surf and rarely venture more than 10 m from shore. When leaving the water, they tend to ride in with the swell and then swiftly crawl up the rocks. If they do not find their territory immediately, they touch the rocks with their tongues and carry scent to a receptor in the roof of the mouth. When they locate their own scent, they follow it to their territory, where they rest on the rocks, lying almost motionless above the high tide.

Sea Snakes

Snakes are descendants of lizards that have lost their limbs as an adaptation to maintain a burrowing lifestyle, a lifestyle that was later abandoned by many species. Although most species of snakes are terrestrial or arboreal, there are about 50 species that live in the marine environment, each one bearing venomous fangs. Sea snakes are less tied to the land than other marine reptiles, with only about half of the species coming onto land to lay their eggs. Like the ancient ichthyosaurs, females of the remaining species are ovoviviparous, retaining their eggs within their body until they hatch. The young that emerge are able to swim and feed immediately.

Although all sea snakes breathe air, some species can remain submerged for several hours. The animal's single lung reaches almost to its tail, and its trachea has become modified to absorb oxygen, thus acting as an accessory lung. Sea snakes can also exchange gases through their skin when they are under water. These adaptations allow the sea snakes to absorb large amounts of oxygen in a very efficient manner. Sea snakes are able to lower their metabolic rate so that they consume less oxygen when submerged, and some species may even be able to extract oxygen from swallowed water. Specialized valves in the snake's nostrils prevent water from entering when they are submerged.

Most sea snakes remain close to the shore, but the yellow-bellied sea snake (*Palamis platurus*) is pelagic, feeding on surface fish. On several occasions, it has been sighted hundreds of miles from land. This species has migrated east and west from the coast of Asia and can be found off the east coast of Africa and the west coast of tropical America. Most of the other species are found in warm coastal waters from the Persian Gulf to Japan and east to Samoa.

The mangrove swamp of Sundarbans at the apex of western Bay of Bengal is the homeland of several reptiles (Table 2.16) that are either aquatic or semi-aquatic in nature. Many species are well adapted in the upper estuarine stretch, where the salinity reaches zero during monsoon.

Table 2.16 List of reptilian fauna reported from Indian Sundarbans region

Species	Family	Habitat
Lissemys punctata (Bonaterre)	Trionychidae	Aquatic
Varanus bengalensis (Daudin)	Varanidae	Wetland associate
Varanus flavescens (Gray)	Varanidae	Wetland associate
Enhydris enhydris (Schneider) smooth water snake	Colubridae	Aquatic
Xenochropis piscator (Schneider) checkered keelback	Colubridae	Aquatic
Cerberus rhynchops (Schneider) dog-faced water snake	Colubridae	Aquatic
Naja naja kaouthia (Lacepede) monocellate cobra	Elapidae	Wetland associate

2.2.5 Marine Mammals

Marine mammals spend their life mostly or entirely in water and possess fins that adapt them for their aquatic lifestyle. This is the most advanced vertebrate group in the marine and estuarine system. They have an insulating body covering of hair to maintain their constant body temperature (homeothermic) and have mammary glands to suckle their young ones. They depend completely on food of marine origin. Marine mammals are placental mammals, which retain their young ones inside their body until they are ready to be born. This specialized organ, placenta is present only during pregnancy. Although marine mammals produce fewer offsprings than other faunal species of seas and estuaries, most of the young survive to adulthood because they receive a great deal of parental care.

Mammals are adapted both to the terrestrial environment and also in the sea. The presence of hair (fur) of some and the layers of blubber in others effectively reduce the loss of body heat to the surrounding water. Like other warm-blooded animals, mammals expend about 10 times more energy, and hence, they need more food to support their high metabolic rate. Being homoeothermic, marine mammals are active feeders around the clock and are adapted to wide range of habitats. For example, baleen whales feed closer to the base of the food chain, while sea otters and toothed whales are second order or higher consumers.

Most marine mammals are quite intelligent compared to other marine animals. This trait, combined with their generally friendly nature, makes them very popular with the common people. Unfortunately, marine mammals share another common characteristic as the bodies of many contain materials that are commercially valuable to humans. As a result, they have been hunted in large numbers over centuries. During the last 50 years, international conservation measures helped to reduce the decline in population and the number of some species is now increasing gradually.

About 4,300 species of marine mammals have been documented. The three living groups of marine mammals are the porpoises, dolphins and whales of order **Cetacea**; the seals, sea lions, walruses and sea otters of order **Carnivora**; and the manatees and dugongs of order **Sirenia**. Each of these orders arose independently from land ancestors. They exhibit the mammalian traits of being endothermic, breathing air, giving birth to living young that they suckle with milk from mammary glands and having hair at sometime in their lives. All marine mammals share four common features:

- Presence of *streamlined body shape* with limbs that is adapted for swimming in the aquatic phase. A slippery skin or hair covering reduces drag.
- Possess the power of *generating internal body heat* from a high metabolic rate and conserve this heat with layers of insulating fat, as in most cases fur. Their large size gives them a favourable surface-to-volume ratio; with less surface area per unit volume, they lose less heat through the skin.
- Possess a highly efficient *respiratory system*, which is modified to collect and retain large quantities of oxygen. The biochemistry of blood and muscle is optimized for the retention of oxygen during sleep, prolonged dives, etc.
- Reduction of the freshwater requirement through a number of *osmotic adaptations*. Minimal intake of sea water, coupled with their kidney's abilities to excrete concentrated and highly saline urine, permits them to meet their water needs with the metabolic water derived from the oxidation of food.

Order Cetacea

Cetaceans include the whales, which are the most beautiful denizens of the ocean and represent the present-day survivors of the ancient stock that reverted to the seas. The magnificence of whales is epitomized in the fact that whales are the largest animals ever to inhabit the earth. The blue whale for instance by reaching a length of 100 ft and the weight of 200 tonnes (as water can support its great weight) is 4 times heavier than the dinosaur that roamed the earth earlier. There are 79 living species of cetaceans which are thought to have evolved from hoofed land mammals related to today's horses and sheep, whose descendants spend most of their time in productive shallow waters searching for food. Modern whales range in size from 2.8 m (6 ft) to 33 m (110 ft) in length and weigh up to 100,000 kg (110 tonnes). Their paddle-shaped forelimbs are used primarily for steering, and their hind limbs are reduced to vestigial bones that do not protrude from the body. Mainly horizontal tail flukes moving up and down by powerful muscles at the animal's posterior end propel them.

Cetaceans are broadly divided into two sub-orders. Suborder **Odontoceti** includes the **toothed whales**, which are active predators and possess teeth to subdue their prey. Toothed whales have a high brain-weight-to-body-weight ratio and much of their brain tissue is involved in formulating and receiving the sounds on which they depend for feeding and socializing. Smaller whales in this group are the **killer whales** (the largest of all dolphins). The largest toothed whale is the sperm whale, which is 18 m in length and can dive at least 1,140 m (3,740 ft). It feeds mainly on large squids. Toothed whales search for prey using **echolocation**, which is otherwise the biological equivalent of sonar. Some common examples of toothed whales are beluga whale, pilot whale, pygmy sperm whale and false killer whale. The largest of the group is the sperm whale, which feeds primarily on the squids. The best-known toothed whales are the dolphins that perform in shows at numerous aquariums and oceanariums. The harbour porpoise is one of the smallest cetaceans and is a familiar site to both boaters and beach combers. The largest of the dolphins is the killer whale or orca. These are the only cetaceans to feed on warm-blooded animals mainly seals and penguins.

The ears of toothed whales are modified to receive a wide range of water vibrations. This adaptation improves the animal's ability to hear underwater and refines its hearing for the purpose of echolocation. Echolocation is similar to sonar.

Dolphins are strongly social animals. They exhibit problem-solving skills, have long periods to mature with many learning experiences and are capable of intraspecies and interspecies cooperation.

The term dolphin is generally used for small cetaceans having a slender body and beak-like snout, and porpoise for animals having a stocky body and blunt snout. Dolphin and porpoise are common names of two subtly different groups of small odontocete whales. Porpoise, as a term, refers to smaller members of the group, which have spade-shaped teeth, a triangular dorsal fin and a smooth front end tapering to a point. Black finless porpoise (*Neomeris phocaenoides*) is often sighted in the brackish waters of Indian Sundarbans. Dolphins are usually larger and have an extended bottlelike jaw filled with sharp round teeth. The small jumping whales in most oceanarium shows are dolphins and killer whales. However, this terminology has not been strictly followed in naming the species.

Habitat wise, the cetaceans of the Indo-Malayan region fall under three categories: (i) the river dolphin, (ii) coastal and estuarine forms and (iii) marine species which live in deep-waters. The latter may be resident of tropical waters or those, which mainly live in cold waters but migrate seasonally towards the tropics such as *Balaenoptera musculus*, *Balaenoptera physalus*.

The common species of dolphin found in the Gangetic stretch adjacent to coastal Bay of Bengal are Gangetic dolphin (*Platanista gangetica*) and Irrawaddy dolphin (*Orcaella brevirostris*). Gangetic oolphin is restricted in fresh water zone, whereas Irrawaddy dolphin is widely visible in the brackish water in and around the deltaic Sundarbans. The identifying characters, food preference, distribution and conservation status of these species are listed in Table 2.17.

The Ganges river (Hooghly estuary) is noted for pouring and draining huge freshwater and silt in the Bay of Bengal. This stretch is the homeland of *Platanista gangetica* (Gangetic dolphin), which is locally known as the '**susu**'. The species was once common in many of the rivers in India, Nepal and Bangladesh. However, it has disappeared from much of its former habitat over the last 100 years and its future is uncertain. There may be as few as 4,000–5,000 left. Dams and barrages are major threats to the future of this

Table 2.17 Identifying characters of dolphins distributed in Indian coastal waters

Species	Description	Distribution	Food habits	Status
Orcaella brevirostris (Irrawaddy dolphin)	It is of the size of Ganges river dolphin with bulging forehead, downwardly pointed mouth, everted lips, dorsal fin and ellipsoid flippers. Body bluish grey above, paler below, total length 2.0–2.75 m; newborn about 0.85 m, maximum age around 30 years, tail is horizontally flattened and notched in the middle	This species is mostly found along the coastal waters from East coast of India through Myanmar, Thailand and Indonesia to northern Australia; particularly found in larger rivers such as Ganga, Krishna, Irrawaddy and Mekong (Marsh et al. 1989). In India, this species is found in Visakhapatnam (Andhra Pradesh), Chilika Lake (Odisha) and Hooghly river (West Bengal)	Primarily takes fishes, also crustaceans and cephalopods	Protected from killing under Schedule II of Indian Wildlife (Protection) Act and from international trade under Appendix II of CITES
Platanista gangetica (Gangetic dolphin)	It is freshwater species weighing up to 90 kg. Its size varies between 2.0 m and 2.5 m, males are smaller than females. Maximum age is about 28 years. Females give birth within October to March with a peak in December and January after a gestation period of 1 year	It is typically found in freshwater zones along the Gangetic stretch of India. It is also found in the Brahmaputra, Karnaphuli and Meghna river from the foot of the Himalayas downstream to the upper limits of the tidal zone in India	Feeds mainly on fishes and aquatic invertebrates	It is included in IUCN Red Data Book, Appendix 1 of CITES and also in the Schedule 1 of Indian Wildlife Protection Act, 1972.

unique species, as they isolate the dolphins, preventing them from swimming freely up and down the rivers. Polluted waters flowing into the river from a number of tributaries are severely straining the habitat of these shy creatures. Irrigation canals in the Ganga river system is another major threat to the existence of dolphins. The drawing of water from the riverine stretch may be a boon to the agriculturists, but are basically curse for the unfortunate dolphins in the Ganga. As water is whisked away from barrages for irrigation, large tracts of rivers become too shallow for the dolphins to survive. (Care 4 Nature; An Ecological Footprint; WWF-India publication, May 4, 2004; http://www.care4nature.org/wildindia/dolphin/conservation.htm). The varied factors posing threats to Gangetic dolphin may be grouped into following points:

- Accidental killing through entanglement in fishing gear, most often nylon gill nets.
- Direct harvest generally for dolphin oil used as a fish attractant and for medicinal purposes.
- Water development projects (e.g. extension of irrigation facilities, construction of barrages, high dams and embankments).
- Increasing level of chemical pollution, such as from industrial discharges and the use of pesticides.
- Increasing levels of other forms of pollution, such as municipal sewage discharge and noise from vessel traffic, wastes discharged from shrimp farms.
- Overexploitation of prey, mainly due to the widespread use of non-selective fishing gear during the period of migrations and early juvenile growth.

The Gangetic dolphin distributed in all the rivers associated with Bay of Bengal has declined in numbers compared to historical levels. Conservation-related measures have already been initiated to check further decline of their population. In this context, WWF-India established a Dolphin Action Group in May 1997, to strengthen the ongoing efforts for the protection of the Indian River dolphin. Few approaches adopted to conserve the species are listed here:

- Strengthening community awareness and participation.
- Encouraging community ownership and management of fisheries.
- Regular monitoring of the status of dolphins and environmental conditions in the aquatic stretch of river Ganga and adjacent estuaries connecting Bay of Bengal.
- Promoting oil synthesis from the fish scraps as an alternative of using dolphin oil as fish attractant.
- Educating government agencies, local fisherman and other stakeholders associated with River Ganga and adjacent estuaries. The industries situated on the bank of the river and coastal industries also come under this scheme.
- Conducting focused conservation efforts in areas where dolphins are found in greatest abundance.

Irrawaddy River dolphin is widely distributed in the brackish water area and estuarine sector of Bay of Bengal, where the salinity fluctuates between 5 and 30‰. The maximum age is around 30 years. The species is rarely found in isolated condition and are sighted in groups of 3–10 individuals. Males are generally larger than females. Current population of this species in Chilika Lake (Odisha) in India is isolated and in danger of extermination (Dhandapani 1992). About 20 dolphins are estimated to be present in the Lake though only 5 were sighted during 1985 and 1987. The species is common in lower parts of Irrawaddy but exact estimate is not known (Leatherwood et al. 1984). About 100–150 dolphins are reported from Mahakam River, Pola River and Semayang Lake in E. Borneo (Kalimantan).

In India and Myanmar, the oil of the species is considered as a cure for rheumatism, when applied externally. Meat is utilized for human consumption. In the coastal waters of Bay of Bengal, hunting for oil as well as catching of dolphins accidentally in gill nets appears to be the main threat to the survival of this species. Pollution of water through industrial wastes, sewage, etc., silting of the lakes and rivers and introduction of prawn fishery in mass scale in Chilika Lake is further adversely effecting the population of Irrawaddy dolphin. The Indian Wildlife Protection

Act (1972) has stated about the protection of the species, and several agencies (e.g. Chilika Development Authority) have undertaken several initiatives to conserve the species through creation and ecorestoration of habitats.

Suborder **Mysticeti** includes the whalebone or baleen whales, who have no teeth. They are not deep divers and are mainly filter feeders and feed actively on krill, few metres below the surface. They have baleen plates, which help in screening plankton engulfed along with sea water. The baleen consists of keratin fibres that are fused together and function to strain the plankton, mainly krill on which the animal feeds from the water. Baleen whales include right whales, rorquals and grey whales. The largest animal is the blue whale, which requires 3 metric tonnes of krill per day during the feeding season. Some species migrate annually from polar to tropical waters and back. Some common examples of baleen whales are humpback whale, bowhead whale, minke whale, right whale, fin whale, sei whale and gray whale.

The Bay of Bengal is a bay that forms the north-eastern part of the Indian Ocean. It occupies an area of 2,172,000 km^2. It is bordered by India and Sri Lanka to the west, Bangladesh to the north, and Myanmar and the southern part of Thailand to the east. Its southern boundary extends as an imaginary line from Dondra Head at the southern end of Sri Lanka to the northern tip of Sumatra. A number of large rivers—Ganges, Brahmaputra, Irrawaddy, Godavari, Mahanadi, Krishna and Cauvery—flow into the Bay of Bengal. The Ayeyarwady River of Myanmar also flows into the bay. This river-influenced bay sustains diverse population of marine mammals.

Schools of dolphins can be seen, whether they are the bottlenose dolphin (*Tursiops truncatus*), pantropical spotted dolphin (*Stenella attenuata*) or the spinner dolphin (*Stenella longirostris*). Tuna and dolphins are usually residing in the same waters. In shallower and warmer coastal waters, the Irrawaddy dolphins (*Orcaella brevirostris*) can be found. A list of marine mammals sighted in and around the bay is highlighted in Table 2.18.

Order Carnivora

This order includes land predators ranging from dogs and cats to bears and weasels, but the **suborder Pinnipedia** (that includes seals, sea lions and walruses) are exclusively marine. Unlike the cetaceans, the gregarious pinnipeds leave the ocean for varying periods of time to mate and raise their young. **Pinnipeds** have four limbs that are modified into flippers. Although they are more at home in the water, they possess the capability to come on the land, mainly to mate, give birth, and moult. Their bodies are spindle-shaped and many species have several layers of fat under the skin to provide insulation. **Seals** are important members of suborder Pinnipedia. Seals have smooth head with no external ear flaps, which enhances further streamlining of the body. They are graceful swimmers and generally their diet consists of small fishes. The rear appendages of hind limbs are fused partially and always point back from the hind end of the body. The elephant seal, named for its long snout and large size, holds the diving depth record for air-breathing vertebrates. Eared seals, which include fur seals and sea lions, have visible external ears. True seals and walruses lack external ears. Eared seals primarily use their front limbs to propel themselves through the water. True seals use their hind limbs and walruses use a combination of both. The hind limbs of eared seals can be rotated at right angles to the body axis and act as legs on land. **Sea lions** have a coarse coat of nothing but hairs, while fur seals have a thick coat of fur. Fur seals were at times relentlessly hunted for their coats, but now they are protected by international law. The populations of sea lions are mainly concentrated in the Pacific coast. California sea lions (*Zalophus californianus*) are found in the near-shore waters along the Pacific coast from Vancouver Island, British Columbia to Baja Mexico. North of southern California, the hauling out grounds are occupied by males only, who migrate north for the winter. The females and their pups remain in California all year. Males often reach 850 pounds and 7 ft in length and females can acquire a weight of 220 pounds and up to 6 ft in length. They are extremely social animals. The main haul-out areas along the Oregon coast are

Table 2.18 List of whales, dolphins and porpoise found in and around Bay of Bengal

Species	Common name	Distribution	Status
Neophocaena phocaenoides	Black finless Porpoise	Coastal waters and estuaries throughout the Indian subcontinent	Schedule I of Wildlife Protection Act and Appendix I of CITES
Steno bredanensis	Rough-toothed dolphin	Tropical and warm temperate waters; Nicobar Island, Sri Lanka and Java	Schedule II of Wildlife Protection Act and Appendix II of CITES
Stenella attenuata	Pantropical spotted dolphin	Tropical and subtropical waters both coastal and offshore; Indian Ocean from Sri Lanka, Sundarbans, etc.	Schedule II of Wildlife Protection Act and Appendix II of CITES
Delphinus delphis	Common dolphin	Warm temperate to tropical waters, both coastal and offshore; Indian Ocean, Bay of Bengal (Madras coast), Sri Lanka, Maldives Island Andaman Island, etc.	Schedule II of Wildlife Protection Act and Appendix II of CITES
Peponocephala electra	Melon-headed whale	Tropical and subtropical waters, mainly in equatorial waters; Mekran coast of Indian Ocean	Schedule II of Wildlife Protection Act and Appendix II of CITES
Feresa attenuata	Pigmy killer whale	Tropical and subtropical waters of the world; recorded from Trincomalee, Sri Lanka	Schedule II of Wildlife Protection Act and Appendix II of CITES
Pseudorca crassidens	False killer whale	Tropical and warm temperate oceans; In India, Calicut, Trivandrum, Gulf of Cambay, Cape Comorin and Port Blair	Schedule II of Wildlife Protection Act and Appendix II of CITES
Globicephala macrorhyncha	Short-finned pilot whale	Warm temperate and tropical waters; Indian Ocean and Bay of Bengal	Schedule II of Wildlife Protection Act and Appendix II of CITES
Kogia breviceps	Pygmy sperm whale	Throughout tropical and warm temperate oceans; eastern coast of India	Schedule II of Wildlife Protection Act and Appendix II of CITES
Kogia simus	Dwarf sperm whale	All tropical and warm temperate oceans; eastern and western coasts of India	Schedule II of Wildlife Protection Act and Appendix II of CITES
Megaptera novaeangliae	Humpback whale	Polar to tropical waters: Reported from India (Travancore), Sri Lanka (off Colombo and Gulf of Mannar)	Schedule II of Wildlife Protection Act and Appendix I of CITES
Balaenoptera musculus	Blue whale	Prefers cold waters and open seas; northern Indian Ocean, recorded around Sri Lanka, east and west coast of India (Mangalore, Cochin, Tuticorin, Calicut, etc.)	Schedule II of Wildlife Protection Act and Appendix I of CITES
Balaenoptera physalus	Fin whale	Migrating from temperate feeding ground in summer towards the tropics in winter; reported from Indian Ocean, Sri Lanka, eastern (Calcutta) and western coasts of India (Bombay)	Schedule II of Wildlife Protection Act and Appendix I of CITES
Balaenoptera edeni	Bryde's whale	Tropical and warm temperate waters; Indian Ocean and Bay of Bengal	Schedule II of Wildlife Protection Act and Appendix I of CITES

Rogue Reef, Three Arch Rocks, Cascade Head and Shell Island of Simpson reef. Stellar sea lions (*Eumetopias jubatus*) are found in the Pacific Ocean from Japan to southern California. This species has a tendency to remain offshore or haul-out in unpopulated areas. Stellar sea lions are much larger and lighter in colour than California sea lions. The males weigh up to 2,200 pounds and can be 8–11 ft, while the females are comparatively smaller weighing around 600–800 pounds and growing up to 6–8 ft long. Stellar sea lions roar rather than bark and are much larger and lighter in colour than California sea lions. Presence of a thick neck representing a lion's mane is a striking characteristic of the species. The breeding grounds of stellar sea lions occur along the north Pacific Rim from Ario Nuevo Island in Central California to the Kuril Islands, north of Japan, with the greatest concentration in the Gulf of Alaska and the Aleutian Islands.

Walruses are much larger than either seals or sea lions and may reach an average weight of 1,800 kg. Walruses resemble true seals—they too have no external ears. But the true seal and the eared seal have 18 teeth, while walruses have 34. Walruses can walk on their four fins unlike other pinnipeds, which have to drag their body around. Despite having a heavy body, a walrus can move as fast as man. The Pacific walrus inhabits the coast along the Berring Sea and its neighbouring waters, while the ones in the Atlantic are found in the Arctic from the Kara Sea till the Hudson Bay. Walruses have tusks, which can grow as long as 68 cm and is seen among both males and females. The thicker and longer the tusk, the higher is the bull's rank. Walruses also use their tusks to pull themselves out of water and crack breathing holes in the ice during harsh, cold winters. Walruses usually feed on fishes, crustaceans, echinoderms and molluscs. At times, these animals were slaughtered for their ivory tusks but now law protects them.

The suborder **Fissipedia** includes only one ideal marine representative known as the **sea otter**. The otters rarely exceed 120 cm in length, eat voraciously, playful and intelligent. Their daily diet includes molluscs, echinoderms, crustaceans, which constitute 20 % of their body weight. **Sea otters** inhabit the northern Pacific Ocean. Instead, of thick layers of blubber, this animal has a thick coat of fur to keep it warm. Sea otters mainly stay close to shores, favouring areas near coral reefs and kelp beds. They are on the verge of extinction on account of hunting, because of their valuable fur, and are now coming back after international protective measures were enacted. *Lutra lutra* (Eurasian otter) has been documented in Indian Sundarbans.

Order Sirenia

The bulky lethargic small-brained dugongs and manatees collectively called sirenians are the only herbivorous marine mammals. They spend their life grazing on seagrasses, marine algae and estuarine plants in coastal temperate and tropical waters of North America, Asia and Africa. Some species also live in fresh water. The largest sirenians reach 4.5 m (15 ft) in length and weigh 680 kg. Sirenians have been hunted extensively, and only 10,000 individuals are now thought to exist worldwide.

Manatees and dugongs represent sirenians. These animals were widely distributed at times in the past, but they are now restricted to coastal areas and estuaries of the tropics. The primary differences between manatees and dugongs are in anatomy and habitat. Manatees inhabit both the sea and inland rivers and lakes. Dugongs are strictly marine mammals living in coastal areas, where they can feed on shallow water grasses. The head of the dugong is larger and the flippers are shorter than those of the manatee. The dugong's tail is notched, whereas the manatee's tail is rounded.

There are three species of **manatee**. The northern manatee (*Trichechis manatus*) is found from the south-eastern United States to northern South America. The Brazilian manatee (*Trichechis inunguis*) is a freshwater species endemic to the Amazon and Orinoco rivers. The African manatee (*Trichechis senegalensis*) is found in coastal habitats, rivers and lakes of western Africa. All three species of manatee have been extensively hunted and law in Florida now protects the northern species.

Manatees mate and give birth under water and the male remains with his mate even after the breeding season. The females give birth to a single calf after 11-month gestation period. Manatees are strict vegetarians and consume large amounts of shallow water plants. A single manatee may consume at least 27 kg (60 pounds) of aquatic plants per day. When manatees eat, they guide the water plants to their mouths with their flippers. This observation may account for the association with mermaids.

Dugongs are interesting marine mammals. They live up to 70 years and grow up to a length of 3 m. They are commonly called as 'sea cows' since they graze only on seagrass meadows. Modified according to the medium, in which they live, dugongs have got a streamlined body with a massive head and a small mouth with nostrils situated on top of the head for surface breathing. They have got a whale-like fluked tail for swimming. A pair of clawed limbs in the front is used for balancing and turning. Body is grey or bronze grey in colour dorsally and white or flesh coloured ventrally. Adult males possess tusk-like incisors useful for tilling the grass meadows while feeding. Dugongs mostly prefer to dwell at depths below 1 fathom where seagrasses grow in plenty. They feed predominantly on seagrass species of the genera *Halophila* and *Halodule*, which are pioneer species that are low in fibre, low in available nitrogen and very easily digestible. An average adult consumes an estimated biomass of 25 kg a day. When seagrass is scarce, marine algae are also consumed. Decline in dugong numbers due to habitat loss was documented in India during 1964 cyclone. At Palk Bay, large quantities of sand brought by floods were deported over seagrass and algal beds and completely destroyed them. This led to the near total absence of turtles and dugongs in the late 1960s and early 1970s. In this area, *Cymodacea serrulata* and *C. isoetifolida* form the major food items of dugong. The same phenomenon occurred in 1992 at Hewey Bay, Australia, where the seagrass bed was lost due to flood leading to the death of dugongs. Dugongs are related more to elephants than to other marine mammals. Dugongs are found in waters around South-East Asia, Africa and Australia. Presently the growth of human populations has posed an increasing hunting pressure on the dugong in the southern Pacific. Another close relative, Stellar's cow, which was present along the Pacific coast from Mexico to Japan, was commercially hunted to extinction by 1768. Dugongs are also being given the same kind of barbarous treatment even after more than two hundred years of development in human civilization.

It is interesting to note that sea mammals such as sea lion, whale and sea otter thrive and depend on the blue carbon domain such as kelp community. Thus, carbon flows from seaweeds to members of higher trophic level in the marine ecosystem (Fig. 2.27).

2.3 Decomposer Community

The marine and estuarine environments are the nursery and survival place for a large variety of bacterial strains. The abundance of microbes in the neritic zone of this ecosystem may be related to the discharge of huge amount of untreated sewage into this system, run-off from the adjacent forest or mangrove ecosystem containing litter and detritus, discharge from the aquacultural farms, etc. In the Indian subcontinent, a very important picture has been obtained regarding the trend of bacterial load while approaching from the inshore to the offshore region. The bacterial population is very high and variable all along the nearshore coastal waters, which, however, decreases towards the offshore region. This is primarily because of grossly inadequate sanitation in coastal areas and discharge of untreated sewage and other effluents through land drainage (Table 2.19). In Indian Sundarbans region, a case study conducted by the present authors during 1990–2012 in three stations, viz. Harinbari (inshore region; marked as 1 in Fig. 2.28), Chemaguri (midshore region; marked as 2 in Fig. 2.28) and Jambu Island (offshore region; marked as 3 in Fig. 2.28), exhibited unique spatial variation of nitrate and phosphate level. The gradual decrease of nitrate and phosphate level (which are primarily liberated from sewage)

Fig. 2.27 Marine mammals in the kelp bed (diagrammatic)

while approaching from the inshore to the offshore waters (Figs. 2.29, 2.30, 2.31, 2.32, 2.33 and 2.34) speaks in favour of severe anthropogenic pressure along the nearshore waters.

The coliform groups of bacteria are used as an indicator to bacterial contamination of water. These groups of bacteria consisting of 16 families and 256 species originate from the intestinal tracts of human and other warm-blooded animals and therefore may occur in both terrestrial and marine environments (Bhattacharyya 1996).

Mangrove vegetation is very common in the coastal and estuarine regions of the tropical countries. In fact mangroves are woody plants that grow at the interface between land and sea in tropical and subtropical latitudes (Fig. 2.35) and are characterized by the presence of pneumatophores, salt glands, lateral roots and cryptoviviparous germination.

These plants, and the associated microbes, fungi, plants and animals constitute the mangrove forest community or mangal (Kathiresan and Bingham 2001). The mangrove ecosystem provides a unique ecological environment for diverse bacterial communities. The bacteria occupy a number of niches and are fundamental to the functioning of these habitats. They are particularly important in controlling the chemical environment of the mangal. For example, sulphate-reducing bacteria (e.g. *Desulfovibrio*, *Desulfotomaculu*, *Desulfosarcina* and *Desulfococcus*) are the primary decomposers in anoxic mangrove sediments (Chandrika et al. 1990; Loka-Bharathi et al. 1991). These bacteria largely control iron, phosphorous

Table 2.19 Statewise annual mean profile of salinity and bacterial load along inshore, coastal and offshore stations

State	Length of stretch (km)	Zone	Salinity (psu)	Bacteria no./ml
Gujarat	1663	Inshore	13.33	360
		Coastal	20.60	180
		Offshore	26.23	40
Maharashtra and Goa	720	Inshore	31.72	210
		Coastal	32.73	160
		Offshore	33.35	60
Karnataka	290	Inshore	31.77	210
		Coastal	33.62	180
		Offshore	33.98	37
Kerela	560	Inshore	15.07	437
		Coastal	32.52	72
		Offshore	33.72	51
Tamilnadu	884	Inshore	33.10	340
		Coastal	33.42	120
		Offshore	34.60	50
Andhra Pradesh	930	Inshore	17.80	280
		Coastal	26.30	160
		Offshore	28.60	40
Odisha	430	Inshore	11.89	180
		Coastal	22.39	60
		Offshore	23.59	30
West Bengal	220	Inshore	15.03	290
		Coastal	16.19	90
		Offshore	18.99	25

Source Coastal Pollution in the seas around India (1996)

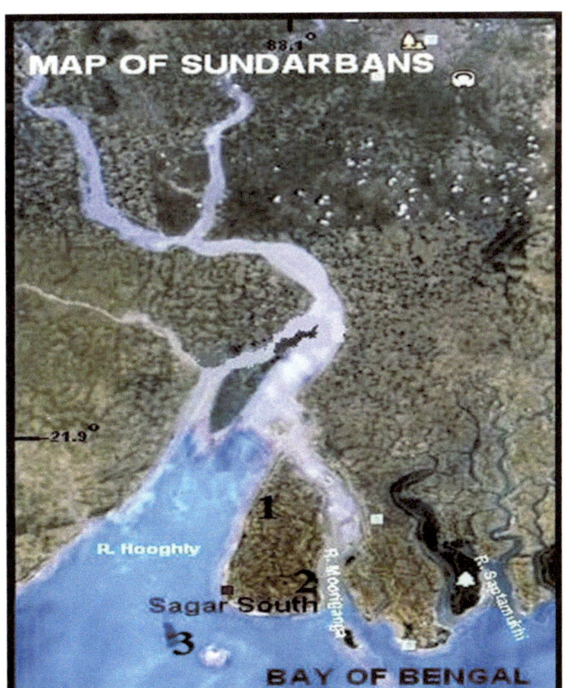

Fig. 2.28 Map showing three selected stations in the inshore, midshore and offshore region of Indian Sundarbans

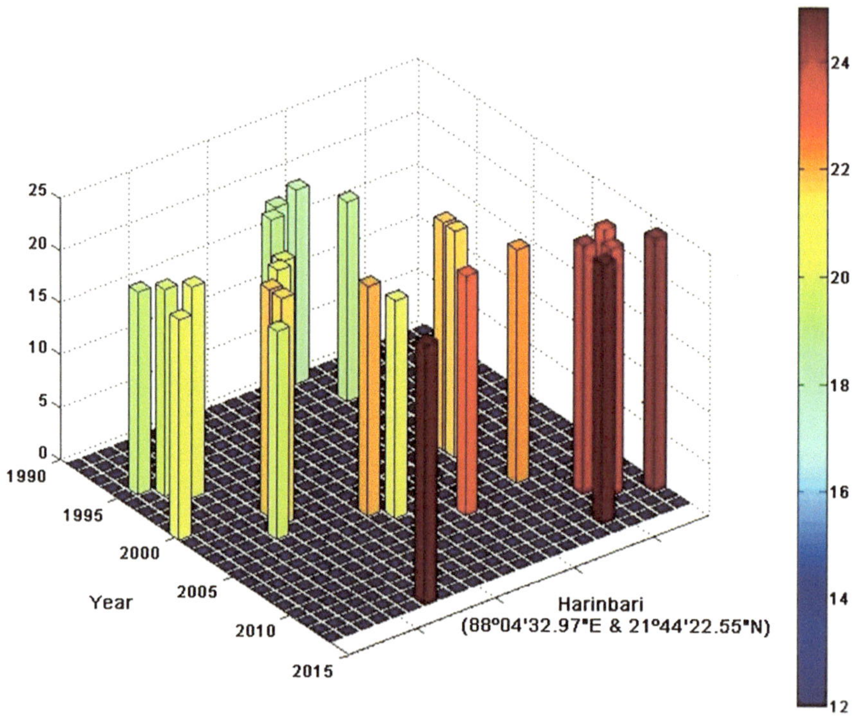

Fig. 2.29 Nitrate level (in µg at l^{-1}) in the inshore region of Indian Sundarbans (graph designed by Dr. Nibedita Mukhopadhyay, Environmentalist)

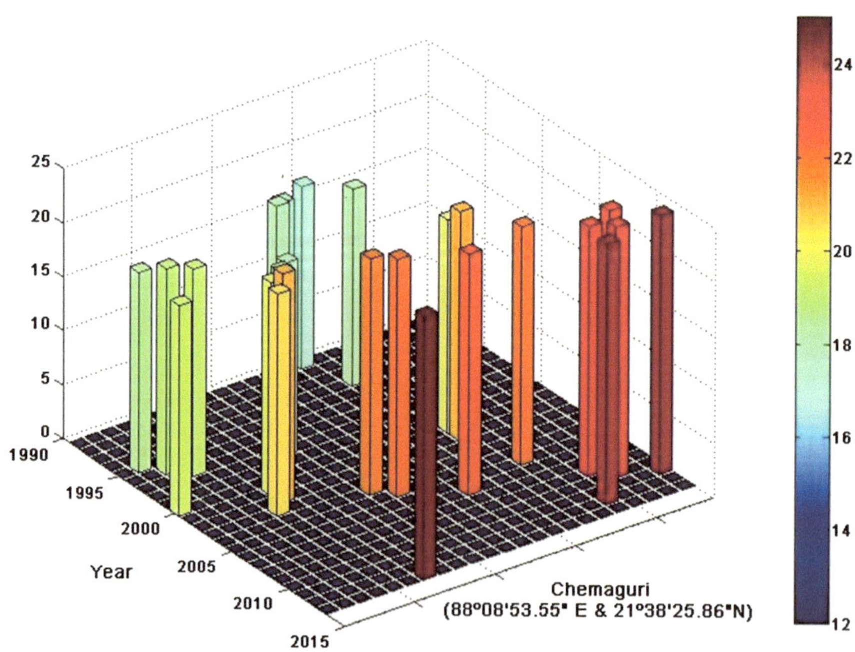

Fig. 2.30 Nitrate level (in µg at l^{-1}) in the midshore region of Indian Sundarbans (graph designed by Dr. Nibedita Mukhopadhyay, Environmentalist)

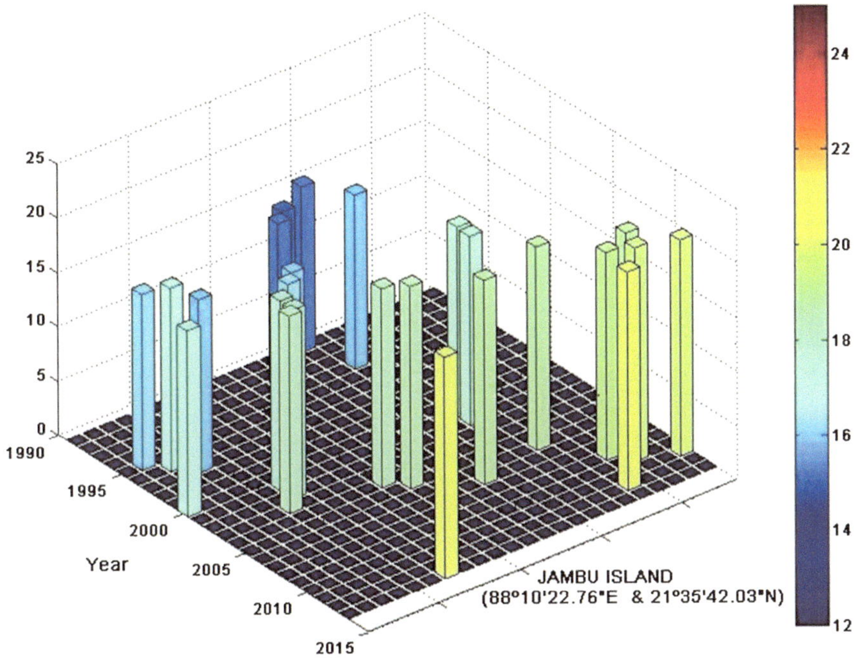

Fig. 2.31 Nitrate level (in μg at l^{-1}) in the offshore region of Indian Sundarbans (graph designed by Dr. Nibedita Mukhopadhyay, Environmentalist)

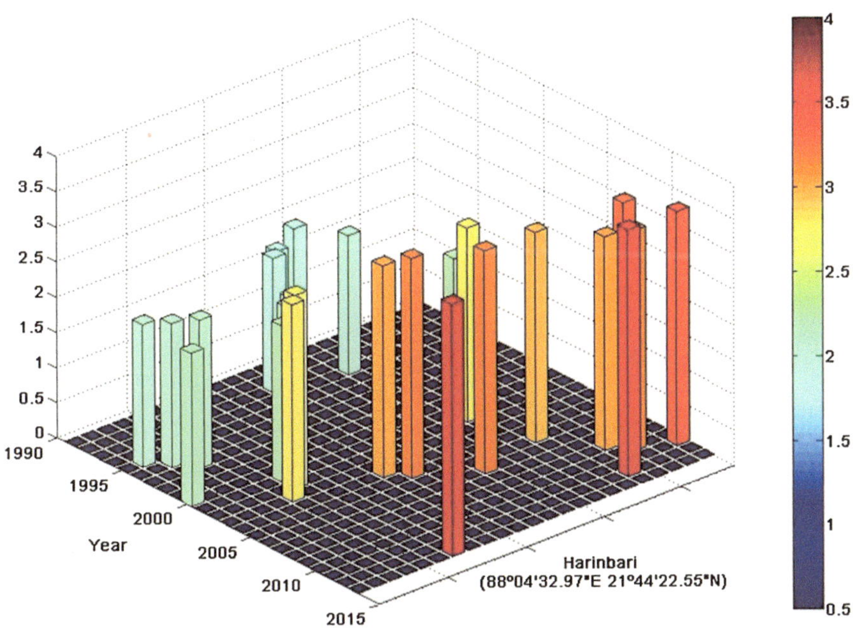

Fig. 2.32 Phosphate level (in μg at l^{-1}) in the inshore region of Indian Sundarbans (graph designed by Dr. Nibedita Mukhopadhyay, Environmentalist)

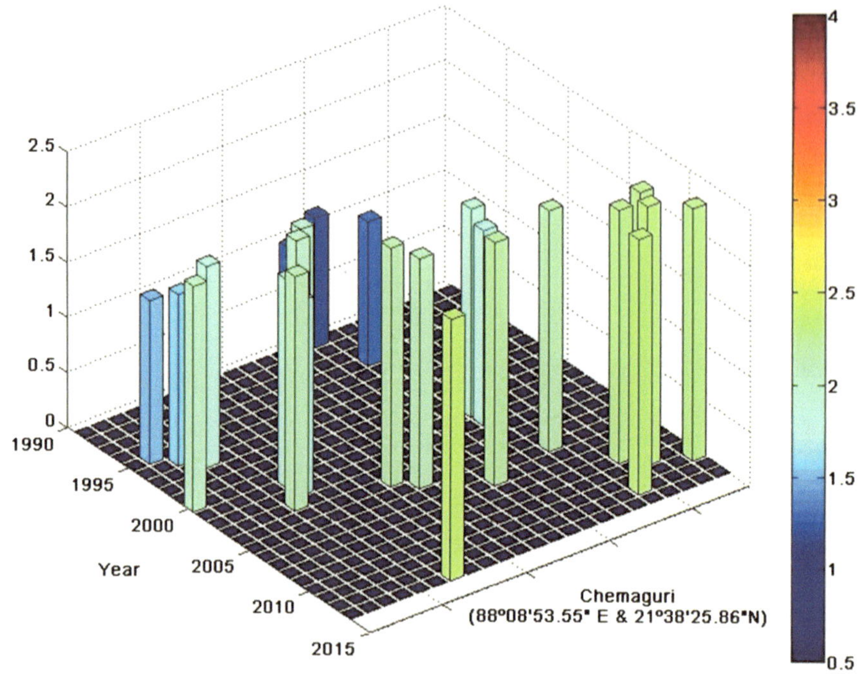

Fig. 2.33 Phosphate level (in μg at l^{-1}) in the midshore region of Indian Sundarbans (graph designed by Dr. Nibedita Mukhopadhyay, Environmentalist)

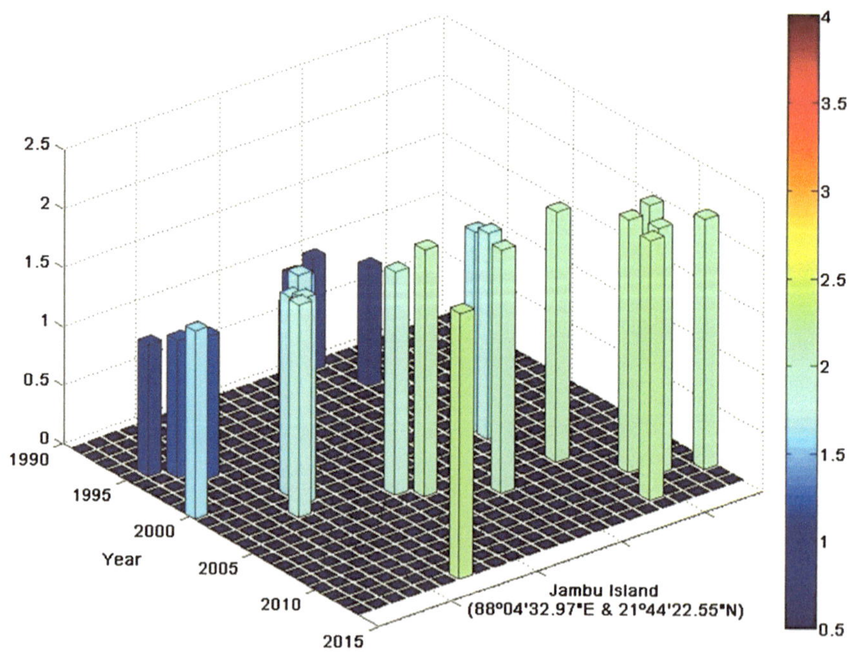

Fig. 2.34 Phosphate level (in μg at l^{-1}) in the offshore region of Indian Sundarbans (graph designed by Dr. Nibedita Mukhopadhyay, Environmentalist)

Fig. 2.35 Mangrove mudflats and vegetation are the survival ground for wide spectrum of microbial strains

and sulphur dynamics and contribute to soil vegetation patterns (Sherman et al. 1998). Methanogenic bacteria are seasonally abundant in sediments where *Avicennia* species dominate (Ramamurthy et al. 1990; Mohanraju and Natrajan 1992). Subsurface bacterial communities (along with epibenthic microalgae) may sequester nutrients and hold them within nutrient-limited mangrove mud (Alongi et al. 1993; Rivera-Monroy and Twilley 1996).

A comparative study done in two selected stations of Indian Sundarbans showed unique variation of microbial population, which may be attributed to the presence of mangrove flora. It was found that the total bacterial density was high in the soil collected from mangrove belt rather than the non-mangrove habitat. Seasonal variation of microbial load was also recorded in both the stations. Abundance of microbes in Zobell's marine agar was recorded almost three times greater in pre-monsoon at mangrove-dominated area in comparison to non-mangrove zone (Tables 2.20 and 2.21). The bacterial density in the other media (nutrient agar) also exhibited the same trend.

The high microbial density in the monsoon season in both the mangrove and non-mangrove zones as revealed in Tables 2.20 and 2.21 may be attributed to the presence of excessive nutrient concentrations in the ambient aquatic phase due to run-off from the adjacent landmasses and the supporting vegetation matter (Banerjee et al. 2001).

The marine and estuarine environments, the mangrove ecosystem and the brackish water aquacultural farms have also become the dwelling spots of variety of pathogenic organisms. These organisms known as coliform bacteria are discharged from the human intestine and their

Table 2.20 Seasonal distribution of different types of microbes at mangrove-dominated zone during March 1998 to February 1999

Medium	Nutrient agar	Zobell's marine agar	Proteolytic medium
Season			
Pre-monsoon (March–June)	13	20	08
Monsoon (July–October)	19	45	10
Post-monsoon (November–February)	08	13	05
Dilution factor 10^3/l ml			

Table 2.21 Seasonal distribution of different types of microbes at non-mangrove-dominated zone during March 1998 to February 1999

Medium	Nutrient agar	Zobell's marine agar	Proteolytic medium
Season			
Pre-monsoon (March–June)	05	07	04
Monsoon (July–October)	11	09	06
Post-monsoon (November–February)	03	06	02
Dilution factor 10^3/l ml			

presence indicates the possibility of the presence of pathogenic organisms. The coliform group comprises all of the facultative and aerobic Gram-negative, non-spore-forming rod-shaped bacteria that ferment lactose with gas formation within 48 h at 37 °C. The *Coliform* bacteria include in genera *Escherichia*, *Citrobacter*, *Enterobacter* and *Klebsiella*. Studies on the faecal coliform load in the Hooghly estuarine complex of West Bengal revealed a range between 55.29 and 183.06 mg/l, which is not a healthy picture of the estuarine ecosystem.

Mangals of the brackish water system are home to a group of fungi called 'manglicolous fungi'. These organisms are vitally important to nutrient cycling in these habitats (Hyde and Lee 1995; Kohlmeyer et al. 1995). Kohlmeyer and Kohlmeyer (1979) were the first to review this group. They recognized 43 species of higher fungi, including 23 Ascomycetes, 17 Deuteromycetes and 3 Basidiomycetes. Hyde (1990a) listed 120 species from 29 mangrove forests around the world. These included 87 Ascomycetes, 31 Deuteromycetes and 2 Basidiomycetes.

Work in individual habitats has revealed surprisingly diverse fungal communities (e.g. Hyde 1990b; Hyde 1996). Chinnaraj (1993a) identified 63 species of higher fungi in mangrove samples from Andaman and Nicobar Islands alone. Similar samples from Lakshadweep Island yielded 32 species (Chinnaraj 1992), and 39 species were found in mangrove samples from the Maldives (Chinnaraj 1993b). Ravikumar and Vittal (1996) found 48 fungal species in decomposing *Rhizophora* debris in Pichavaram, South India. On the Indian Ocean coast of South Africa, Steinke and Jones (1993) identified 93 species of marine fungi, including 55 from mangrove wood (particularly *Avicennia marina*). Table 2.22 lists some of the fungal species identified in these studies.

Table 2.22 Fungal species isolated from mangrove

Species
Aigialus striatispora
Aniptodera longispora
Aniptodera salsuginosa
Calathella mangrovei
Cryptovalsa halosarceicola
Eutypa bathurstensis
Falciformispora lignatilis
Halophytophthora kandilae
Halophytophthora kandilae
Halophytophthora vesicular
Halophytophthora spinosa
Halosarpheia munuta
Hapsidascus hadrus
Hypoxylon oceanicum
Julella avicenniae
Khuskia oryzae
Lophiostoma asiana
M. ramunculicola
Massarina armatispora
Massarina velatospora
Payosphaeria minuta
Pedumispora sp.
Phomopsis mangrovei
Saccardoella sp.
Trematospaeria lineolatispora

2.3.1 Taxonomy of Micro-organisms in the Blue Soup

The marine environment is the dwelling place of a diverse group of micro-organisms such as bacteria, filamentous fungi, yeasts, microalgae and protozoa. All these groups of marine micro-organisms have different divisions and subdivisions as stated here in tabular forms (Tables 2.23, 2.24 and 2.25).

2.3.2 Habitats of Marine Micro-organisms

Marine micro-organisms inhabit various types of habitats. They are distributed at the surface of the sea as **neuston** (also known as **pleuston**) or at the photic zone of the pelagic region as plankton or at the epibiotic habitats (attached communities). They are also present inside the tissues of other marine organisms (endobiotic habitats). Many marine microbes are also distributed on the seabed.

Approximately 5,000 species of nekton swim freely through the neritic and pelagic zones of the ocean. The important members of nekton are fishes, reptiles and mammals. These animals also produce organic debris referred to as **seston** which contribute significantly to nutrition of the micro-organisms.

Epibiotic habitats may be inanimate (such as biofouling communities) or animate surfaces on which attached communities occur. The endobiotic usually denotes the environment within the tissues of other larger organisms. Here, the relationship with the host may be beneficial (mutualism), detrimental to the host (parasitism) or may cause diseases (pathogenesis).

The intertidal zone of the marine environment is rich in seagrass, salt marsh grass, seaweeds, etc., which offer unique site for dwelling of marine microbes. Evidences suggest that grasses pose a narrower range of micro-organisms than seaweeds. Sieburth et al. (1974) observed that cord grass *Spartina alterniflora* is colonized initially by fungi notably *Sphaerulina pedicellata*, whereas eel grass (*Zostrea marina*) mostly possess the pinnate diatom, *Cocconeis scutellum*. On seaweeds diatoms, yeast and bacteria thrive luxuriantly.

Microbes are also available in plenty in the rhizosphere of mangroves. The rhizosphere is the narrow region of soil that is directly influenced by root secretions and associated soil microorganisms. Soil which is not part of the rhizosphere is known as bulk soil. The rhizosphere contains several strains of bacteria that feed on sloughed-off plant cells, termed rhizodeposition, and the proteins and sugars released by roots. Protozoa and nematodes that graze on bacteria are also more abundant in the rhizosphere. Thus, much of the nutrient cycling and disease suppression needed by plants occurs immediately adjacent to roots. We enumerated microorganisms in the rhizosphere and non-rhizosphere soil samples of mangroves and mangrove associate species and observed that the rhizosphere soil sample contained higher microbial populations compared to non-rhizosphere soil (Figs. 2.36, 2.37 and 2.38).

Most of the bacteria in mangrove rhizosphere belong to Gram-negative type. Gram-positive bacteria constituted less than 15 % of the total bacterial population in the rhizosphere zone of Indian Sundarban mangroves. *Arthrobacter* and endospore-producing forms *Bacillus* and *Clostridium* (Family: Bacillaceae) have also been isolated. Majority of bacteria belong to the families Pseudomonadaceae and Vibrionaceae. Most rhizosphere zone occupied by bacteria is aerobic or facultatively anaerobic. Nitrogen-fixing bacteria such as members of the genera *Azospirillum*, *Azotobacter*, *Rhizobium*, *Clostridium* and *Klebsiella* were isolated from the sediments, rhizosphere and root surfaces of various mangrove species, where the soil pH is slightly acidic in nature.

The deep-sea environment is also an important site of marine micro-organisms. Turner (1979) observed bacteria on the surface of faecal pellets and concluded that many of the deep-sea micro-organisms have their origin at the surface layer of the ocean. This view has been supported to some extent by the results of experiments which demonstrated an enhanced rate of

Table 2.23 Bacterial genera indigenous to the marine environment

Category	Family	Genus
Phototrophs	Chromatiaceae	*Chromatium, Thiospirillum, Thiocystis, Thiocapsa*
	Thiocapsaceae	*Chlorobium, Prosthecochloris, chloroherpeton*
	Related genus	*Ectothiorhodopsira*
	Ectothiorhodopiraceae	*Rhodocyclus,*
	Rhodospirillaceae	*Rhodomicrobium,*
		Rhodopseudomonas
		Rhodospirillum
		Erythrobacter
	Related genus (Chroococcacean)	*Synechocystis*
		Synechococcus
	(Nostocacean)	*Nostoc*
	(Oscillatorian)	*Lyngbya, Oscillatoria*
		Plectonema, Spirulina
		Trichodesmium
	(Pleurocapsean)	*Dermocarpa*
	(Rivularian)	*Calotrhix, Dichothrix*
		Richelia, Haliarachne
		Dactyliococcopsis, katagnymene, Nodularia, Pegalothrix
Other phototrophic prokaryotes	Prochloraceae	*Prochloron*
Gliding bacteria	Cytophagaceae	*Cytophaga, Flexibacter*
		Flexithrix, Herpetisiphon, Saprospira
		Sprocytophaga, Microscilla
	Related organisms	*Beggiatoa, Thioploca*
	Beggiatoaceae	*Leucothrix, Thiothrix*
	Leucotrichaceae	
Budding and appendaged bacteria	Caulobacteraceae	*Caulobacter*
	Hyphomicrobiaceae	*Hyphomicrobium*
		Hyphomonas
		Pedomicrobium
	Plantomycetaceae	*Planctomycetes, Pirella,*
		Prosthecomicrobium
Aerobic/microaerophilic non-motile/ motile, helical/vibroid gram-negative bacteria	Spirillaceae	*Bdellovibrio*
		Oceanospirillum
	Spiromaceae	*Flectobacillus*

(continued)

Table 2.23 (continued)

Category	Family	Genus
Gram-negative aerobic rods and cocci	Halobacteriaceae	*Halococcus*, *Halobacterium*,
		Methylophaga
	Methylococcaceae	*Actinetobacter*
	Neisseriaceae	*Pseudomonas*
	Pseudomonadaceae	*Alcaligenes*, *Alteromonas*
	Alcaligenaceae	*Chromobacterium*, *Deleya*
		Flavobacterium
		Janthinobacterium
		Marinomonas, *Paracoccus*
		Shewanella, *Halomonas*
Facultatively aerobic rods	Enterobacteriaceae	*Serratia*
	Vibrionaceae	*Photobacterium*, *Vibrio*
		Listonella
Gram-negative anaerobic rods and cocci	Haloanaerobiceae	*Halobacteroides*
	Desulfurococcaceae	*Desulfobacter*
		Desulfobulbus
		Desulfosarcina
		Desulfuromonas
		Desulfovibrio
Gram-negative chemolithotrophs (ammonia-or-nitrite-oxidizing bacteria)	Nitrobacteraceae	*Nitobacter*, *Nitrococcus*, *Nitrosococcus*, *Nitrospina*, *Nitromonas*, *Nitrosospira*, *Nitrospira*
(Sulphur bacteria)	–	*Macromonas*, *Thiobacillus*, *Thiomicrospora*, *Thiovulum*, *Thiobacterium*, *Achromatium*
	Achromatiaceae	
Methane bacteria	Methanobacteriaceae	*Methanobacterium*
		Methanospirillum
	Methanococcaceae	*Methanococcus*
	Methanomicrobiaceae	*Methanococcoides*
		Methanogenium
		Methanomicrobium
	Methanoplanaceae	*Methanoplanus*
	Methanosarcinaceae	*Methanosarcina*
	Micrococcaceae	*Micrococcus*
		Staphylococcus
	Planococcaceae	*Marinococcus*
Endospore-forming rods and cocci	Bacillaceae	*Bacillus*
	Clostridiaceae	*Clostridium*
Actinomycetes and related bacteria	Actinomycetaceae	*Actinomycetes*
	Micromonosporoceae	*Micromonospora*
	Nocardiaceae	*Nocardia*, *Rghodococcus*
	Mycobacteriaceae	*Mycobacterium*
	Streptomycetaceae	*Streptomyces*
		Coryneforms (*Arthobacter*, *Brevibacterium*, *Corneybacterium*, *Curtobacterium*)
Spirochaetes	Spirochataceae	*Crispispira*, *Sprirochaeta*

Table 2.24 Filamentous fungi recovered from the marine environment

Subdivision	Class	Family	Genus
Ascomycotina	Discomycetes	Orbiliaceae	*Orbilia*
	Plectomycetes	Eurotiaceae	*Amylocarpus*, *Eiona*
	Pyrenomycetes		*Laboulbenia*
	Incertae sedis	Laboulbeniaceae	*Buergenerula*
	Loculoascomycetes	Physosporellaceae	*Spathulospora*
		Spathulosporaceae	*Gnomonia*
		Diaporthaceae	*Aniptodera*, *Bathyacus*
		Halosphaeriaceae	*Carbosphaerella*,
			Ceriosporopsis,
			Chadefaudia, *Corollospora*, *Haligena*, *Halosarpheia*, *Halosphaeria*, *Lignincola*, *Lindra*, *Lulworthia*, *Nais*, *Nautosphaeria*, *Trailia*, *Halonectria*, *Nectriella*
		Hypocreaceae	*Haloguignardia*, *Phycomelaina*, *Biconiosporella*, *Zopfiella*
		Polystigmataceae	*Abyssomyces*, *Chaetosphaeria*
		Sodariaceae	*Pontogeneia*
			Pharcidia, *Turgidosculum*
		Sphariaceae	*Oceanitis*, *Ophiobolus*, *Savoryella*, *Torpedospora*, *Herpotrichiella*
		Verrucariaceae	*Leiophloea*
			Didymella, *Mycosphaerella*
	Incertae sedis		*Bankegyia*, *Kymadiscus*, *Didymosphaeria*, *Halotthia*, *Helicascus*
		Herpotrichiellaceae	*Keissleriella*, *Leptosphaeria*
	Loculoascomycetes	Mycoporaceae	*Manglicola*, *Massarina*
		Mycosphaerellaceae	*Microthelia*, *Paraliomyces*, *Phaeosphaeria*, *Pleospora*
		Patellariaceae	*Pontoporeia*, *Thalassoascus*
		Pleosporaceae	*Trematosphaeria*, *Crinigera*, *Orcadia*, *Sphaerulina*
	Incertae sedis		
Basidiomycotina	Gasteromtcetes	Melanogastraceae	*Nia*
	Hymenomycetes	Corticiaceae	*Digitatispora*
		Incertae sedis	*Halosyphina*
	Teliomycetes	Tilletiaceae	*Melanotaenium*
Deuteromycotina	Hyphomycetes	Agronomytcetaceae	*Papulospora*
		Moniliaceae	*Blodgettia*, *Botryophilalophora*, *Clavatospora*, *Varicosporina*
		Dermatiaceae	*Asteromyces*, *Cirrenelia*, *Cladosporium*, *Clavariopsis*, *Cremasteria*, *Dendryphiella*, *Dictoyosporium*, *Derchslera*, *Humicola*, *Monodictys*, *Orbimyces*, *Periconia*, *Sporidesmium*, *Stemphylium*, *Trichocladium*, *Zalerion*, *Allescheriella*, *Tubercularia*, *Dinemasporium*, *Sphaceloma*, *Ascochyta*, *Ascochytula*
		Tuberculariaceae	*Camarosporium*, *Coniothyrium*, *Cytospora*
		Excipulaceae	*Diplodia*, *Macrophoma*
		Melanconiaceae	*Phialophorophoma*, *Phoma*
		Sphaerioidaceae	*Rhabdosphora*, *Robillardo*, *Septoria*, *Stagonospora*
	Coelomycetes		

Based on Kohlmeyer and Kohlmeyer (1979), Kohlmeyer (1986)

Table 2.25 Marine yeasts

Subdivision	Family	Genus
Ascomycotina	Metschnikowiaceae	*Metschnikowia*[a]
	Saccharomycetaceae	*Debaryomyces, Hanseniaspora, Hansenula, Klayveromyces*[a], *Pichia*[a], *Saccharomyces*
Basidiomycotina	Sporobolomycetaceae	*Leucosporidium*[a], *Rhodosporidium*[a], *Sporobolomyces*
Deutermycotina	Torulopsidaceae (yeast-like cells)	*Candida*[a], *Cryptococcus*[a], *Kloleckera, Rhodotorula*[a], *Sterigmatomyces*[a], *Torulopsis*[a], *Trichosporon, Aureobasidium/ Pullularia*

[a] Contains obligate marine species. After Kohlmeyer and Kohlmeyer (1979)

metabolic activity of marine micro-organisms with a reduction of pressure (Jannach and Wirsen 1982). This helps to draw an inference that the activity of marine micro-organisms is more in the shallow water and decreases with the increase of depth and pressure.

Micro-organisms are also distributed in the deep-sea sediments. Deming and Colwell (1985) used epifluroscence microscopy to determine the vertical distribution of bacteria in deep-sea sediments. Thus, using core samples collected at depths exceeding 4,000 m, it was recorded that bacterial populations at the surface layer of sediment amounted to 4.65×10^8 bacteria/gm dry weight. However, there was a doubling in numbers to 8.29×10^8 bacteria/gm dry weight at a sediment depth of 3 cm followed by progressively decline to 1.7×10^7 bacteria gm dry weight in a core sample at 15 cm from the surface of the sediment. Parallel results were obtained in a second core collected from a similar depth. Higher counts of approximately 3.07×10^{10} bacteria/gm dry weight were recorded from faecal pellets. These counts were 9-folds to 72-folds higher than in the underlying surface sediment (Deming 1985).

An outline search performed by us on the habitat of marine microbes (http://www.geocites.com/RainForest/Vines/4301/microbe/html) revealed the estuarine environment highly favourable for the survival of diverse strains of microbes. This is because the constantly changing environmental parameters can create a wide diversity of ecological niches in this brackish water ecosystem (Atlas 1998). Estuaries have high nutrients and high photon energy and are the most productive ecosystem for photosynthetic aerobes. The range of saline concentrations creates three types of niches: fresh water, brackish water and saline water. Each niche is occupied by organisms that are adapted for those conditions. This form of ecological portioning reduces exploitative competition and enhances growth of different types of microbial communities (Campbell 1993). Similarly, the continental shelf and the coral reefs are areas of high productivity due to high nutrients and photon energy, but without the extreme salinity gradient (Atlas 1998).

Conversely, the ocean zones are not as highly productive except for the pleuston layer where there is adequate light for microbes that are dominant primary producers. The pelagic offshore zone does not have enough nutrients at the surface to support significant microbial growth. The primary producers lyse and sink to the deep benthic zone. The benthic zone, rich in nutrients, does not have enough light energy to support primary productivity. Other forms of energy deep in the hadal trenches, combined with fresh outpourings of chemical nutrients from the molten core, provide the congenial environmental conditions for islands of deep undersea communities of rare bacteria and peculiar species (National Geographic 1979). The deep-sea vents occur in the ocean floor where the ocean crustal plates spread apart and cause plumes of hot lava to erupt into the ocean. The high concentrations of electron-rich elemental compounds are very congenial for the growth and survival of eubacteria such as the chemoautotrophs (Campbell 1993). Each type of bacteria has special adaptations that enable the organism to obtain

Fig. 2.36 Rhizosphere of *Avicennia marina* with soil clinging on the roots

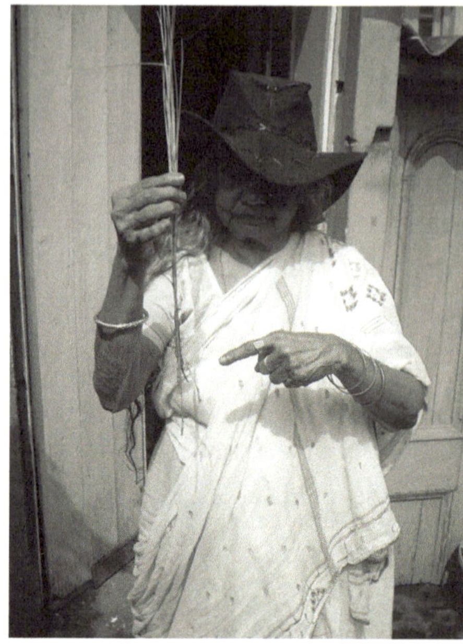

Fig. 2.38 Rhizosphere of *Portrersia coarctata*, a common salt marsh grass of lower Gangetic delta region

Fig. 2.37 Rhizosphere of mangrove associates with soil clinging on the roots

metabolic energy and to withstand the extreme environmental conditions of high pressure and temperature (over 100 °C).

Moving away from the deep-sea vents, the concentration of microbial population drops dramatically (Atlas 1998). There are heterotrophic bacteria in the sea floor sediments which feed on photoautotrophic cyanobacteria that drift down attached to sediment particles.

2.3.3 Role of Marine Micro-organisms in the Carbon Cycle

In sea water, there are 34,500 billion tonnes of carbon, the cycling of which occurs in a steady state (Hobbie and Melillo 1984). Carbon dioxide fixation to form organic molecules is performed by producers such as algae, cyanobacteria and various green and purple photosynthetic bacteria. Global estimates of important fluxes or transfers between reservoirs are shown in Fig. 2.39. Net primary production (NPP = gross photosynthesis-respiration) is approximately equal in terrestrial and marine environments. Approximately 20 % of the ocean NPP occurs in the coastal ocean; 80 % of this is deposited in surface sediments. Turnover or residence times for the reservoirs range from

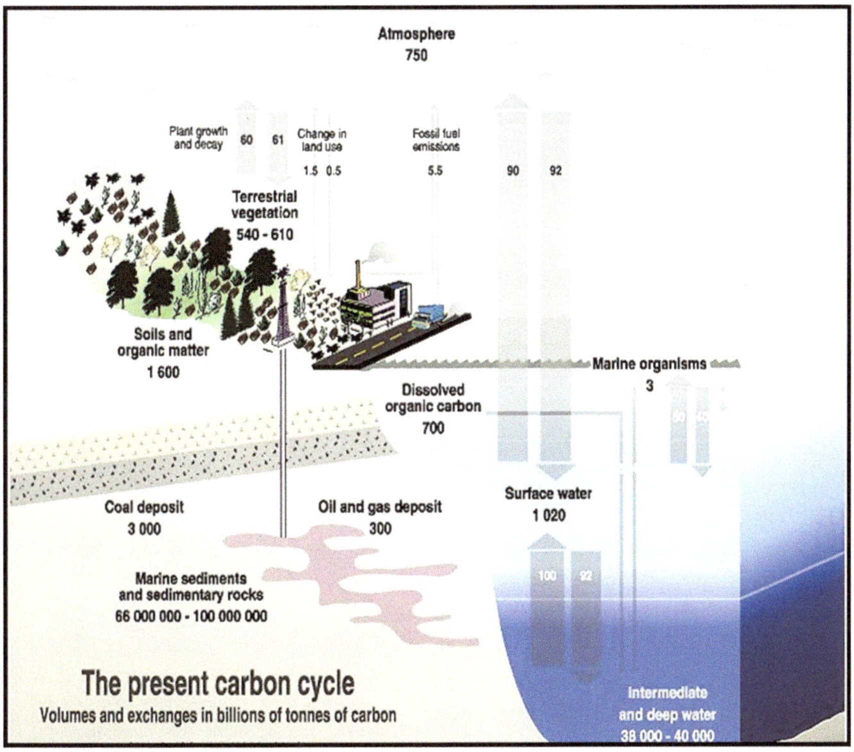

Fig. 2.39 Carbon reservoirs and fluxes. *Sources* http://www.grida.no/publications/vg/climate/page/3066.aspx—Center for climatic research, Institute for environmental studies, University of Wisconsin at Madison; Okanagan University College in Canada; World Watch, November–December, 1998; Climate Change, 1995; The Science of Climate Change, contribution of Working Group 1 to the second assessment report of the Intergovernmental Panel on Climate Change; UNEP and WMO, Cambridge Press University, 1996; www.un.org/earthwatch/about/docs/ewwp4wp6.htm

$>10^6$ year for kerogen in the sediment reservoir to 10^3–10^5 year for peats and soil carbon, to about three years for atmospheric carbon dioxide and less than one year for ocean biomass. Because of its small size and relatively slow equilibration with the ocean reservoir, the atmospheric carbon reservoir is presently out of balance. The difference between atmospheric sources (deforestation and combustion) and sinks annual atmospheric increment and the difference between ocean influx and efflux) is the 'missing link' of 1.8 Gt Cy^{-1}. The major long-term sink for carbon is burial in deep-sea sediments. This removal of a small portion (0.1 %) of annual NPP is responsible for oxygen in the Earth's atmosphere. Protection of photosynthetically fixed organic carbon from oxidation by photosynthetic oxygen (respiration) has permitted accumulation of oxygen in the atmosphere and ocean over geological time. The carbon cycle is completed by weathering of uplifted marine shales or by combustion of fossil fuels. Conversely, methanogens reduce carbon dioxide anaerobically to form methane, which in turn is used by comparatively few organisms. Heterotrophs are also important players in the carbon cycle, by generating carbon dioxide through the activity of respiration. The process of decomposition of plant materials (particularly coastal litter generated from mangroves, salt marsh, seagrass, seaweeds, etc.) also plays a major role in completing the carbon cycle.

The flow and exchange of carbon dioxide between two types of vegetation in the marine and estuarine environments is a unique feature of marine carbon cycle. Seagrasses are closely associated with mangrove habitats in many parts of the world's marine ecosystem. In the Andaman Sea, there are three mangrove-associated seagrasses, *Thalassia hemprichii*, *Enhalus acoroides* and *Halophila ovalis* (Poovachiranon and Changsang 1994). Intertidal mangrove areas in the Gazi Bay, Kenya, are colonized by *Thalassia hemprichii*, *Halophila ovalis* and *Halodule wrightii* (Coppenjans et al. 1992). *Halophila baccarii* occurs on intertidal mudflats of the Indian mangals (Jagtap 1991). Tussenbrock (1995) found that seagrass growth, biomass and primary production were all higher in the vicinity of mangrove discharges than they were in other habitats. Respiratory carbon dioxide derived from mangrove particulate organic matter (POM) could be a carbon source for seagrass and could promote faster growth.

In the present *era*, great emphasis has been given on the study of methanogenesis, which may be regarded as the final step in the process of mineralization of organic matter. It has been recorded that oceanic water is supersaturated with methane, attributed to its in situ production (Seiler and Schmidt 1974).

Methanogens are strict anaerobes, being inactivated by the presence of oxygen. The organisms are abundant below Eh values of −200 mV (Mah et al. 1977) and are unable to compete with sulphate reducers until the sulphate has been depleted. In marine sediments, only the upper few millimetres are oxygenated. Below this layer for a few centimetres, NO_3^- serves as the electron acceptor. Then, below this level for a further few centimetres, SO_4^{2-} is the electron acceptor and, in the deeper sediment, methanogenesis proceeds. The energy yield also decreases from oxygen to methane. Moreover, there is a decrease in the range of energy substrates which can be utilized, i. e., this is unique picture of decline in nutritional versatility. Organic carbon may be oxidized to carbon dioxide in the anaerobic layers of methanogens (such as *Methanococcus vanielii*), which have been recovered from the anaerobic environment of the ocean floor. The methane generated by methanogens is undoubtedly re-oxidized to carbon dioxide at the sediment surface, before assimilation by **chemolithotrophs**. Complete mineralization inevitably involves many different organisms (Jorgensen 1980).

References

Aizpuru M, Achard F, Blasco F (2000) Global assessment of cover change of the mangrove forests using satellite imagery at medium to high resolution. EEC research project no. 15017-1999-05 FIEF ISP FR—Joint Research Centre, Ispra

Alongi DM, Christoffersen P, Tirendi F (1993) The influence of forest type on microbial-nutrient relationships in tropical mangrove sediments. J Exp Mar Biol Ecol 171:201–223

Atlas R (1998) Microbial ecology: fundamentals and applications. Benjamin Cummings, Menlo Park, p 694

Azocar A, Rada F, Orozco A (1992) Water relations and gas exchange in two mangrove species with contrasting mechanisms of salt regulation. Ecotropicos 5:11–19

Balsamo RA, Thomson WW (1995) Salt effects on membranes of the hypodermis and mesophyll cells of *Avicennia germinans* (Avicenniaceae): a freeze-fracture study. Am J Bot 82:435–440

Banerjee K, Mitra A, Bhattacharyya DP (2001) Abundance of microbial population in deltaic Sundarbans, West Bengal. Indian J Environ Ecoplantation 5(3):471–473

Bhattacharyya RN (1996) Coastal pollution in the seas around India: an overview. In: The manual of integrated coastal zone management. Departement of Environment, Government of West Bengal, pp 183–194

Bunt JS (1992) Introduction. In: Robertson AI, Alongi DM (eds) Tropical mangrove ecosystem. American Geophysical Union, Washington, DC, pp 1–6

Campbell N (1993) Biology, 3rd edn. Benjamin Cummings, Redwood City, p 1190

Chandrika V, Nair PVR, Khambadhkar LR (1990) Distribution of phototrophic thionic bacteria in the anaerobic and micro-aerophilic strata of mangrove ecosystem of Cochin. J Mar Biol Assoc India 32(1–2):77–84

Chaudhuri AB, Choudhury A (1994) Mangroves of the Sundarbans, vol 1. IUCN—The World Conservation Union, India, pp 165

Chinnaraj S (1992) Higher marine fungi of Lakshwadweep Islands and a note on *Quintaria lignatilis*. Crytogamie Mycol 13(40):313–319

Chinnaraj S (1993a) Higher marine fungi from mangroves of Andaman and Nicobar Islands. Sydowia 45 (10):109–115

Chinnaraj S (1993b) Manglicolous fungi from atolls of Maldives, Indian Ocean. Indian J Mar Sci 22(2):141–142

Clough BF (1993) Constraints on the growth, propagation and utilization of mangroves in arid regions. In: Lieth H, Al Masoom A (eds) Towards the rational use of high salinity tolerant plants, vol 1. Kluwer Academic Publishers, Amsterdam, pp 341–352

Chakraborty P (1996) 'Map-info' in coastal zone management. In: Ghosh AK and Sanyal P (eds) Integrated Coastal Zone Management—A Manual. Published by Department of Environment, Government of West Bengal, pp 44–49

Coppejans E, Beeekman H, Wit MDe (1992) The seagrass and associated macroalgal vegetation of Gazi Bay

(Kenya). In: Jaccarini V, Alartens E (eds) The ecology of mangrove and related ecosystems. Hydrobiologia 247:59–75

Deming JW (1985) Bacterial growth in deep-sea sediment trap and boxcore samples. Mar Ecol Prog Ser 25:305–312

Deming JW, Colwell RR (1985) Observations of barophilic microbial activity in samples of sediment and intercepted particulates from the Demerara Abyssal Plain. Appl Environ Microbiol 50:1002–1006

Dhandapani P (1992) Status of Irrawaddy river dolphin *Orcaella brevirostris* in Chilka Lake. J Mar Biol Assoc India 34(1–2):90–93

Diop ES (1993) Conservation and sustainable utilization of mangrove forests in Latin America and Africa regions. Part II—Africa, pp 245–261. Mangrove ecosystems technical reports, vol 3. ITTO/ISME project PD 114/90. ISME, Okinawa, p 262

Drennan P, Pammenter NW (1982) Physiology of salt secretion in the mangrove *Avicennia marina* (Forsk.) Vierh. New Phytol 91:1000–1005

Duke NC (1992) Mangrove floristics and biogeography. In: Robertson AI, DM Alongi (eds) Tropical mangrove ecosystems. American Geophysical Union, Washington, DC, p 100

FAO, UNEP (1981) Tropical forest resources assessment project. Forest resources of tropical Africa. Part II: country briefs FAO, UNEP, 586 p

FAO (1994) Mangrove forest management guidelines. FAO forestry paper no. 117, Rome

Fisher P, Spalding MD (1993) Protected areas with mangrove habitat. Draft Report World Conservation Centre, Cambridge, 60 p

Groombridge B (1992) Global biodiversity: status of the earth's living resources. WCMC/The National History Museum/IUCN/UNEP/WWF/WRI. Chapman & Hall, London, 594 p

Gouda R, Panigrahy RC (1996) Ecology of phytoplankton in coastal water of Gopalpur, East coast of India. Indian J Mar Sci 2:13–18

Hobbie JE, Melillo JM (1984) Comparative carbon and energy flow in ecosystems. In: Klug MJ, Reddy CA (eds) Current perspective in microbial ecology. American Society in Microbiology, Washington, DC, pp 389–393

Hyde KD (1990a) A comparison of the intertidal mycota of five mangrove tree species. Asian Mar Biol 7:93–108

Hyde KD (1990b) A new marine ascomycete from Brunei, *Aniptodera longispora* sp. nov. from intertidal mangrove wood. Bot Mar 33(4):335–338

Hyde KD (1996) Biodiversity of micro-fungi in north Queensland. Aust Syst Bot 9(2):261–271

Hyde KD, Lee SY (1995) Ecology of mangrove fungi and their role in nutrient cycling: what gaps occur in our knowledge? Hydrobiologia 295:107–118

ITTO/ISME (1993) The world of mangroves. Part I. Japan, pp 1–63

Jagtap TG (1991) Distribution of seagrasses along the Indian coast. Aquat Bot 40(4):379–386

Jannasch HW, Wirsen CO (1982) Microbial activities in undecompressed and decompressed deep sea water samples. Appl Environ Microbiol 43:1116–1124

Jorgensen BB (1980) Mineralization and the bacterial cycling of carbon, nitrogen and sulfur in marine sediments. In: Ellwood DC, Hedger JN (eds) Contemporary microbial ecology. Academic Press, New York

Kathiresan K, Bingham BL (2001) Biology of mangrove ecosystems. Adv Mar Biol 40:81–251

Kohlmeyer J (1986) Taxonomic studies of the marine Ascomycotina. In: Moss ST (ed) The biology of marine fungi. Cambridge University Press, Cambridge, pp 199–210

Kohlmeyer J, Kohlmeyer E (1979) Marine Mycology. Academic Press, New York, p 690

Kohlmeyer J, Bebout B, Volkmann-Kohlmeyer B (1995) Decomposition of mangrove wood by marine fungi and teredinids in Belize. Mar Ecol 16:27–39

Lacerda LD, Carvalho CEV, Rezende CE, Pfeiffer WC (1993) Mercury in sediments from the Paraiba *do Sul* river continental shelf, SE Brazil. Mar Pollut Bull 26:220–222

Lalli GM, Parsons TR (1997) Energy flow and nutrient cycling. In: Biological oceanography: an introduction, 2nd edn. Open University, pp 112–146

Lear R, Turner T (1977) Mangroves of Australia. University of Qld Press, Brisbane

Leatherwood S, Bowles AE, Krygier E, Hall JD, Ingell S (1984) Killer whales (*Orcinus orca*) in Southeast Alaska, Prince William Sound, and Shelikof Strait: a review of available information. Rep Int Whaling Comm 34:521–530

Loka-Bharathi PA, Oak S, Chandramohan D (1991) Sulfate reducing bacteria from mangrove swamps. Their ecology and physiology. Oceanol Acta 14 (2):163–171

Macnae W (1968) A general account of the fauna and flora of mangrove swamps and forests in the Indo-West-Pacific region. Adv Mar Biol 6:73–270

Mah RA, Ward DM, Baresi L, Glass TL (1977) Biogenesis of methane. Annu Rev Microbiol 31:309–341

Marsh H, Lloze R, Heinsohn GE, Kasuya T (1989) Irrawaddy dolphin—*Orcaella brevirostris* (Gray, 1866). In: Ridgway SH, Harrison SR (eds) Handbook of marine mammals river dolphins and the larger toothed whales, vol 4. Academic Press, London, pp 101–118

Mepham RH, Mepham JS (1984) The flora of tidal forests—a rationalization of the use of the term 'mangrove'. S Afr J Bot 51:75–99

Mitra A, Banerjee K (2005). In: Banerjee SR (eds) Living resources of the sea; focus Indian sundarbans. WWF India, South 24 Parganas, p 96

Mohanraju R, Natrajan R (1992) Methanogenic bacteria in mangrove sediments. Hydrobiologia 247:187–193

National Geographic (1979) Strange world without sun. Natl Geogr Mag 156(5):680–705

Poovachiranon S, Changsang H (1994) Community structure and biomass of seagrass beds in the Andaman Sea and mangrove-associated seagrass beds. Res Bull Phuket Mar Biol Centre 59:53–64

Ramamurthy T, Raju RM, Natarajan R (1990) Distribution and ecology of methanogenic bacteria in mangrove sediments of Pichavaram, east coast of India. Indian J Mar Sci 19(4):269–273

Ravikumar DR, Vittal BPR (1996) Fungal diversity on decomposing biomass of mangrove plant *Rhizophora* in Pichavaram estuary, east coast of India. Indian J Mar Sci 25:142–144

Reimold JR, Queen WH (1974) Ecology of halophytes. Elsevier Science and Technology Books, New York. ISBN-100125864507

Rivera-Monroy VH, Twilley RR (1996) The relative role of denitrification and immobilization in the fate of inorganic nitrogen in mangrove sediments (Terminos Lagoon, Mexico). Limnol Oceanogr 41 (2):284–296

Santhanam R, Srinivasan A (1998) Distribution and biomass of copepods in the coral reef areas of Tuticorin, South India. Paper presented in the conference on coastal zone Canada '98, Victoria, BC, 30 Aug–3 Sept 1998

Scholander PF, Hammel HT, Hemmingsen EA, Bradstreet ED (1964) Hydrostatic pressure and osmotic potential in leaves of mangroves and some other plants. Proc Natl Acad Sci USA 52:119–125

Scholander PF (1968) How mangroves desalinate water. Physiol Plant 21:251–261

Scholander PF, Hammel HT, Hemmingsen E, Garey W (1962) Salt balance in mangroves. Plant Physiol 37:722–729

Schwamborn R, Saint-Paul U (1996) Mangroves—forgotton forests? Nat Res Dev 43(44):13–36

Saenger P, Hegerl EJ, Davie JDS (1983) Global status of mangrove ecosystems. Commission on ecology papers no. 3. IUCN. Gland, Switzerland, 88 p

Seiler W, Schmidt U (1974) Dissolved nonconservative gases in seawater. In: Goldberg ED (eds) The sea, vol 5. Wiley, New York, pp 219–243

Sherman RE, Fahey TJ, Howarth RW (1998) Soil plant interactions in neotropical mangroves forests: iron, phosphorous and sulfur dynamics. Oceologia 115:533–556

Sieburth JMcN, Brooks RD, Gessner RV, Thomas CD, Tuttle JL (1974) Microbial colonization of marine plant surfaces as observed by scanning electron microscopy. In: Colwell RR, Monta RY (eds) Effect of the ocean environment on microbial activities. University Park Press, Baltimore, pp 418–443

Spalding M, Blasco F, Field C (1997) World mangrove atlas. The International Society for Mangrove Ecosystems, Okinawa, p 178

Steinke TD, Jones EBG (1993) Marine and mangrove fungi from the Indian coast of South Africa. S Afr J Bot 59(4):385–390

Tang DL, Kawamura H (2001) Long-term series satellite ocean colour products on the Asian waters. In: Proceedings of the 11th PAMS/JECSS workshop. Hanrimwon Publishing (CD-ROM: 0112-PO3), Seoul, pp 49–52

Tateoka T (1965) Porteresia, a new genus of Graminae. Bull Natl Sci Museum (Tokyo) 8:405–406

Tomlinson PB (1986) The botany of mangroves. Cambridge University Press, London

Turner JT (1979) Microbial attachment to copepod fecal pellets and its possible ecological significance. Trans Am Microsc Soc 98:131–135

Tussenbrock BI (1995) *Thalassia testudinum* leaf dynamics in a Mexican Caribbean coral reef lagoon. Mar Biol 122(1):33–40

Twilley RR, Chen RH, Hargis T (1992) Carbon sinks in mass and their implications to carbon budget of Tropical coastal ecosystems. Water Air Soil Pollut 64:265–288

UNEP (2002) Global distribution of coral, mangrove and sea grass diversity. http://maps.grida.no/go/graphic/distributionofcoralmangrove-and-seagrass-diversity. Retrieved 26 May 2007

UNESCO (1973) International classification and mapping of vegetation. United Nations Education, Scientific and Cultural Organisation, Paris

Van Eijk M (1939) Analyse der Wirkung des NaCl auf die Entwicklung Sukhulenz and Transpiration bei *Salicornia herbacea*, Sowie untersuchungen uber den Einfluss der Salzauf Nahme auf die Wurzelatmung bei *Aster tripolium*. Recueil des Travaux. Botaniques Neerlandais 36:559–657

Wilkie ML, Fortuna S (2003) Status and trends in mangrove area extent worldwide. Forest resources assessment working paper no. 63. FAO, Rome. Available at www.fao.org/forestry/mangroves/statistics

Internet References

http://www.geocites.com/RainForest/Vines/4301/microbe/html

http://www.care4nature.org/wildindia/dolphin/conservation.htm

http://www.epa.qld.gov.au/nature_conservation/habitats/mangroves_andwetlands/mangroves

http://oceanlink.island.net/ask/pollution.html

http://www.grida.no/publications/vg/climate/page/3066.aspx

http://www.partnersvoorwater.nl/wp-content/uploads/2012/07/MoUBangladeshDeltaPlan2100-22May20120001.pdf

www.un.org/earthwatch/about/docs/ewwp4wp6.htm

3 Blue Carbon Assessment

...Blue carbon of the planet Earth
Can save mankind from oxygen dearth....
The Authors

3.1 Remote Sensing Technology

Remote sensing may be defined as the collection and interpretation of information about an object, area or event without being in physical contact with the object. Aircraft and satellites are the common platforms for remote sensing of the earth and its natural resources. Aerial photography in the visible portion of the electromagnetic (EM) wavelength was the initial stage of remote sensing, but technological developments have enabled the acquisition of information at other wavelengths including near infrared (NIR), thermal infrared and microwave. Collection of information over a large numbers of wavelength bands is referred to as multispectral or hyperspectral data. The development and deployment of manned and unmanned satellites has enhanced the collection of remotely sensed data and offers an inexpensive way to obtain information over large areas. The capability of remote sensing to identify and monitor vegetation has expanded greatly over the last few years. Today, the satellite imageries (output of the remote sensing technology) are used to scan the health of coastal vegetation, their degradation rate and restoration. Thus, in the blue carbon domain, the utility of remote sensing is immense not only to understand the magnitude of threat, but also to monitor the gradual steps of ecorestoration.

Nowadays, the phytopigments (pigments of phytoplankton) in the waterbodies are also monitored through remote sensing. Launched by ISRO's Polar Satellite Launch Vehicle (PSLV) from SHAR Centre, Sriharikota, IRS-P4 (OCEANSAT) is the first satellite primarily built for ocean applications. The 1,050-kg satellite is placed in a polar sun-synchronous orbit of 720 km height. IRS-P4 has on-board Ocean Colour Monitor (OCM) and a Multifrequency Scanning Microwave Radiometer (MSMR). OCM is a solid-state camera operating in eight narrow spectral bands. The camera is used to collect data on chlorophyll concentration, detect and monitor phytoplankton blooms and obtain data on atmospheric aerosols and suspended sediments in the water. MSMR, which operates in four microwave frequencies both in vertical polarization and horizontal polarization, is used to collect data on sea surface temperature, wind speed, cloud water content and water vapour content in the atmosphere above the ocean. IRS-P4 vastly augments the IRS satellite system of ISRO, which at present comprises four satellites, IRS-1B, IRS-1C, IRS-P3 and IRS-1D, and extend remote sensing applications to several newer areas. The salient features of IRS-P4 along with its scanner specifications are listed in Tables 3.1 and 3.2.

The productivity of aquatic system is presently monitored through satellite imageries. Time series analysis of satellite data reflects accurately the variation of phytoplankton community and associated phytopigments on which with the distribution and population of fish population depends. This is the basics of predicting potential fishing zone (PFZ).

Some common satellites used for scanning objects from distances are listed here.

- GOES
  5 spectral bands 1–41 km spatial resolution geostationary

A. Mitra and S. Zaman, *Blue Carbon Reservoir of the Blue Planet*,
DOI 10.1007/978-81-322-2107-4_3, © Springer India 2015

Table 3.1 Salient features of IRS-P4

Payloads	
Ocean colour monitor	
Swath	1,420 km
Field of view	360 m
Spectral bands	8 (from 400 to 885 nm)
Multifrequency scanning microwave radiometer	
Frequency	6.6, 10.65, 18 and 21 GHz
Swath	1,360 km

Table 3.2 Scanner specification of IRS-P4

Salient features	
Orbit	Polar sun-synchronous circular
Altitude	720 km
Inclination	98.28°
Period	99.31 min
Local time of equator crossing	12 noon
Repetitivity cycle	2 days
Size	2.8 m × 1.98 m × 2.57 m
Mass at lift-off	1,050 kg
Length when fully deployed	11.67 m
Altitude and orbit control	Three-axis body-stabilized using reaction wheels, magnetic torquers and hydrazine thrusters
Power	9.6 m^2 solar array generating 750 W; two 21 Ah Ni–Cd batteries
Mission life	5 years

- NOAA AVHRR
  5 spectral bands 1.1 km spatial resolution 1-day repeat cycle
- Landsat TM
  7 spectral bands 30 m spatial resolution 16-day repeat cycle
- MODIS
  Multi-spectral bands 250–1,000 m spatial resolution (band dependent) 1-day repeat cycle
- IKONOS
  4 spectral bands 4 m spatial resolution 5-day repeat cycle.

Remote sensing has a wide range of applications in many different fields:

- Coastal applications: monitor shoreline changes, track sediment transport and map coastal features. Data can be used for coastal zone management and erosion prevention.
- Ocean applications: monitor ocean circulation and current systems, measure ocean temperature and wave heights and track sea ice. Data can be used to understand the health of ocean water and manage marine biodiversity.
- Hazard assessment: track hurricanes, earthquakes, erosion and flooding. Data can be used to assess the impacts of a natural disaster and create preparedness strategies to be used before and after a hazardous event.
- Natural resource management: monitor land use, map wetlands and wildlife habitats. Data can be used to minimize the damage that urban growth has on the environment and help formulate policies to conserve natural resources.

For understanding the basics of remote sensing and its working, few terminologies are essential to understand, which are discussed here in brief.

3.1.1 Electromagnetic Energy

The EM spectrum is the continuous range of EM radiation, extending from gamma rays (highest frequency and shortest wavelength) to radio waves (lowest frequency and longest wavelength) and including visible light. The EM spectrum can be divided into seven different regions such as gamma rays, X-rays, ultraviolet, visible light, infrared, microwaves and radio waves (Fig. 3.1).

Remote sensing involves the measurement of energy in many parts of the EM spectrum. The major regions of interest in satellite sensing are visible light, reflected and emitted infrared and the microwave regions. The measurement of this radiation takes place in what are known as spectral bands.

3.1.2 Spectral Band

A spectral band is defined as a discrete interval of the EM spectrum. For example, the wavelength range of 0.4–0.5 μm (μm = micrometers or

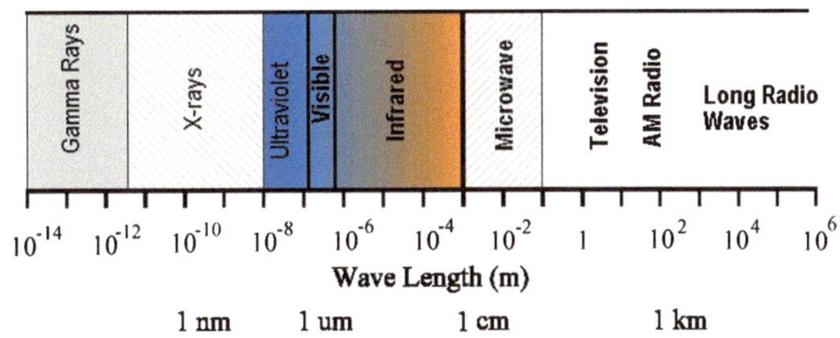

Fig. 3.1 Divisions of EM spectrum

10^{-6} m) is one spectral band. Satellite sensors have been designed to measure responses within particular spectral bands to enable the discrimination of the major Earth's surface materials. Researchers or scientists choose a particular spectral band for data collection depending on what they wish to examine. The design of satellite sensors is based on the absorption characteristics of Earth's surface materials across all the measurable parts in the EM spectrum.

3.1.3 Spectral Signature

When the solar radiation reaches the surface of the Earth, some of the energy at specific wavelengths is absorbed and the rest of the energy is reflected by the surface material. The only two exceptions to this situation are whether the surface of a body is a perfect reflector or a true black body. The occurrence of such materials in the planet Earth is extremely rare. In the visible region of the EM spectrum, the feature we describe as the colour of the object is the visible light that is not absorbed (but reflected) by that object. In the case of a green leaf, for example, the blue and red wavelengths are absorbed by the leaf, while the green wavelength is reflected and detected by our eyes. In remote sensing, a detector measures the EM radiation that is reflected back from the Earth's surface materials. These measurements can help to distinguish the type of land covering. Soil, water and vegetation have clearly different patterns of reflectance and absorption over different wavelengths. The reflectance of radiation from one type of surface material, such as soil, varies over the range of wavelengths in the EM spectrum. This is known as the spectral signature of the material. All Earth's surface features, including minerals, vegetation, dry soil, sand, water and snow, have unique (specific) spectral reflectance signatures.

The reflectance of **clear water** is generally low. However, the reflectance is maximum at the blue end of the spectrum and decreases as wavelength increases. Hence, water appears dark bluish to the visible eye. **Turbid water** has some sediment suspension that increases the reflectance in the red end of the spectrum and would be brownish in appearance. The reflectance of **bare soil** generally depends on its composition. In the example shown, the reflectance increases monotonically with increasing wavelength. Hence, it should appear yellowish red to the eye.

Figure 3.2 shows the typical reflectance spectra of water, bare soil and two types of vegetation.

When solar radiation hits a target surface, it may be **transmitted**, **absorbed** or **reflected**. Different materials reflect and absorb differently at different wavelengths. The reflectance spectrum of a material is a plot of the fraction of radiation reflected as a function of the incident wavelength and serves as a unique signature for the material. In principle, a material can be identified from its spectral reflectance signature if the sensing system has sufficient spectral resolution to distinguish its spectrum from those of other materials. This premise provides the basis for multispectral remote sensing.

Vegetation has a unique spectral signature that enables it to be distinguished readily from other types of land cover in an optical/near-infrared image. The reflectance is low in both the blue and red regions of the spectrum, due to absorption by chlorophyll for photosynthesis. It has a

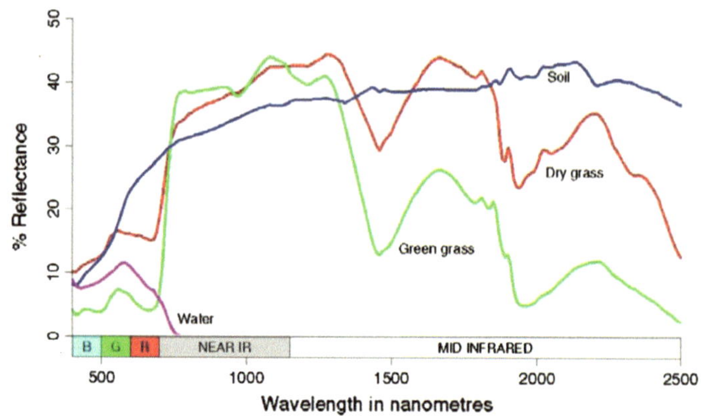

Fig. 3.2 Typical reflectance spectra

peak at the green region. In the NIR region, the reflectance is much higher than that in the visible band due to the cellular structure in the leaves. Hence, vegetation can be identified by the high NIR but generally low visible reflectance. This property has been used in early reconnaissance missions during war times for 'camouflage detection'.

The shape of the reflectance spectrum can be used for identification of vegetation type (Fig. 3.3). For example, the reflectance spectra of dry grass and green grass in the previous figures can be distinguished although they exhibit the general characteristics of high NIR but low visible reflectance. Dry grass has higher reflectance in the visible region, but lower reflectance in the NIR region. For the same vegetation type, the reflectance spectrum also depends on other factors such as the leaf moisture content and health of the plants. These properties enable vegetation condition to be monitored using remotely sensed images.

3.1.4 Sensors and Platforms

A sensor is a device that measures and records EM energy. Sensors can be divided into two groups (Fig. 3.4). **Passive sensors** depend on an external source of energy, usually the sun. The most common passive sensor is the photographic camera. **Active sensors** have their own source of energy, e.g., a radar gun. These sensors send out a signal and measure the amount reflected back.

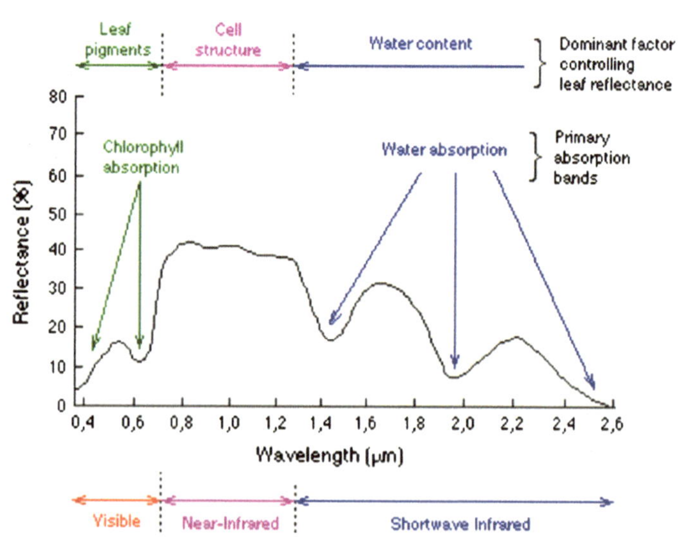

Fig. 3.3 Typical reflectance spectra of vegetation

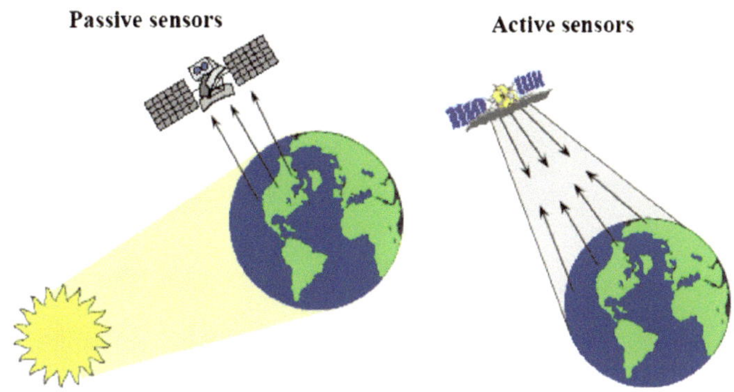

Fig. 3.4 Passive and active sensors

Active sensors are more controlled because they do not depend upon varying illumination conditions. Remote sensors collect data by detecting the energy that is reflected from Earth. These sensors can be on satellites or mounted on aircraft. Remote sensors can be either passive or active. Passive sensors respond to external stimuli. They record radiation that is reflected from Earth's surface, usually from the sun. Because of this, passive sensors can only be used to collect data during daylight hours.

In contrast, active sensors use internal stimuli to collect data about Earth. For example, a laser beam remote sensing system projects a laser onto the surface of Earth and measures the time that it takes for the laser to reflect back to its sensor.

3.1.5 Orbits and Swaths

The path followed by a satellite is referred to as its orbit. Satellites which view the same portion of the Earth's surface at all times have geostationary orbits. Weather and communication satellites commonly have these types of orbits. Many satellites are designed to follow a north–south orbit which, in conjunction with the Earth's rotation (west–east), allows them to cover most of the earth's surface over a period of time. These are near-polar orbits. Many of these satellites orbits are also sun-synchronous such that they cover each area of the world at a constant local time of day. Near-polar orbits also mean that the satellite travels northward on one side of the earth and the southward on the second half of its orbit. These are called ascending and descending passes. As a satellite revolves around the Earth, the sensor sees a certain portion of the Earth's surface. The area imaged is referred to as the swath. The surface directly below the satellite is called the nadir point.

3.1.6 Satellite Sensor Characteristics

The main function of most satellite sensors is to collect information about the reflected radiation along a pathway, also known as the field of view (FOV), as the satellite orbits the Earth. The smallest area of ground that is sampled is called the instantaneous IFOV. The IFOV is also described as the pixel size of the sensor. This sampling or measurement occurs in one or many spectral bands of the EM spectrum. The data collected by each satellite sensor can be described in terms of spatial, spectral and temporal resolution.

Spatial Resolution

The spatial resolution (also known as ground resolution) is the ground area imaged for the IFOV of the sensing device. Spatial resolution may also be described as the ground surface area that forms one pixel in the satellite image. The IFOV or ground resolution of the Landsat Thematic Mapper (TM) sensor, for example, is 30 m. The ground resolution of weather satellite sensors is often larger than a square kilometre. There are satellites that collect data at less than 1 m ground

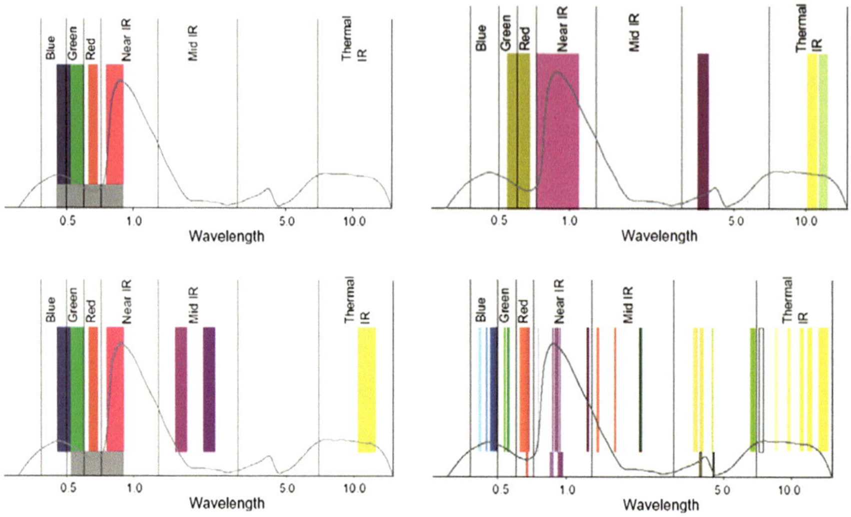

Fig. 3.5 Multispectral images

resolution, but these are classified military satellites or very expensive commercial systems.

Temporal Resolution

Temporal resolution is a measure of the repeat cycle or frequency with which a sensor revisits the same part of the Earth's surface. The frequency will vary from several times per day, for a typical weather satellite, to 8–20 times a year for a moderate ground resolution satellite, such as Landsat TM. The frequency characteristics will be determined by the design of the satellite sensor and its orbit pattern.

Spectral Resolution

The spectral resolution of a sensor system is the number and width of spectral bands in the sensing device. The simplest form of spectral resolution is a sensor with one band only, which senses visible light. An image from this sensor would be similar in appearance to a black-and-white photograph from an aircraft. A sensor with three spectral bands in the visible region of the EM spectrum would collect similar information to that of the human vision system. The Landsat TM sensor has seven spectral bands located in the visible and near- to mid-infrared parts of the spectrum.

A panchromatic image consists of only one band. It is usually displayed as a greyscale image, i.e., the displayed brightness of a particular pixel is proportional to the pixel digital number which is related to the intensity of solar radiation reflected by the targets in the pixel and detected by the detector. Thus, a panchromatic image may be similarly interpreted as a black-and-white aerial photograph of the area, though at a lower resolution.

Multispectral and hyperspectral images (Fig. 3.5) consist of several bands of data. For visual display, each band of the image may be displayed one band at a time as a greyscale image or in combination of three bands at a time as a colour composite image. Interpretation of a multispectral colour composite image will require the knowledge of the spectral reflectance signature of the targets in the scene.

3.1.7 Geodesy, Geodetic Data and Map Projections

Geodesy is the branch of science concerned with the determination of the size and shape of the Earth. Geodesy involves the processing of survey measurements on the curved surface of the Earth, as well as the analysis of gravity measurements.

Knowing the exact location of a pixel on the Earth's surface (its spatial location) is an essential component of remote sensing. It requires a detailed knowledge of the size and the shape of the Earth.

The Earth is not a simple sphere. Topographic features such as mountain ranges and deep oceans disturb the surface of the Earth. The ideal reference model for the Earth's shape is one that can represent these irregularities and identify the position of features through a coordinate system. It should also be easy to use.

The 'flat Earth' model is not appropriate when mapping larger areas. It does not take into account the curvature of the Earth. A 'curved Earth' model more closely represents the shape of the Earth. A spheroid best represents the shape of the Earth because it is significantly wider at the equator than around the poles (unlike a simple sphere). A spheroid (also known as an ellipsoid) represents the equator as an elliptical shape, rather than a round circle. Surveying and navigation calculations can be performed over a large area when a spheroid is used as a curved Earth reference model.

The surface of the sea is not uniform. The Earth's gravitational field shapes it. The rocks that make up the Earth's interior vary in density and distribution, causing anomalies in the gravitational field. These, in turn, cause irregularities in the sea surface. A mathematical model of the sea surface can be formulated; however, it is very complex and not useful for finding geographic positions on a spheroid reference model.

3.1.8 Satellite/Sensor: Exclusive for Monitoring Marine Ecosystem and Blue Carbon

The Nimbus-7 satellite, launched in 1978, carried the first sensor, the Coastal Zone Colour Scanner (CZCS), specifically intended for monitoring the Earth's oceans and water bodies. The primary objective of this sensor was to observe ocean colour and temperature, particularly in coastal zones, with sufficient spatial and spectral resolution to detect pollutants in the upper levels of the ocean and to determine the nature of materials suspended in the water column. The Nimbus satellite was placed in a sun-synchronous, near-polar orbit at an altitude of 955 km. Equator crossing times were local noon for ascending passes and local midnight for descending passes.

The repeat cycle of the satellite allowed for global coverage every 6 days or every 83 orbits. The CZCS sensor consisted of six spectral bands in the visible, near-IR and thermal portions of the spectrum each collecting data at a spatial resolution of 825 m at nadir over a 1,566-km swath width.

The first Marine Observation Satellite (MOS-1) was launched by Japan in February 1987 and was followed by its successor, MOS-1b, in February of 1990. These satellites carry three different sensors: a four-channel Multispectral Electronic Self-Scanning Radiometer (MESSR—50 m resolution), a four-channel Visible and Thermal Infrared Radiometer (VTIR—900 to 2,700 m resolution) and a two-channel Microwave Scanning Radiometer (MSR), in the microwave portion of the spectrum. The MESSR bands are quite similar in spectral range to the Landsat MSS sensor and are thus useful for land applications in addition to observations of marine environments. The MOS systems orbit at altitudes around 900 km and have revisit periods of 17 days.

The SeaWiFS (sea-viewing wide field of view sensor) on-board SeaStar spacecraft is an advanced sensor designed for ocean monitoring. It consists of eight spectral bands of very narrow wavelength ranges (see accompanying table) tailored for very specific detection and monitoring of various ocean phenomena including ocean primary production and phytoplankton processes, ocean influences on climate processes (heat storage and aerosol formation) and monitoring of the cycles of carbon, sulphur and nitrogen. The orbit altitude is 705 km with a local equatorial crossing time of 12 noon. Two combinations of spatial resolution and swath width are available for each band: a higher resolution mode of 1.1 km (at nadir) over a swath of 2,800 km and a lower resolution mode of 4.5 km (at nadir) over a swath of 1,500 km.

Today, the technology of remote sensing is used to generate a land-use map of the mangrove forest and its surrounding. Two steps are usually followed for generation of the land-use map.

Step 1: A maximum likelihood supervised classification is carried using training areas chosen according to extensive field knowledge, but without any specific reference to the grip sample points.

Step 2: In this step, the raw result of the supervised classification is checked to visual interpretation of the satellite image and field visit. We present here the application of remote sensing technology to generate a land-use map of mangrove ecosystem located around the Gautami-Godavari estuary (Fig. 3.6). The study area extends a subset between 16° 30′ N–17° 05′N and 82° 30′E–82° 25′E and includes Kakinada Bay and the Coringa wildlife sanctuary where the most important stretch of mangrove is found and also the mangrove forest situated south of Gautami-Godavari River.

Several different landscapes compose the area, with paddy fields and coconut tree plantations in the West and South, mangrove forests, aquaculture ponds for shrimps farming spreading into mangrove forest, salt pans, casuarinas plantation along the beach and on Hope Island, village and urban areas (Kakinada and Yanam).

The present study was undertaken to assess the three-decade spatial changes in vegetation dynamics using multitemporal satellite data and geographical information system (GIS) considering the importance of Godavari delta and mangroves.

For the convenience of the reader, the approach of this study is explained here in points.

Field Survey

During April 2006, field survey was carried out in the study area to investigate the vegetation and other land cover. For ground collection of data, false colour composite satellite images and survey of India topographical maps (65H and 65L) were used. GPS was used for ground truth data collection at various locations before the image classification.

Data Used

Temporal satellite images of Landsat MSS (Landsat Multispectral Scanner-path/row: 152/48; date: 8 January 1977), Landsat TM (Landsat Thematic Mapper-path/row: 141/49; date: 12 October 1988), Landsat ETM+ (Landsat Enhanced Thematic Mapper+ path/row: 141/48, 141/49; date: 8 December 2000) and IRS P6 LISS III (Indian Remote Sensing P6 Linear Imaging Self Scanner-path/row: 103/61; date: 13 December 2005) was used to carry out the study. The orthorectified Landsat data were downloaded from GLCF Web site (http://glcf.umiacs.umd.edu/). ERDAS Imagine 9.0 software was used for processing the images.

To obtain the same spatial resolution, the Landsat TM, ETM+ and IRS P6 LISS III images was geometrically corrected in relation to the MSS image (since Landsat multispectral scanner (MSS) image downloaded at 57 m resolution from http://www.glcf.umiacs.umd.edu), i.e., from 30 to 57 m (in case of IRS LISS III from 23.5 to 57 m).

Classification

Mapping of vegetation and land cover of the study area is assessed by supervised digital classification method. This has been the most frequent method for remotely sensed data classification. The samples of known identity were used to classify pixels of unknown identity in supervised classification. To represent the typical spectral information of the land cover classes (dense mangrove, open mangroves, plantations, agricultural lands, built-up area, aquaculture, sand, mudflats and water bodies), training sites in the images are generated (Fig. 3.7). Using maximum likelihood in ERDAS Imagine 9.0 software, the classification was run on the images after the selection of training sites. Signature separability analysis was carried out on the training signatures for better classification prior to the classification.

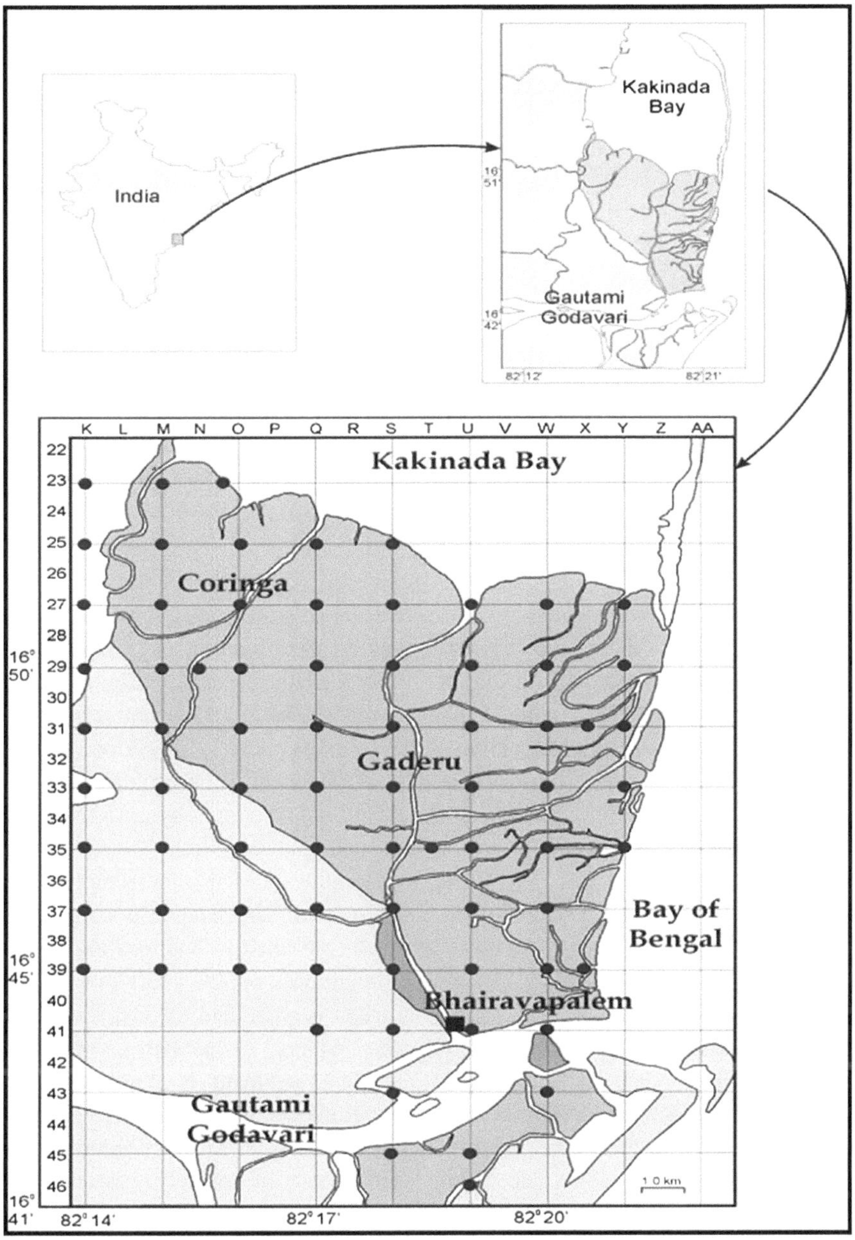

Fig. 3.6 Mangrove areas in Gautami-Godavari river (*dark shaded* area refers to mangroves)

Output

From land cover map of the year 2005, it may be seen that water bodies (sea/rivers/streams/canals/reservoir/tank) in the study area constitute 44.1 % of the area. Another major feature is most of the inland area is under intense agriculture (20.2 %). In 1977, the forest cover (dense mangroves and open mangroves) accounted for about 194.8 km^2 (Table 3.3; Fig. 3.8).

The land-use map exhibits dense mangroves (represented by shinning green colour), open mangroves (represented by dull green colour),

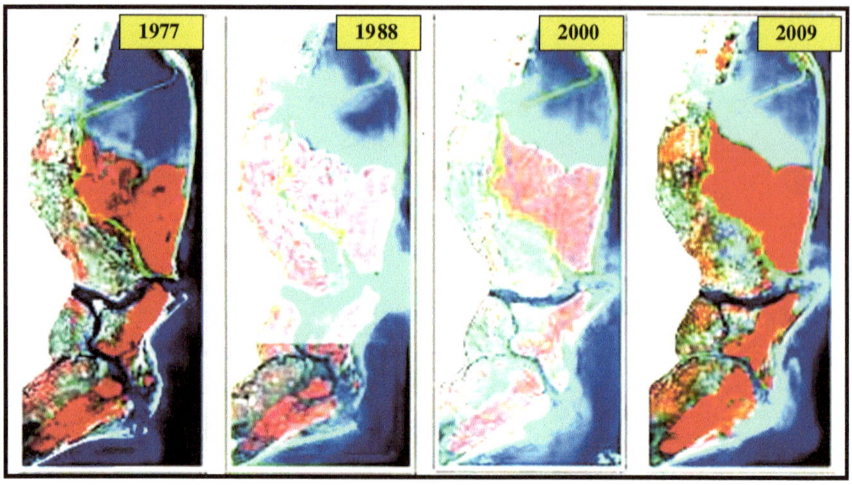

Fig. 3.7 Multitemporal false colour composite images of Godavari mangroves

aquaculture (represented by sky blue colour), mudflats (represented by dull blue colour), plantation (represented by brownish tinge), agriculture (represented by yellow colour) and built-up area (represented by pink colour) with water surrounding the zone (deep blue colour) (Fig. 3.8). The aerial extent of vegetation and other land cover in the Godavari mangroves is represented in Table 3.3.

3.2 Chemical Method

In 1980, Froelich showed that organic carbon and inorganic carbon cannot be quantified with high accuracy and precision by difference upon combustion methods because the thermal decomposition ranges of naturally occurring organic carbon and inorganic carbon overlap. It was also observed by Froelich that when sediments are treated with aqueous acid to remove inorganic carbon, from 5 to 45 % of the original organic carbon can be solubilized and lost (or counted as inorganic carbon) if the spent acid is discarded without analysis. To overcome these problems, he devised an alternate analytical method for organic carbon that accomplishes selective removal of inorganic carbon by acidification, but accounts for organic carbon lost by dissolution. This procedure requires organic carbon determinations of both the solids left after acidification with an elemental (CHN) analyser and the spent acid solution with a dissolved organic carbon (DOC) analyser.

A more recently developed method (Weliky et al. 1983) also discriminates inorganic carbon from organic carbon by acidification, but has the advantage that both forms are determined sequentially in the same sample using only one instrument. The sample is weighed into a small flask connected in line to a molecular sieve carbon dioxide trap and a carbon analyser. Inorganic carbon is quantified by acidification with H_3PO_4 and boiling to evolve carbon dioxide. The residual sediment is then boiled with acidic dichromate solution to evolve organic carbon as carbon dioxide. This procedure is applicable to sediments of widely varying composition including calcareous oozes that contain high levels of acid-soluble organic matter (Froelich 1980).

Both of these procedures provide substantial improvements in accuracy. However, Froelich's procedure is laborious and requires both a DOC and solid carbon analyser. The method of Weliky et al. (1983) relies on wet oxidation for organic carbon analysis, a technique which is slower and not as dependably complete as high temperature combustion. In addition, total nitrogen values are not obtained. Both procedures require relatively

Table 3.3 Aerial extent of vegetation and other land cover in Godavari mangroves: study area (area in km^2)

Land cover class	1977		1988		2000		2005	
	Area	Area (%)	Area	Area (%)	Area	Area (%)	Area	Area (%)
Dense mangrove	20.6	2.1	7.9	0.8	8.4	0.9	10.3	1.1
Open mangrove	174.2	17.9	162.3	16.7	176.8	18.2	175.7	18.1
Subtotal	**194.8**	**20.0**	**170.2**	**17.5**	**185.2**	**19.1**	**186.1**	**19.1**
Plantations	10.6	1.1	11.4	1.2	17.4	1.8	18.2	1.9
Agriculture	208.8	21.5	200.8	20.7	181.9	18.7	196.2	20.2
Built-up area	14.2	1.5	19.5	2.0	20.7	2.1	22.4	2.3
Aquaculture	0.0	0.0	17.6	1.8	54.7	5.6	46.5	4.8
Water bodies	438.2	45.1	456.0	46.9	434.0	44.7	428.8	44.1
Mudflat	85.5	8.8	77.1	7.9	56.5	5.8	60.1	6.2
Sand	19.8	2.0	19.2	2.0	21.3	2.2	13.6	1.4
Grand total	**972.0**	**100.0**	**972.0**	**100.0**	**972.0**	**100.0**	**972.0**	**100.0**

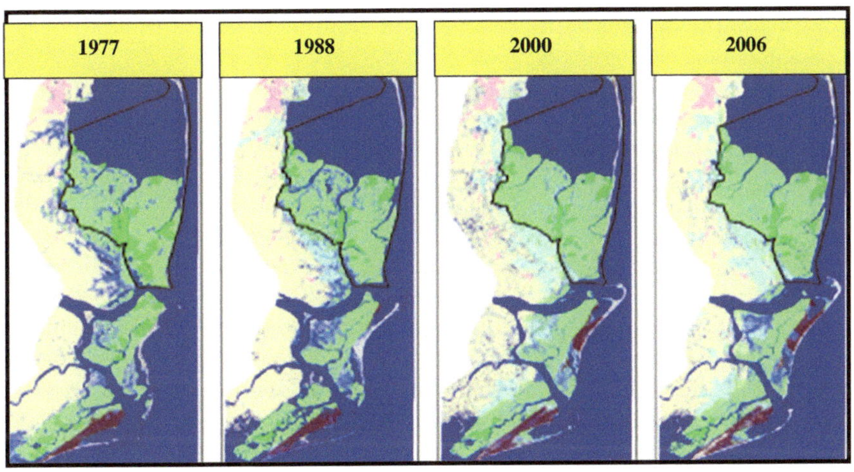

Fig. 3.8 Classified vegetation and land cover maps of Godavari mangroves

large (50–200 mg) samples, are somewhat tedious and cannot be easily automated.

Hedges and Stern (1984) described two methodologies for the analysis of organic carbon, total nitrogen and inorganic carbon in sediments, sediment trap materials and plankton. Both methods discriminate organic carbon and inorganic carbon by acidification, avoid losses of acid-soluble organic and inorganic matter and require only high temperature combustion. The two methods differ primarily in whether inorganic carbon is removed by aqueous- or vapour-phase acidification.

(A) **Aqueous Acidification Procedure**

In the aqueous acidification procedure, the wet sample is homogenized and a subsample equivalent to 50–100 mg dry weight is weighed into a pre-weighed 20-ml glass scintillation vial. The sample is dried overnight at 50 °C and reweighed after cooling to obtain the initial dry weight (W_o) and the wt% water. If inorganic carbon is to be determined (by difference), the dried sample is powdered with an agate mortar and pestle and 1–10 mg subsamples are weighed out in duplicate into solvent-rinsed tin sample boats for elemental analysis. If this is done, W_o

must be redetermined for the residual dry sediment in the scintillation vial before proceeding.

The dry sample is then treated with an excess (≥2 ml) of organic-free 1 N HCl (Froelich 1980). The acidified slurry is agitated in an ultrasonic bath at room temperature for 5 min and then dried overnight at 50 °C. After removal from the oven, the vial of dried sample is allowed to sit open for at least 24 h. This waiting period is necessary because hydroscopic salts (e.g. calcium chloride) in HCl-treated samples absorb water vapour upon first exposure to laboratory air, rapidly gaining from a few percentage to >100 % of their original weight before coming to a relatively constant mass. Aqueous HCl acidification followed by drying without gravimetric correction before analysis of a weighed subsample can lead to substantial error in calculated elemental compositions.

After equilibration, the sample vial is reweighed to obtain the final weight (W_f). The same sample is immediately homogenized and duplicate 1–10 mg subsamples are weighed into tin sample boats for elemental analysis. The weighings of W_f and the subsamples for CHN analysis should be made sequentially for individual samples with minimum elapsed time because changes in water content between weighings will affect the accuracy of the final result. This interval is usually <15 min, and results produce no significant weight change for well-equilibrated samples.

(B) **Vapour Acidification Procedure**

The vapour acidification method should not be applied to samples containing >50 wt% calcium carbonate ($CaCO_3$) and is described here specifically for use with a Carlo Erba CHN analyser. A small sample (25–50 mg) is oven-dried overnight at 50 °C. If inorganic carbon is to be determined by difference, a pair of powdered subsamples (1–10 mg) is weighed out into tin sample boats for total carbon analysis. For organic carbon analysis, a second pair of subsamples is prepared by weighing into silver sample boats (supplied by Carlo Erba), which are more resistant to acid attack than tin. The samples in the open silver boats are transferred to a Teflon plate drilled to hold approximately 100 boats in numbered 'wells'. The loaded plate is then enclosed in a glass desiccator along with a small beaker containing concentrated HCl. The samples are exposed to HCl vapour for 24–48 h at room temperature, then removed and heated in an oven for 1 h at 50 °C to drive off residual hydrochloric acid and water. The silver cups are then crimped closed and transferred without reweighing to the CHN analyser. It is important that untreated and acidified samples be handled separately because residual acid vapours in the HCl-treated samples can react with carbonates in untreated counterparts.

Both analytical procedures were designed specifically for use with a CHN analyser linked to an electronic integrator. However, they should be readily adaptable to other elemental analysers capable of handling small samples.

3.2.1 CHN Analyser: An Overview of the Instrument

Carbon, hydrogen and nitrogen are important components of living organisms and non-living substances (such as sediment, coal and wood). The composition of these elements needs to be assessed with precision and high accuracy for analysing the object and living matter in details. The proportion of carbon, hydrogen and nitrogen can be analysed through CHN analyser in which the complete oxidation of the matter is carried out. Several companies manufacture CHN analyser. In 1989, 2,400 CHN Elemental Analyzer was introduced by Perkin Elmer. The principle of measurement is very simple. The sample to be tested is weighed in a tin capsule. The required amount ranges between 2 and 3 mg. In case the carbon content is less, the required quantity may be up to 10 mg. The sample is placed in the autosampler. The tin capsule enclosing the sample reaches the reactor chamber where excess oxygen is introduced. At about 990 °C, the sample is mineralized. Formation of carbon monoxide is possible at this temperature even in the presence of excess oxygen. The complete oxidation is reached through a tungsten trioxide catalyst, which is passed by the gaseous reaction products. The resulting mixture consists of carbon dioxide, water and nitrogen dioxide. However, some excess oxygen passes the catalyst. The resulting mixture now flows through a silica tube packed with copper granules. In this zone, the temperature is around 500 °C, where the remaining oxygen is bound and nitric/nitrous oxides are reduced. The resulting gas stream now includes the analytically important species carbon dioxide, water and nitrogen. Eventually included sulphur dioxide or hydrohalogenides are absorbed at appropriate temperature. In CHN analyser, high-purity helium is used as carrier gas finally and the gas mixture is brought to a definite pressure/volume state and is passed through a gas chromatographer system. Separation of the species is done by so-called zone chromatography. In this technology, a staircase-type signal is registered. Step height is proportional to the substance amount in the mixture. In CHN analyser, calibration is done by elemental analysis of standard substances, whose percentage of C, H and N are known. This is supplied by the manufacturer of the instrument.

While assessing the percentage of C, H and N, the researchers must keep in mind about various problem and interferences that may occur during operation, e.g., the weighing of oily or fluid substances is impossible in tin capsules and, for this purpose, aluminium pans with a lid are available. These pans are tightly closed by cold welding to prevent loss of sample by minimizing evaporation. Phosphorous can interfere in the process of mineralization, leading to the formation of $P_2O_5{\cdot}xH_2O{\cdot}yC$. This effect can be controlled by the addition of vanadium pentoxide. The analysis of numerous fluorine containing compounds may cause errors in the hydrogen result.

3.2.2 Analysis Mode Options in CHN Analyser

Analysis of elements can be done through several modes that are available in the instrument.

1. **CHN Mode**: The CHN mode is the most universal of the analysis modes because of the combination of the reagent design and the optimize combustion control parameters. Interfering elements (halogens) are removed.
2. **CHNS Mode**: The CHNS mode is designed to handle conventional organics. This mode is specifically designed to include sulphur, which reduces the universality of the CHN capability. This includes the capability for limiting the range of sample types and sample size (1–2 mg is recommended). Metal cations are excluded. Special precautionary measures are to be taken in calibration and blank detection for lower levels of sulphur.
3. **Oxygen Mode**: The oxygen mode excludes compounds containing phosphorous, fluorine, silicon and metal cations. Samples containing mineral matter must be demineralized prior to analysis. The user may choose any or all modes. The CHN analyser may be upgraded at any time to add additional mode capability to suit the needs of the laboratory.

3.2.3 Precision

The analyser optimizes the time of analysis for each analysis mode and minimizes maintenance time, changeover in analysis modes and downtime. There is an in-built microprocessor diagnostics, which continually monitor instrument conditions. Analysis time in each mode of operation has been optimized for efficiency and precision. The analysis time is six minutes in CHN mode, eight minutes in CHNS mode and 4 min in oxygen mode. The unique steady-state measurement of all signals improves precision in all modes of operation. Modern organic elemental analysis consists of a series of automated steps: combustion and reduction, homogenization of product gases, separation and detection. Combustion (complete) is the most critical step to the success of the process and ultimately affects the accuracy and precision of the final result in terms of the weight percentage of the elements or the element to be measured in the sample. The modern analyser not only provides for advanced combustion conditions of temperature, time and available oxygen (or pyrolysis gas in the case of oxygen mode), but also enhanced combustion features under user directed microprocessor control to further optimize the combustion step for the widest variety of samples.

3.3 Case Studies

3.3.1 Assessment of Blue Carbon Through Remote Sensing

The blue carbon can also be assessed through remote sensing. A case study from critical habitat information system for Coringa mangrove in Andhra Pradesh (India) (DOD 2001) reveals that mangroves have been severely depleted in this region. The Coringa mangrove situated about 150 km South of Visakhapatnam (16° 44′–16° 53′ N and 82° 14′–82° 22′E) is in the south of Kakinada Bay (Fig. 3.9).

The mangrove ecosystem receives freshwater from four sources, namely (i) Coringa River (ii) Gaderu River, (iii) Distributaries of Gautami-Godavari River and (iv) neritic waters from Kakinada Bay.

The major threats operating on mangroves of Coringa are tabulated in brief (Table 3.4).

The assessment of the health of the mangroves was scanned through satellite data of 1988 IRS 1A and 1998 IRC 1C using ERDAS Image Processing Software along with ground truth verification. Land-Use Maps of 1988 and 1998 were analysed to detect the variation using the overlay facilities of GIS. Table 3.5 (Fig. 3.12) reveals the major land-use/land cover features and their respective area in 1988 and 1998 within Coringa Reserve Forest and its extension.

Fig. 3.9 Location of Coringa mangrove ecosystem

Fig. 3.10 *Sonneratia apetala*: a common mangrove floral species in deltaic Sundarbans

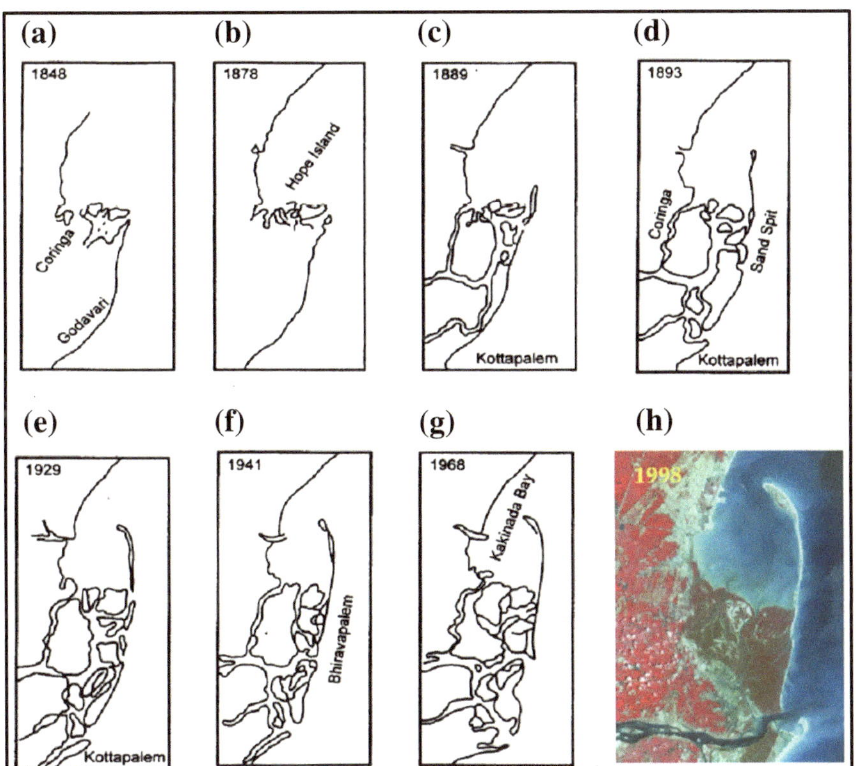

Fig. 3.11 Sand-spit formation and growth

Table 3.4 Threats, description and conservation

Threats	Description	Conservation
1. Shoreline changes	A sand spit has formed near Kakinada Bay over a period of 100 years (Fig. 3.11). During 1848–1852, major confluence of Godavari River shifted its course from Coringa Bay to about 5 miles South East. In addition to this, the construction of dam at Dowleswaram (1846–1852) and deforestation in Godavari Basin greatly reduced the flushing capacity of the river and accelerated the sediment transport. These phenomena resulted in silting up of the Coringa creeks. This silting caused the opening of the major branch of Godavari River in open sea in Hope Island and the increased sediment input resulted in a formation of a shoal during 1851. Littoral currents (flowing towards northern direction) tend to deposit the littoral drift materials, while tidal currents (perpendicular to shore) tend to remove these materials. Since tidal currents became weak (because of the large littoral bay mouth), littoral currents dominated the deposition and the shoal developed into a sand spit. With the increased growth rate of the sand spit northward, the bay entrance (between sand spit and mainland) reduced causing strong tidal currents which is the main cause of development of a hypersaline environment	Dredging of the bay entrance may reduce the magnitude of tidal currents (preferably in terms of amplitude) due to which restoration of ecological condition will occur and the hypersaline condition may be managed
2. Mangrove felling	Felling of mangroves for collecting timber, fuel wood, etc., results in the loss of mangroves. As many as 16 adjacent villages depend on mangroves as a source for firewood, some of the mangrove areas close to these villages exhibit denuded mangrove vegetation	People's participation through the model of joint mangrove management (JMM) will help to increase the awareness of the local villagers regarding the importance of mangroves. Also, plantation programme should be taken through proper institutional framework (preferably Forest Department) involving the local people
3. Hypersaline condition	In some pockets, natural degradation of mangrove occurs due to hypersaline condition because of reduced tidal flushing	Rainwater harvesting and channelization of the harvested rainwater through fish bone model may help diffuse the freshwater in the mangrove ecosystem by increasing the dilution factor of the system
4. Conversion of mangrove area to aquaculture	In Coringa mangrove ecosystem, the area under aquaculture has increased from 669 ha to 4,669 ha during a span of 10 years (1988–1998) which is an increase of 597.90 %. The overlay analysis reveals that about 639 ha of mangrove area has been converted into aquaculture	The best way to preserve the mangrove ecosystem in its pristine form is to reduce the anthropogenic pressure arising out of livelihood. Shrimp culture, being a lucrative livelihood in mangrove ecosystem, can only be diverted if some alternative livelihood is provided to the shrimp culturists in order to meet the financial need. These alternative livelihoods may encompass several non-conventional approaches such as oyster culture, seaweed culture, crab fattening and mangrove floral based fish feed preparation

(continued)

Table 3.4 (continued)

Threats	Description	Conservation
5. Pollution	The Gautami-Godavari mangrove estuarine ecosystem is severely polluted in terms of heavy metals. The region is significantly polluted with heavy metals such as Zn, Cu, Mn, Fe, Co, Ni and Cr. It is also reported that some of the main industries discharge their effluents directly into the aquatic system (about 6500 kL/day) which ultimately reach the Coringa River	Awareness generation programme involving the industrialists, fishermen, Forest Department staffs, pollution control authorities and other stakeholders may be arranged on regular basis to ensure the treatment of effluents before their discharge from the point sources. Also, high-end researches should be initiated to monitor the role of mangrove flora and other associated species in the process of bioremediation
6. Biodiversity loss	In Coringa mangrove ecosystem, fifteen species of true mangroves (dominant species are *Avicennia marina, A. officinalis, Exoecaria agallocha, Sonneratia apetala, Aegiceros corniculatum* and *Rhizophora apiculata*) and six mangrove associate species are very common (Checklist of Coringa mangroves ***Vide* Annexure**). *S. casualaris* is reported to be at the verge of extinction and *A. marina* is the most abundant in the region because of its wide range of salinity tolerance	Loss of biodiversity can be prevented only by the awareness of the local people and by providing alternative livelihood befitted with the local environment and situation. A very interesting research in this context has been published by Pramanick et al. (2014) in which the researchers prepared edible products from fruits of *Sonneratia apetala* (Fig. 3.10), which is not only nutritionally balanced, but also can serve as a unique source of livelihood for the villagers of mangrove fringe zones
	Molluscan shell burning is common in the area, and it is reported that about 3,600 tonnes of molluscs are removed annually from Kakinada Bay and Coringa mudflats for lime production	
7. Wild harvest of tiger prawn seeds	Collection of post-larval stage of *Penaeus monodon* is done to feed the shrimp farms in and around the Coringa mangrove ecosystem. During this collection, a variety of juveniles of other species are caught in the net and destroyed as they are non-remunerative to the seed collectors	Development of shrimp hatcheries at subsidized cost (through Government intervention), awareness to the seed collectors on regular basis and providing alternative livelihood to the tiger prawn seed collectors are some of the possible measures for conservation of fish juveniles that are destroyed during the process of collection and segregation of tiger prawn seeds
8. Grazing	Mangrove flora particularly the seedlings are grazed by cows, buffaloes, etc., due to which trees do not attain their maturity. It is reported that in Coringa village, about 2,500 cattle graze on the peripheral areas of the mangroves	Local-level awareness particularly involving the fishermen communities may yield good results

It is noted from this table that during the time span between 1988 and 1998, a significant reduction in mangrove cover is observed within the Coringa Reserve Forest area. This is confirmed with the subsequent increase in degraded areas and aquaculture areas (Table 3.5). Using the overlay facilities of GIS, analysis of 1988 and 1998 land-use/land cover maps was performed (Fig. 3.13).

The changes that have occurred in the whole area are highlighted in Table 3.6. The results clearly show that out of 13,359 ha of mangrove area, an area of 11,567 ha remained unchanged during the span of 10 years. This may be attributed to dense mangrove vegetative cover with minimum anthropogenic pressure. Changes have occurred in the areas as conversion of mangroves to other categories and vice versa. An area of 1,065 ha of mangrove has changed to other categories such as aquaculture, degraded area and mudflat. In some pockets, the mangroves have

Table 3.5 Land-use/land cover in 1988 and 1998

Area	Coringa Reserve Forest (ha)		Increase/decrease (%)	Whole area (ha)		Increase/decrease (%)
Category	1988	1998		1988	1998	
Mangrove	9,623	9,320	−3.15	13,359	12,127	−9.22
Degraded area	686	875	+27.55	1,309	1,292	−1.31
Aquaculture	0.07	0.39	+457.14	669	4,669	+597.90

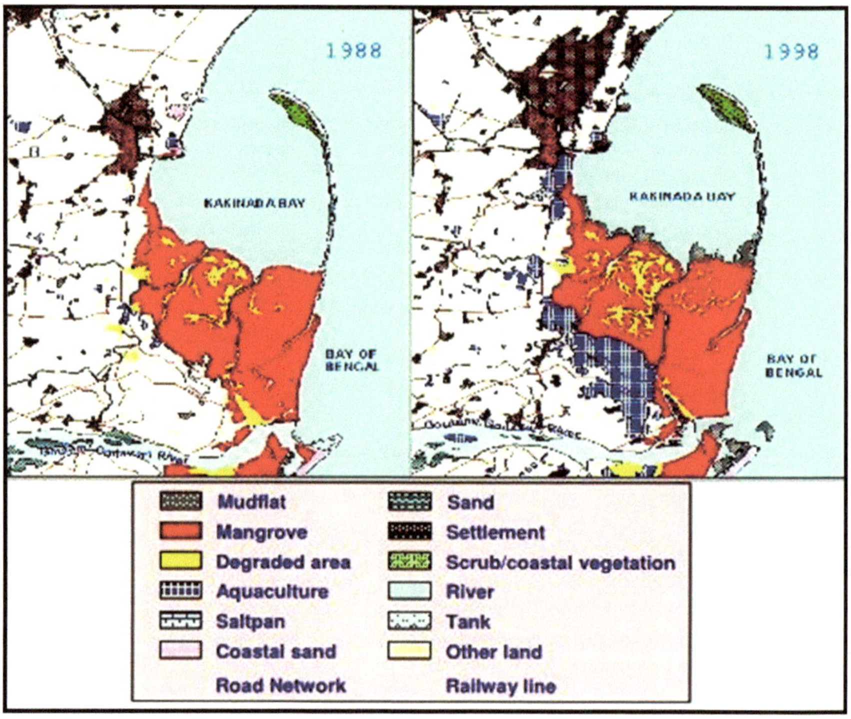

Fig. 3.12 Land use/land cover of Coringa in 1988 and 1998

grown in about 285 ha in the degraded area and other land (Table 3.6). This picture is not specific to Coringa, but is also observed in several other maritime states of India.

Annexure: Check List of Coringa Mangroves

- Family: AVICENNIACEAE
  1. *Avicennia marina* (Forsk.) Vierh.
  2. *Avicennia officinalis* Linn.
  3. *Avicennia alba* Blume.
- Family: ACANTHACEAE
  4. *Acanthus ilicifolius* Linn.
- Family: COMBRETACEAE
  5. *Lumnitzera racemosa* Willd.
- Family: EUPHORBIACEAE
  6. *Excoecaria agallocha* Linn.
- Family: MELIACEAE
  7. *Xylocarpus mekongensis* Pierre.
- Family: MYRSINACEAE
  8. *Aegiceras corniculatum* (Linn.) Blanco.
- Family: RHIZOPHORACEAE
  9. *Rhizophora apiculata*.
  10. *Rhizophora mucronata* Lamk.
  11. *Bruguiera gymnorrhiza* (Linn.) Lamk.

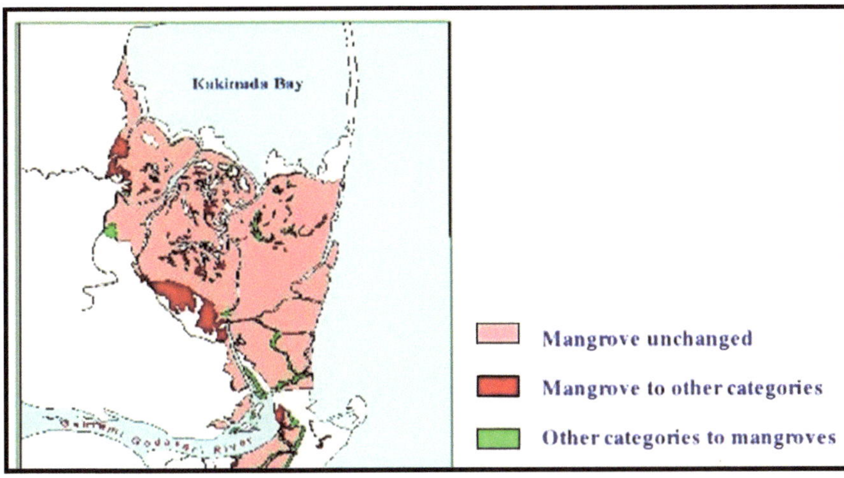

Fig. 3.13 Overlay analysis of 1988 and 1998 land-use maps

Table 3.6 Changes in land use/land cover in study areaĄ

Land use/land cover	Area (ha)
Mangrove remained as mangrove	11,567
Degraded area to mangrove	248
Other land to mangrove	37
Mangrove to aquaculture	639
Mangrove to degraded area	534
Mangrove to other land	213
Mangrove to mudflat	174
Aquaculture remained as aquaculture	625
Other land to aquaculture	2,715
Salt pan to aquaculture	418
Degraded area to aquaculture	159

12. *Bruguiera cylindrica* (Linn.) Bl.
13. *Ceriops decandra* (Griff.) Ding Hou.

- Family: SONNERATIACEAE

14. *Sonneratia apetala* Buch.-Ham.
15. *Sonneratia. caseolaris* (Linn.) Engler.

References

Department of Ocean Development (2001) ICMAM Project Directorate, Chennai, Government of India, Dec 2001

Froelich PN (1980) Analysis of organic carbon in marine sediments. Limnological Oceanogr 25:564–572

Hedges JI, Stern JH (1984) Carbon and nitrogen determinations of carbonate-containing solids. Limnological Oceanogr 29(3):657–663

Pramanick P, Zaman S, Bera D, Raha AK, Mitra A (2014) Mangrove fruit products: a search for alternative livelihood for island dwellers of lower Gangetic Delta. Int J Pharm Res Scholars (IJPRS) 3(I-1):131–137

Weliky K, Suess E, Ungerer CA, Muller PJ, Fischer K (1983) Problems with accurate carbon measurements in marine sediments and particulate matter in seawater: a new approach. Limnological Oceanogr 28:1252–1259

Internet Reference

http://glcf.umiacs.und.edu/

4 Blue Carbon in Floral Community

...Conserve the blue carbon lane
And increase the carbon gain...
The Authors

Forests play a crucial role in regulating the climate of the planet Earth by acting as important storehouses of carbon. Forest plants and soils drive the global carbon cycle by sequestering (storing) carbon dioxide through photosynthesis and releasing it through respiration. When the uptake of carbon dioxide (photosynthesis) exceeds losses via respiration, harvest and management, forests store carbon (C sinks). In an undisturbed forest, ~74 % of the carbon dioxide is stored in live stems and branches, 16 % is stored in roots, and 10 % is stored in soils. The global sink in forest vegetation and soils (Table 4.1) is estimated to be 1,200 Gt of carbon (1 Gt = 10^9 tonnes). This increases at a rate of 1–3 Gt annually.

Several researchers have estimated the carbon stored in forest vegetation in the pre-industrial and present times (Table 4.2).

The planet Earth sustains some 3.9 billion hectares or 39,000,000 km^2 of remaining forests (Brown et al. 2006) which is approximately 30 % of the world's land surface.

The tropical forest zone encompasses 1.76 billion hectares or 17,600,000 km^2 and is divided into six ecofloristic zones: the tropical rain forests, the moist deciduous forests, the dry zone, the very dry zone, the desert zone, the hill forest and mountain forest.

Tropical rain forests are found in areas with more than 2,500 mm of annual rainfall. They are evergreen, luxuriant and rich in animal and plant species. More than half the world's 718.3 million hectares of rain forests are located in two countries: Brazil (41 %) and Indonesia (13 %). Rain forest composition and structure vary with distance from the ocean, distance from rivers, altitude and geographic position.

Moist deciduous forests occur in areas with an annual rainfall of 1,000–2,000 mm. Forest structure varies depending on the amount and distribution of rain, the type of soil and the length of the dry season. Some dominant tree species may lose their leaves towards the end of the dry season. This forest type is generally less diverse than rain forest.

Dry zone forests are found in tropical areas receiving between 500 and 1,000 mm of rainfall per year. They are relatively open and include thornland, shrubland, savannah and other short and sparse woody vegetation. Dry zone forests tend to be fragile and are easily degraded. More than half are in Africa. Dry forest types include oak, mesquite, piñon-juniper, maquis and acacia.

Tropical upland forests are forests above 800 m and include cloud forests (montane rain forests), which are shorter, floristically simpler and more heavily laden with mosses and lichens than lowland rain forests. Tropical upland species are similar to temperate forest species. The upland zone covers the Himalayas, parts of Myanmar, Thailand and Vietnam, the highlands of Mexico, the Andes and the highlands of Ethiopia and mountains around Lake Victoria.

The total mangrove forest area accounts for 0.7 % of the total tropical forest in the world. Most mangroves (42 %) are found in Asia, followed by Africa (20 %), North and Central America (15 %), Oceania (12 %) and South America (15 %).

According to a study conducted by a team of US Forest Service and university scientists,

A. Mitra and S. Zaman, *Blue Carbon Reservoir of the Blue Planet*,
DOI 10.1007/978-81-322-2107-4_4, © Springer India 2015

Table 4.1 Carbon storage totals for global soils

Storage (Gt C)	Reservoir type	Author(s)
1,115 Gt	Soils, present potential ('prehistoric')	Adams et al. (1990)
1,395 Gt (1.)	Peats + soils, present potential	Adams et al. (1990)
1,405 Gt (3.)	Soils, present day	Bazilevich (1974) (s.)
3,000 Gt (4.)	Soils (+peats?), present day	Bohn (1978) (s.)
1,672 Gt (3.)	Soils, present day	Bolin et al. (1979) (s.)
1,477 Gt (5.)	Soils, present day	Buringh (1983) (s.)
1,515 Gt (6.)	Soils (+peat lands?) present day	Schlesinger (1984)
787 Gt	Forest soils only (+fine debris)	Dixon and Krankina (1993)
1,500 Gt (7.)	Soils, in 1989	IPCC (1990) (s.)
1,560 Gt (7.)	Soils, in 'pre-industrial' era	IPCC (1990) (s.)
860 Gt (8.)	Peats, present day	Bohn (1976) (s.)
300 Gt (8.)	Peats	Sjors (1980) (s.)
202 Gt (8.)	Peats	Post et al. (1982)
377 Gt (8.)	Peats	Bohn (1976, 1982a, b)
180–227 Gt (8.)	Peats	Gorham (1990) (s.)
461 Gt (9.)	Subarctic and boreal peat	Gorham and Janssens (1992)
1,576 Gt (10.)	Global soils (present day)	Eswaran et al. (1993)
500 Gt (11.)	Global peats	Markov et al. (1988) (s.)

Table 4.2 Previous global carbon storage estimates for vegetation

Storage (Gt C)	Reservoir type	Author(s)
827 Gt (1.)	Present actual land vegetation	Whittaker and Likens (1975) (s.)
560 Gt (2.)	Present actual land vegetation	Olson et al. (1983)
550 Gt (3.)	Present actual (1980s) land vegetation	IPCC (1990) (s.)
610 Gt (3.)	Pre-industrial (pre-1700) vegetation	IPCC (1990) (s.)
1,080 Gt (4)	Land vegetation, 'prehistoric' times	Bazilevich et al. (1971)
924 Gt	Present potential ('prehistoric') vegetation	Adams et al. (1990)
343 Gt (5.)	Last glacial maximum vegetation	Adams et al. (1990)
350 Gt	Coarse woody debris (present potential)	Harmon et al. (pers. Comm., 1990)
591 Gt (8.)	Present actual land vegetation	Ajtay et al. (1979) (s.)

coastal mangrove forests store more carbon than almost any other forest on Earth. Their findings are published online in the journal Nature Geoscience, (www.nature.com/naturegeoscience.com).

The mangrove forest's ability to store such large amounts of carbon can be attributed, in part, to the deep organic-rich soils in which it thrives (Fig. 4.1).

The soil depth mostly increases with time as the silt of the overlying aquatic phase deposit on the existing soil bed (intertidal mudflats). This happens because the intertidal mudflats suffer total submergence by silty water during high tide (Fig. 4.2).

Mangrove sediment carbon stores were on average five times larger than those typically observed in temperate, boreal and tropical terrestrial forests, on a per-unit-area basis. The mangrove forest's complex root systems, which anchor the plants into underwater sediment, slow down incoming tidal waters allowing organic and inorganic material to settle into the sediment surface. Low oxygen conditions slow decay rates, resulting in much of the carbon accumulating in the soil. In fact, mangroves have more carbon in their soil alone than most tropical forests have in all their biomass and soil combined.

Fig. 4.1 Organic carbon-rich soil of mangrove ecosystem

Fig. 4.2 Complete submergence of intertidal mudflats during high tide

The capacity of mangroves, seagrasses and salt marshes to sequester carbon dioxide from the atmosphere is becoming increasingly recognized at an international level. Of all the biological carbon, also termed as 'green carbon', captured in the world, over half (55 %) is captured by mangroves, seagrasses, salt marshes and other marine living organisms, which are also known more specifically as 'blue carbon'. Mangroves, salt marshes and seagrasses form much of the Earth's blue carbon sinks (http://www.recoftc.org/site/resources/Mangroves-more-Carbon-Rich-and-Important-for-Climate-Change.php).

4.1 Mangroves

Mangrove ecosystems are concentrated along tropical and subtropical coasts and are the only known woody halophytes. A part of their productivity flows into adjacent ecosystems, or conversely, they receive organic materials from estuarine or oceanic ecosystems (Fig. 4.3) (Ong 1993; Kristensen et al. 2008).

The biological and ecological aspects of mangroves have been studied worldwide (Lugo and Snedaker 1974; Boto and Wellington 1984; Hutchings and Saenger 1985; Twilley et al. 1986; Odum and Mclvor 1990; Twilley 1995; Cox and Allen 1999). Most researchers deal with patterns of primary productivity, nutrient cycle and detrital export in riverine mangroves, whereas fringe and scrub mangrove have been studied to a lesser extent (Twilley et al. 1992; Alongi et al. 1992; Lee 1995). The field survey of mangrove biomass and productivity is rather difficult due to muddy soil conditions and the heavy weight of the wood. The peculiar tree form of mangroves, especially their unusual roots, has long attracted the attention of botanists and ecologists (Tomlinson 1986). The survival of mangroves in soft muddy substratum is basically because of the support of their root system. To maintain a bottom-heavy tree form (Ong et al. 2004) or a low ratio of top biomass to root biomass (T/R ratio; Komiyama et al. 2000), mangroves might allocate a great deal of biomass to their roots. This phenomenon may produce peculiar conditions for ecosystem processes in root zones, owing to the anaerobic conditions.

Over the years, forest ecologists have developed various methods to estimate the biomass of forests. Three important methods are usually adopted for estimating forest biomass: the harvest method, the mean-tree method and the allometric method. In a mature mangrove forest, the total weight of an individual tree often reaches several tons (Komiyama et al. 2005). Therefore, the harvest method cannot be easily used in mature forests and in itself is not reproducible because all trees must be destructively harvested. The mean-tree method is utilized only in forests with a homogeneous tree size distribution, such as plantations. The allometric

Fig. 4.3 Mangrove forests supply nutrients to the adjacent estuarine water

method estimates the whole or partial weight of a tree from measurable tree dimensions, including trunk diameter and height, using allometric equations. This is a non-destructive method and is thus useful for estimating temporal changes in forest biomass by means of subsequent measurements. However, the site- and species-specific dependencies of allometric equations pose a problem to researchers because tree weight measurement in mangrove forests is labour intensive. Based on studies of forest biomass using the allometric method and other characters, Kira and Shidei (1967) summarized the so-called summation method for estimating the net primary production (NPP) of forests. In this method, the rates of growth increment, death and consumption by herbivores are summed to obtain the NPP. The gross primary production (GPP) of forests can then be calculated by adding the rate of metabolic respiration to the NPP. Recently, interest has grown in the study of carbon fluxes of an entire ecosystem, which includes carbon emissions from soil respiration. Net ecosystem production (NEP) is a sophisticated criterion to judge carbon fixation from the NPP and the rate of soil respiration. One method for estimating the NEP is through the eddy covariance. Essentially, this consists of taking rapid measurements of the vertical component of air velocity and the concentration of carbon dioxide/water vapour in the air above forest canopies and taking their covariance. However, this method requires large equipment in mangrove forests, high-priced instruments and complex computation.

Allometric equations for mangroves have been developed for several decades to estimate biomass and subsequent growth. Most studies have used allometric equations for single-stemmed trees, but mangroves sometimes have multistemmed tree forms, as often seen in *Rhizophora*, *Avicennia* and *Excoecaria* species (Clough et al. 1997; Dahdouh Guebas and Koedam 2006). Clough et al. (1997) showed that the allometric relationship can be used for trunks in a multistemmed tree.

Moreover, for dwarf mangrove trees, allometric relationships have been used to estimate the biomass (Ross et al. 2001). For studies on single-stemmed trees published from 1984 to 2000, Saenger (2002) cited 43 allometric equations on AGB. His review and subsequent studies by Tam and Wong (1995), Ong et al. (2004), Comley and McGuinness (2005) and Soares and Schaeffer-Novelli (2005) provide a good overall survey of the relevant literature. They found that species-specific trait of allometry (i.e., the allometric equation) is significantly different among mangrove tree species. Clough et al. (1997) found different relationships in different sites, although Ong et al. (2004) reported similar equations applied to two different sites for *Rhizophora apiculata*. This issue is important for practical uses of allometric equations. If the equations are segregated by species and site, then different expressions can be established for each site (Table 4.3).

On both the species- and site-specific issues of allometry, Chave et al. (2005) and Komiyama et al. (2005) proposed the use of a common allometric equation for mangroves. The common allometric equation that Komiyama et al. (2005) proposed is based on the pipe model (Shinozaki et al. 1964) and the static model of plant form (Oohata and Shinozaki 1979). These models predict that the partial weight of the trunk at a certain height physically sustains the weight of the upper tree body, regardless of tree species and locality. By using these two theories, Komiyama et al. (2005) derived equations with trunk diameter and wood density as parameters and found good fits with 104 sample trees comprising 10 mangrove species from Thailand and Indonesia (the data, Tamai et al. 1986; Komiyama et al. 1988, are included in this common equation). The common equation of Chave et al. (2005) was established based on statistical analysis but nevertheless consisted of the same two parameters used by Komiyama et al. (2005) (Table 4.3). These two common equations have the advantage of requiring only two parameters, even though Soares and Schaeffer-Novelli (2005) list a large number of parameters in their allometric equations for mangroves. The measurement of trunk diameter or girth is more practical than other parameters, especially for those working in closed and tall canopies where tree height is difficult to accurately measure. Wood

Table 4.3 Allometric equations for various mangroves based on DBH (cm)

Species	Aboveground tree weight (W_{top} in kg)	References
Avicennia germinans	$W_{top} = 0.140\mathrm{DBH}^{2.40}$	Fromard et al. (1998)[a]
	$r^2 = 0.97$, $n = 45$, Dmax = 4 cm	
	$W_{top} = 0.0942\mathrm{DBH}^{2.54}$	Imbert and Rollet (1989)[a]
	$r^2 = 0.99$, $n = 21$, Dmax: unknown	
Avicennia marina	$W_{top} = 0.308\mathrm{DBH}^{2.11}$	Comley and McGuinness (2005)
	$r^2 = 0.97$, $n = 22$, Dmax = 35 cm	
Laguncularia racemosa	$W_{top} = 0.102\mathrm{DBH}^{2.50}$ $r^2 = 0.97$,	Fromard et al. (1998)[a]
	$n = 70$, Dmax = 10 cm,	
	$W_{top} = 0.209\mathrm{DBH}^{2.24}$ $r^2 = 0.99$,	Imbert and Rollet (1989)[a]
	$n = 17$, Dmax: unknown	
Rhizophora apiculata	$W_{top} = 0.235\mathrm{DBH}^{2.42}$ $r^2 = 0.98$,	Ong et al. (2004)
	$n = 57$, Dmax = 28 cm	
Rhizophora mangle	$W_{top} = 0.178\mathrm{DBH}^{2.47}$ $r^2 = 0.98$,	Imbert and Rollet (1989)[a]
	$n = 17$, Dmax: unknown	
Rhizophora spp.	$W_{top} = 0.128\mathrm{DBH}^{2.60}$ $r^2 = 0.92$,	Fromard et al. (1998)[a]
	$n = 9$, Dmax = 32 cm	
	$W_{top} = 0.105\mathrm{DBH}^{2.68}$ $r^2 = 0.99$,	Clough and Scott (1989)[a]
	$n = 23$, Dmax = 25 cm	
Bruguiera gymnorrhiza	$W_{top} = 0.186\mathrm{DBH}^{2.31}$ $r^2 = 0.99$,	Clough and Scott (1989)[a]
	$n = 17$, Dmax = 25 cm	
Bruguiera parviflora	$W_{top} = 0.168\mathrm{DBH}^{2.42}$ $r^2 = 0.99$,	Clough and Scott (1989)[a]
	Dmax = 25 cm, $n = 16$	
Ceriops australis	$W_{top} = 0.189\mathrm{DBH}^{2.34}$ $r^2 = 0.99$,	Clough and Scott (1989)[a]
	$n = 26$, Dmax = 20 cm	
Xylocarpus grnatum	$W_{top} = 0.0823\mathrm{DBH}^{2.59}$ $r^2 = 0.99$, $n = 15$, Dmax = 25 cm	Clough and Scott (1989)[a]
Common equation	$W_{top} = 0.251\mathrm{pD}^{2.46}$ $r^2 = 0.98$,	Komiyama et al. (2005)
	$n = 104$, Dmax = 49 cm	
	$W_{top} = 0.168\mathrm{pDBH}^{2.47}$ $r^2 = 0.99$,	Chave et al. (2005)
	$n = 84$, Dmax = 50 cm	

W_{top}: aboveground tree weight
[a] After Saenger (2002)
D_{max}: the upper range of samples

density differs significantly in different mangrove species, but less for individuals within a species (Komiyama et al. 2005).

A number of reports are now available on mangrove biomass from different regions of the world. AGB of 460 t ha^{-1} was reported from a forest dominated by *R. apiculata* in Malaysia (Putz and Chan 1986). AGB of more than 300 t ha^{-1} was documented in mangrove forests in Indonesia (Komiyama et al. 1988) and French Guiana (Fromard et al. 1998). The AGB was less than 100 t ha^{-1} in most secondary forests or concession areas. In high-latitude areas (>24° 23′ N or S), primary forests mostly have an AGB of around 100 t ha^{-1}; however, even at 27° 24′S, an AGB of 341 t ha^{-1} was reported for an *Avicennia marina* forest (Mackey 1993). The lowest AGB reported was 7.9 t ha^{-1} for a *Rhizophora mangle* forest in Florida, USA (Lugo and Snedaker 1974). The canopy height of mangrove forests is generally lower at higher latitudes (Pool et al. 1977; Saenger and Snedaker 1993) which is a justified reason for relatively lower AGB in higher latitudes (Table 4.4).

Table 4.4 Global data of AGB of different mangrove species

Region	Location	Condition or age	Species	ABG (t ha $^{-1}$)	References
Australia	27° 24′S, 153° 8′E	Secondary forest	*A. marina* forest	341.0	Mackey (1993)
Thailand (Ranong Southern)	9°N, 98°E	Primary forest	*Sonneratia* forest	281.2	Komiyama et al. (1987)
Sri Lanka	8° 15′N, 79° 50′E	Fringe	*Avicennia* sp.	193.0	Amarasinghe and Balasubramaniam (1992)
Indonesia (Halmahera)	1° 10′N, 127° 57′E	Primary forest	*Sonneratia* forest	169.1	Komiyama et al. (1987)
Australia	33° 50′S, 151° 9′E	Primary forest	*A. marina* forest	144.5	Briggs (1977)
French Guiana	4° 52′N, 52° 19′E	Matured coastal	*Laguncularia* sp., *Avicennia* sp., *Rhizophora* sp.	315.0	Fromard et al. (1998)
South Africa	29° 48′S, 31° 03′E	–	*Bruguiera gymnorrhiza, A. marina*	94.5	Steinke et al. (1995)
French Guiana	5° 23′N, 52° 50′E	Pioneer stage 1 year	*Avicennia* sp.	35.1	Fromard et al. (1998)
Western Indian Sundarbans	21° 43′ 08.58″ N 88° 10′ 44.55″ E	Natural forest	*Sonneratia apetala*	53.70	This study (average of all selected stations and seasons)
Central Indian Sundarbans	22° 16′ 33.79″ N 88° 48′ 17.60″ E	Natural forest	*Sonneratia apetala*	7.12	This study (average of all selected stations and seasons)

In contrast, at low latitudes, primary or mature mangrove forests generally have high AGB. The AGB is always low in temperate areas and may be related to different climatic conditions, such as temperature, solar radiation, precipitation and frequency of storms.

It is interesting to note that AGB in mangroves is primarily influenced by stem and not by branches and leaves. A study conducted by Mitra et al. (2009) on three dominant mangrove species (*Sonneratia apetala*, *Excoecaria agallocha* and *Avicennia alba*) in western and central Indian Sundarbans clearly confirms that stem biomass (which is a direct function of DBH) is a unique indicator of mangrove AGB unlike branches and leaves that contribute substantially to litter fall and less to permanent biomass (Figs. 4.4, 4.5 and 4.6).

The mangrove ecosystem acts as a unique sink of carbon. The magnitude or the quantum of stored carbon in mangroves is a direct function of stem biomass. The scatter plots computed to evaluate the interrelationships between stored carbon in the same three species (*S. apetala*, *E. agallocha* and *A. alba*) aboveground vegetative parts highlight greatest dependency of stored carbon on stem biomass (Figs. 4.7, 4.8 and 4.9).

Atmospheric carbon dioxide has increased from 280 parts per million by volume (ppmv) in the year 1880 to nearly 370 ppmv in the year 2000 (Houghton et al. 2001). Most atmospheric carbon dioxide resulting from fossil fuels will be absorbed into the ocean, affecting ocean chemistry. Question may arise at this point that whether the rise in carbon dioxide level will increase the potentiality of mangrove system to

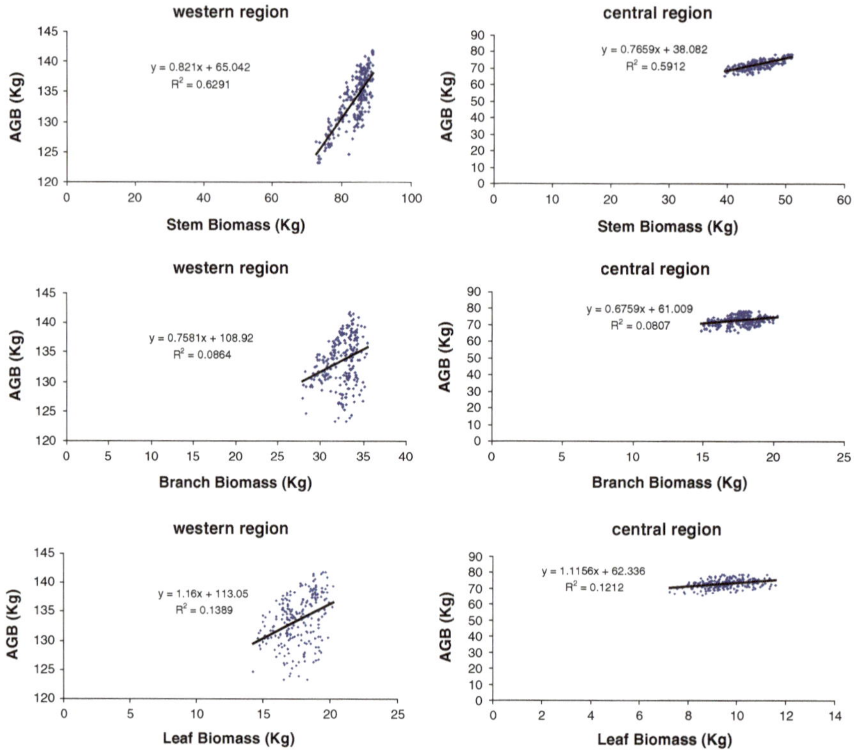

Fig. 4.4 Scatter plots showing the relationship between AGB and stem, branch and leaf biomass of *Sonneratia apetala* in the western and central regions of Indian Sundarbans

store carbon. This is a relevant question, no doubt, but the answer is controlled by several environmental parameters (such as salinity and soil nutrients) that affect the growth and survival of mangroves. The storage potential of mangroves is also salinity specific; for example, *Heritiera fomes* (locally known as Sundari) can survive luxuriantly in complete freshwater environment (Fig. 4.10) environment and can store substantial carbon.

Heritiera fomes is gradually getting extinct owing to increase in salinity in central Indian Sundarbans and can be considered as most sensitive species (Figs. 4.10 and 4.11) in relation to sea-level rise and subsequent increase in salinity.

Buttress roots are seen in several mangrove species such as *Heritiera littoralis and H. fomes* (Fig. 4.12). Buttress roots provide stability to the trees, especially in tropical areas. They can grow up to 10 m in height.

According to UNEP (1994), the efficiency of mangrove water use will be enhanced, and there will be specific species variation in response to elevated carbon dioxide. Due to the increase in water-use efficiency, mangroves in arid regions may benefit because decreased water loss via transpiration will accompany carbon dioxide uptake (Ball and Munns 1992). If salinity increases in arid regions, then this advantage may be lost, because increases in carbon dioxide do not affect mangrove growth when salinity is too high for a species to maintain water uptake.

Increases in carbon dioxide are not likely to cause mangrove canopy photosynthesis to increase significantly (UNEP 1994). However, in an experiment testing, the effects of humidity, salinity and increased carbon dioxide on two Australian mangrove species, *Rhizophora stylosa* and *Rhizophora apiculata*, photosynthesis increase significantly with increased levels of

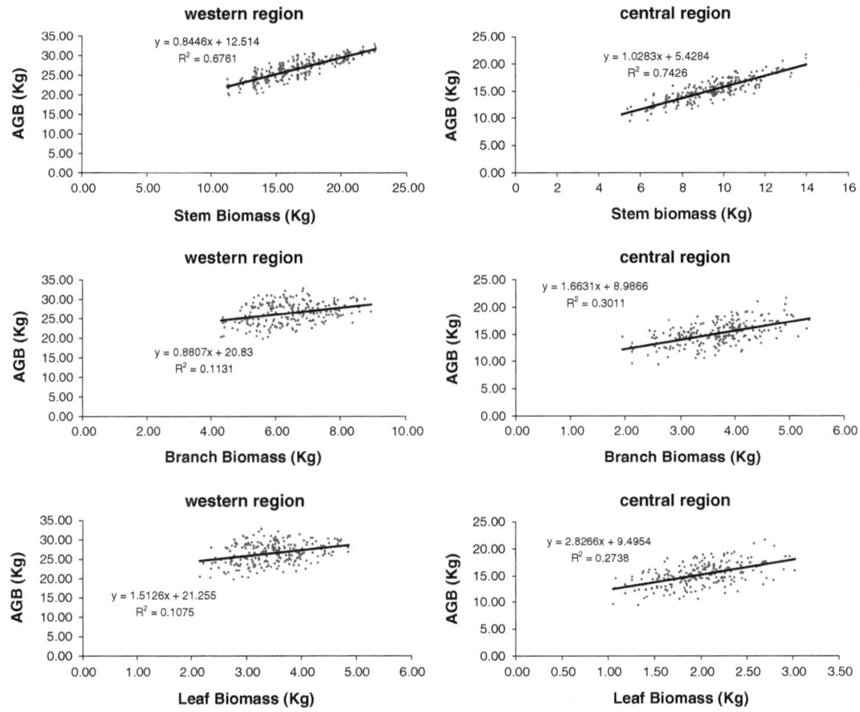

Fig. 4.5 Scatter plots showing the relationship between AGB and stem, branch and leaf biomass of *Excoecaria agallocha* in the western and central regions of Indian Sundarbans

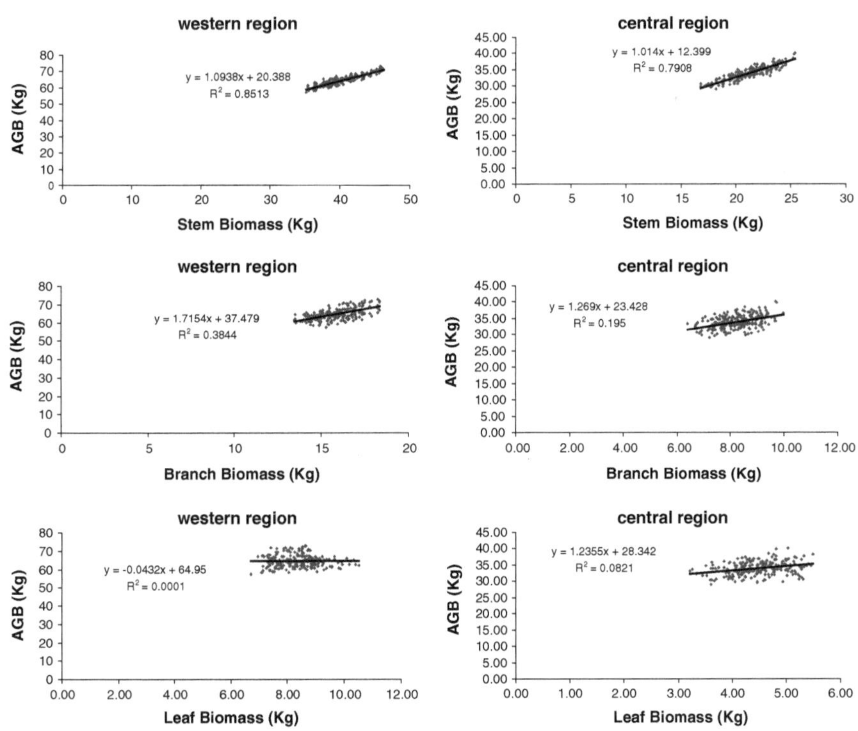

Fig. 4.6 Scatter plots showing the relationship between AGB and stem, branch and leaf biomass of *Avicennia alba* in the western and central regions of Indian Sundarbans

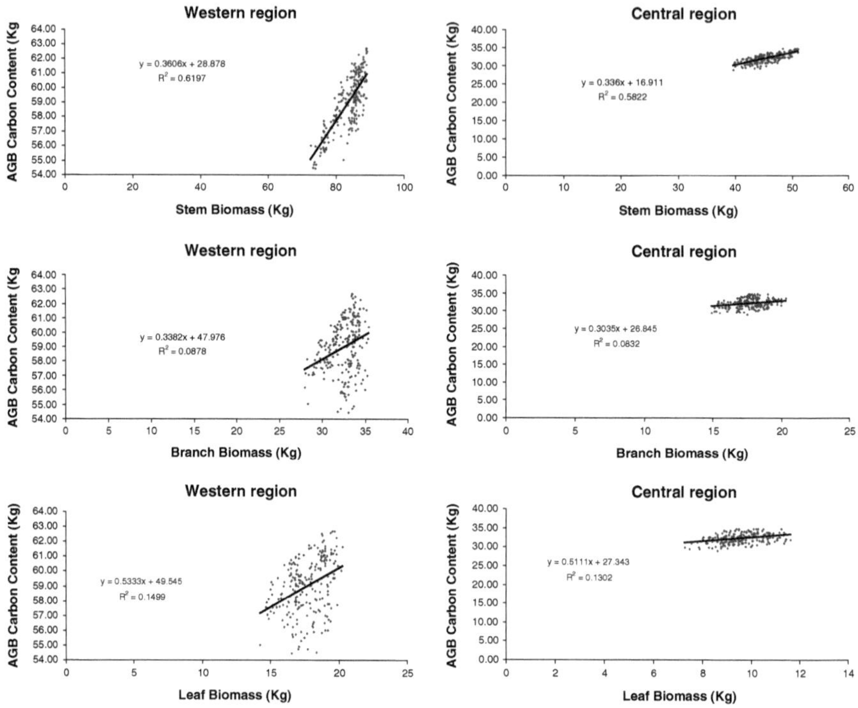

Fig. 4.7 Scatter plots showing the relationship between AGB carbon content and stem, branch and leaf biomass of *Sonneratia apetala* in the western and central regions of Indian Sundarbans

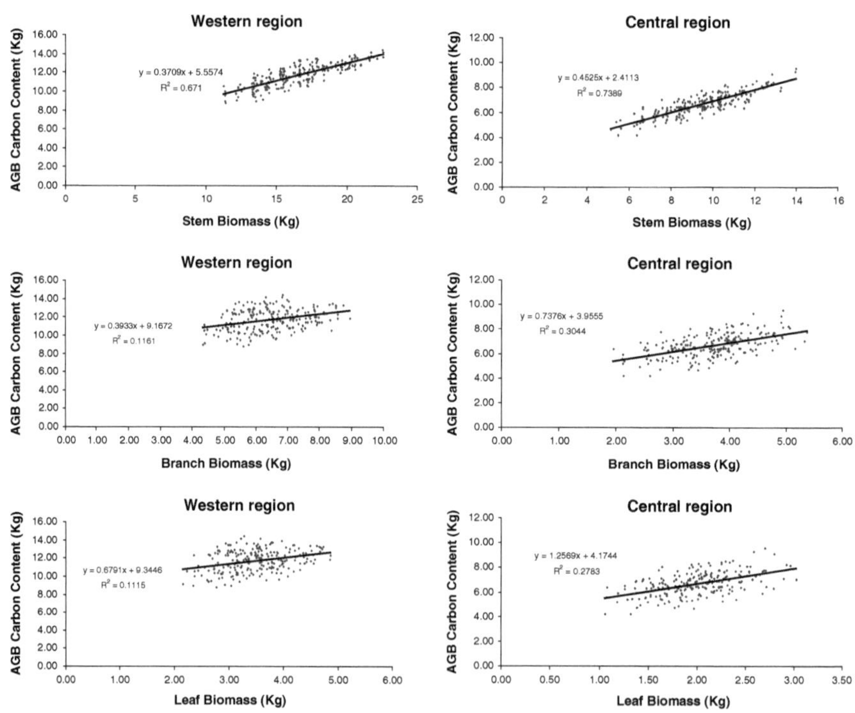

Fig. 4.8 Scatter plots showing the relationship between AGB carbon content and stem, branch and leaf biomass of *Excoecaria agallocha* in the western and central regions of Indian Sundarbans

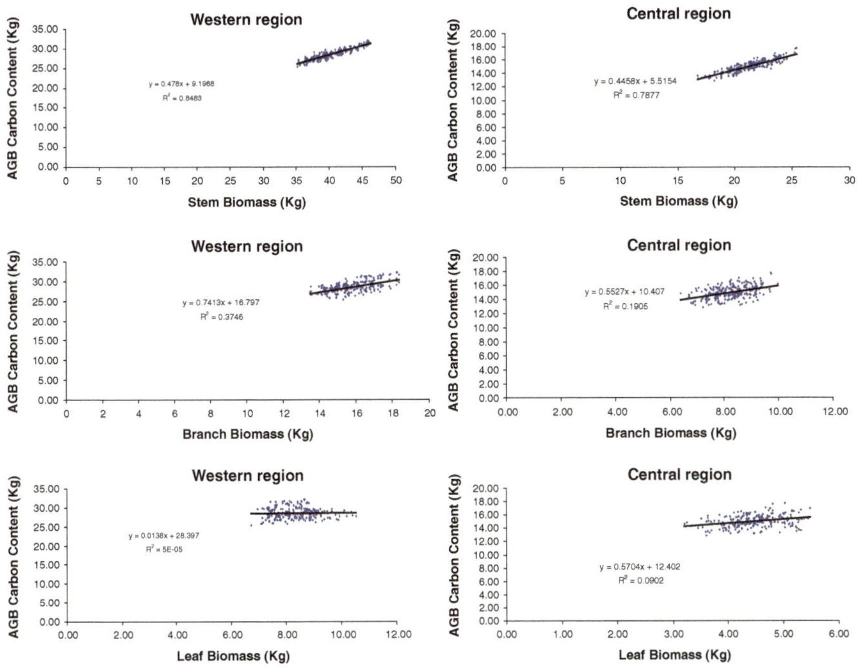

Fig. 4.9 Scatter plots showing the relationship between AGB carbon content and stem, branch and leaf biomass of *Avicennia alba* in the western and central regions of Indian Sundarbans

Fig. 4.10 Dr. Sufia Zaman with the *Heritiera fomes* seedlings growing normally in the freshwater environment of the city of Kolkata (India)

Fig. 4.11 Freshwater-loving Sundari (*Hertiera fomes*) is gradually vanishing from the Matla riverine zone: a threat posed by rising salinity

carbon dioxide (Ball et al. 1997). The mangroves were grown in glasshouses for 14 weeks with different combinations of atmospheric carbon dioxide (340 and 700 ppm), relative humidity (43 and 86 %) and salinity (25 and 75 % of sea water) to determine the effects of these variables on their development and growth (Ball et al. 1997). Although *R. stylosa* has a slower relative growth rate and greater salt tolerance than *R. apiculata*, the scientists concluded that elevated carbon dioxide significantly increased rates of net photosynthesis in both mangrove species, but only when grown at the lower salinity level. In addition, while increased carbon dioxide levels did not significantly affect the relative growth rate of either species, the average growth rates of both species increased with atmospheric carbon dioxide enrichment in the lower salt environment (Ball et al. 1997). These scientists postulated that increased levels of carbon dioxide might allow these two mangrove species to expand into areas of greater aridity, thus increasing species diversity in those regions.

Snedaker and Araújo (1998) exposed four mangrove species *R. mangle*, *Avicennia germinans, Laguncularia racemosa* and *Conocarpus erectus* to increased carbon dioxide (361–485 ppm). All four species demonstrated significant decreases in stomatal conductance and transpiration and an increase in instantaneous transpiration efficiency. Only *L. racemosa* demonstrated a significant decrease in net primary

Fig. 4.12 Buttress root of *H. fomes* (marked with *yellow arrows*): a unique feature of the species

productivity when exposed to increased carbon dioxide. Snedaker and Araújo (1998) suggest that increased levels of carbon dioxide on a global scale may result in a competitive disadvantage of *L. racemosa* in mixed mangrove communities relative to the other species whose rates of net primary productivity are not significantly affected by increases in carbon dioxide. The results of this study indicate that global increases in carbon dioxide may result in a competitive advantage of mangroves in arid regions due to their ability to minimize water use during periods of water stress while maintaining relatively high rates of carbon dioxide uptake (Snedaker and Araújo 1998).

Farnsworth et al. (1996) analysed the effects of doubled levels of carbon dioxide on *R. mangle* seedlings. The seedlings demonstrated significant increases in biomass, total stem length, branching activity and total leaf area compared to seedlings grown in normal levels of carbon dioxide. In this study, reproduction of *R. mangle* was achieved after only one year of growth in a high carbon dioxide environment, whereas it typically takes a full two years before they are able to reproduce in the field; thus, elevated carbon dioxide also appeared to accelerate maturation in addition to growth. However, Ellison and Farnsworth (1996) predict that whether increased atmospheric carbon dioxide results in enhanced growth of mangroves, it will likely not be enough to compensate for the negative impacts of sea-level rise.

It has been reported that mangroves are among the most carbon-dense forests in the tropics (sample-wide mean: 1,023 MgC ha^{-1} ±88 s:e:m:) and exceptionally high compared to mean carbon storage of the world's major forest domains (Donato et al. 2011). Estuarine sites contained a mean of 1,074 MgC ha^{-1} (±171 s:e:m:); oceanic sites contained 990 ± 96 MgC ha^{-1}. AGC pools were sizeable (mean 159 MgC ha^{-1}, maximum 435 MgC ha^{-1}), but belowground storage in soils dominated, accounting for 71–98 % and 49–90 % of total storage in estuarine and oceanic sites, respectively. High per-hectare carbon storage coupled with a pantropical distribution (total area ~14 million hectares; FAO 2007; Bouillon et al. 2009), suggests mangroves are a globally important surface carbon reserve.

A pilot project undertaken in the Indian Sundarbans region with the financial assistance of Ministry of Earth Science, Government of India revealed significant spatial differences of stored

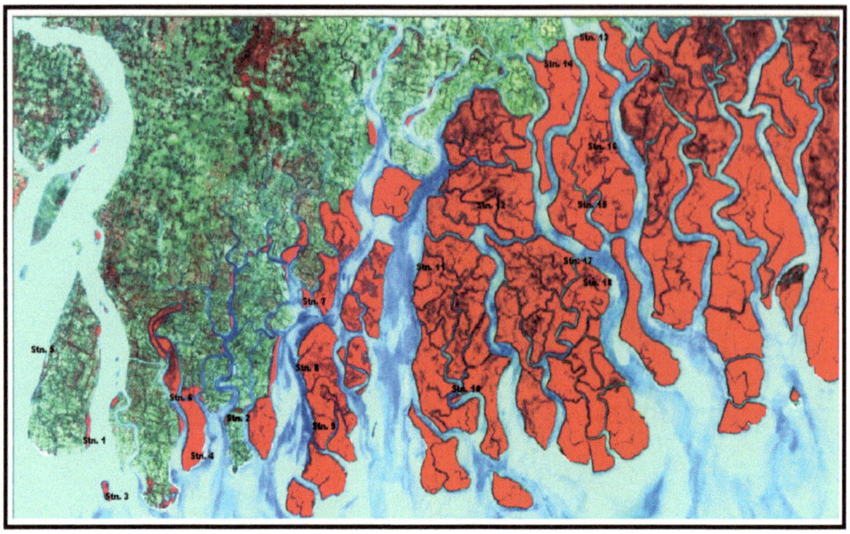

Fig. 4.13 Location of sector-wise sampling stations in Indian Sundarbans; the *red* colour indicates the mangrove vegetation

carbon in the study area owing to variation in mangrove forest area. This study was conducted in three sectors of Indian Sundarbans (Fig. 4.13 and Table 4.5) which are drastically different in terms of salinity (Fig. 4.14). Two interesting findings came out from this study. The first finding is that the three sectors are different in context to AGB and AGC of *Sonneretia apetala, E. agallocha, A. alba, A. officinalis and A. marina* (Table 4.7). Question may arise at this point that why the growth rate and stored carbon exhibited such a variation? Although the answer is clear from Fig. 4.14 that points towards salinity, but still we examined spatial variations of parameters such as

Table 4.5 Sampling stations in the western, central and eastern sectors of Indian Sundarbans

Sectors		Sampling stations	Latitude	Longitude
Western sector	Stn. 1	Chemaguri (W_1)	21° 38′ 25.86″N	88° 08′ 53.55″E
	Stn. 2	Saptamukhi (W_2)	21° 40′ 02.33″N	88° 23′ 27.18″E
	Stn. 3	Jambu Island (W_3)	21° 35′ 42.03″N	88° 10′ 22.76″E
	Stn. 4	Lothian (W_4)	21° 38′ 21.20″N	88° 20′ 29.32″E
	Stn. 5	Harinbari (W_5)	21° 44′ 22.55″N	88° 04′ 32.97″E
	Stn. 6	Prentice Island (W_6)	21° 42′ 47.88″N	88° 17′ 55.05″E
Central sector	Stn. 7	Thakuran Char (C_1)	21° 49′ 53.17″N	88° 31′ 25.57″E
	Stn. 8	Dhulibasani (C_2)	21° 47′ 06.62″N	88° 33′ 48.20″E
	Stn. 9	Chulkathi (C_3)	21° 41′ 53.62″N	88° 34′ 10.31″E
	Stn. 10	Goashaba (C_4)	21° 43′ 50.64″N	88° 46′ 41.44″E
	Stn. 11	Matla (C_5)	21° 53′ 15.30″N	88° 44′ 08.74″E
	Stn. 12	Pirkhali (C_6)	22° 06′ 00.97″N	88° 51′ 06.04″E
Eastern sector	Stn. 13	Arbesi (E_1)	22° 11′ 43.14″N	89° 01′ 09.04″E
	Stn. 14	Jhilla (E_2)	22° 09′ 51.53″N	88° 57′ 57.07″E
	Stn. 15	Harinbhanga (E_3)	21° 57′ 17.85″N	88° 59′ 33.24″E
	Stn. 16	Khatuajhuri (E_4)	22° 03′ 06.55″N	89° 01′ 05.33″E
	Stn. 17	Chamta (E_5)	21° 53′ 18.56″N	88° 57′ 11.40″E
	Stn. 18	Chandkhali (E_6)	21° 51′ 13.59″N	89° 00′ 44.68″E

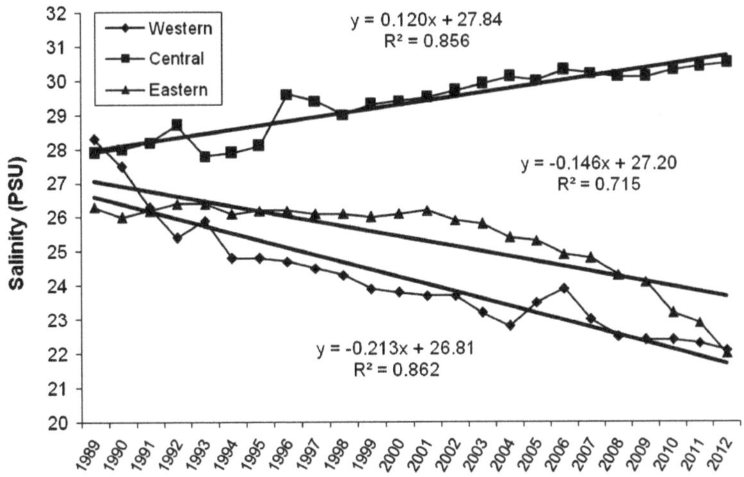

Fig. 4.14 Spatio-temporal variations of salinity in the Indian Sundarbans; each point is the average of six stations and three seasons

Table 4.6 ANOVA of hydrological parameters in Indian Sundarbans

Variable	F_{cal}	F_{crit} ($p < 0.05$)
Salinity		
Between sectors	17,612	2.743
Between years	0.740	1.705
pH		
Between sectors	0.886	4.102
Between stations	0.791	3.325
DO		
Between sectors	0.697	4.102
Between stations	1.128	3.325
Nitrate		
Between sectors	0.552	4.102
Between stations	0.106	3.325
Phosphate		
Between sectors	1.922	4.102
Between stations	0.867	3.325
Silicate		
Between sectors	1.288	4.102
Between stations	1.501	3.325
Soil organic carbon		
Between sectors	0.164	4.102
Between stations	0.342	3.325

Table 4.7 ANOVA of AGB and AGC of selected mangrove species in Indian Sundarbans

Variable	F_{cal}		F_{crit} ($p < 0.05$)
	AGB	AGC	
Sonneratia apetala			
Between sectors	31.366	23.895	4.102
Between stations	1.226	1.232	3.325
Avicennia alba			
Between sectors	5.261	5.068	4.102
Between stations	0.440	0.396	3.325
Excoecaria agallocha			
Between sectors	10.311	9.861	4.102
Between stations	1.327	1.203	3.325
Avicennia officinalis			
Between sectors	5.337	5.166	4.102
Between stations	0.487	0.453	3.325
Avicennia marina			
Between sectors	5.085	4.971	4.102
Between stations	0.889	0.840	3.325

pH, DO, nitrate, phosphate, silicate and soil organic carbon through ANOVA and confirmed that salinity is the only causative factor regulating the spatial variations of AGB and AGC (Table 4.6). The second important finding points towards the resilience of *E. agallocha* and vulnerability of *S. apetala* to salinity. Such responses of the species towards salinity have been confirmed through correlation coefficient values between the AGB, AGC and salinity of the aquatic phase surrounding the species (Table 4.8).

Mangrove plants use C_3 photosynthetic biochemistry. There are several proofs to confirm this statement. Firstly, the leaves lack the anatomical features (such as bundle sheath cells) characteristic of leaves employing C_4 photosynthesis.

Table 4.8 Interrelationships between aboveground biomass and carbon with salinity in the western, central and eastern sector of Indian Sundarbans

Combination	*r* Value (AGB)			*r* Value (AGC)		
	Western sector	Central sector	Eastern sector	Western sector	Central sector	Eastern sector
Salinity × *S. apetala*	−0.8640*	−0.8307*	−0.9234*	−0.8479*	−0.8477*	−0.9754*
Salinity × *A. alba*	−0.0336	−0.2950	0.0401	−0.0232	−0.2842	0.1375
Salinity × *E. agallocha*	0.9183*	0.9714*	0.8811*	0.9340*	0.9654*	0.8246*
Salinity × *A. officinalis*	−0.0396	−0.1679	0.1874	−0.0279	−0.1657	0.3318
Salinity × *A. marina*	−0.0049	−0.0647	−0.2131	0.0038	−0.0714	−0.1237

*$p < 0.01$

Secondly, the carbon isotope composition of leaves shows a range of $\delta^{13}C$ values from −23.2 to −34.3°/oo reported for C_3 plants (Smith and Epstein 1971). Thirdly, the carbon dioxide compensation point is at 25 °C and saturated light intensity ranges from 45 to 60 μl l^{-1}, consistent with the photorespiratory activity typical of C_3 photosynthesis. Fourthly, the photosynthetic rates are maximal at leaf temperature less than 35 °C and generally become light saturated at intensities ranging from 30 to 50 % of full sunlight. Finally, there is no convincing evidence in mangroves of environmentally induced from C_3 to either C_4 or CAM photosynthetic biochemistry. The gas exchange in mangroves is strongly regulated by the amount and properties of the enzyme ribulose-1, 5-biphosphate carboxylase/oxygenase (RuBisCO), which catalyses the combination of carbon dioxide with the acceptor molecule ribulose-1, 5-biphosphate (Fig. 4.15).

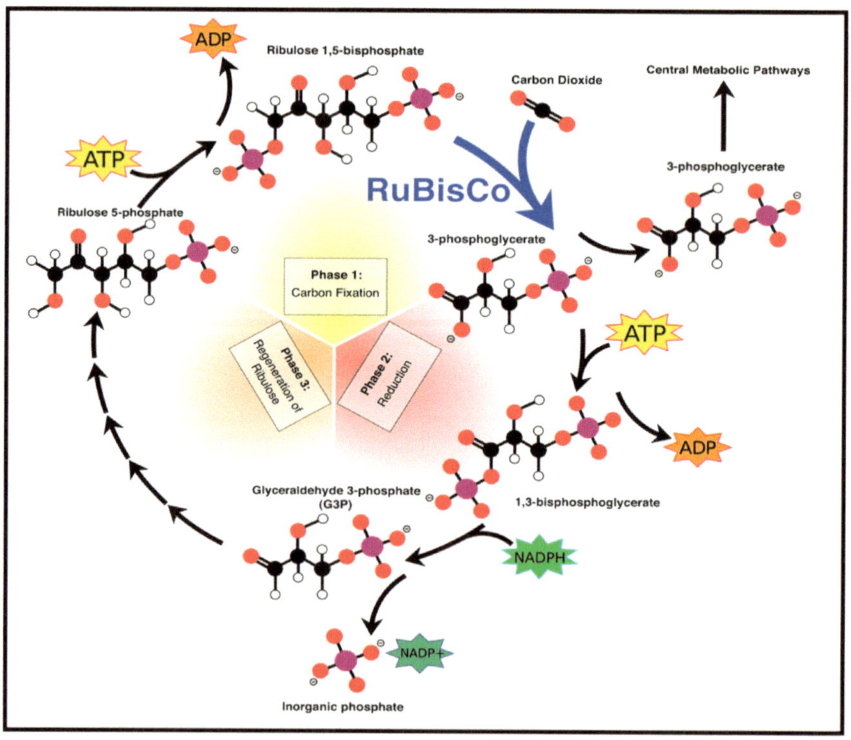

Fig. 4.15 RuBisCo action in mangrove flora

4.2 Seagrass and Salt Marsh Grass

4.2.1 Seagrass

Seagrasses are marine plants that generally inhabit the protected shallow waters of temperature and tropical coastal areas. These plants are not true grasses, and the represent several genera appear to be more closely related to members of the lily family. The most extensive areas of seagrasses are found in the tropics. They are concentrated in two major reasons of the indo-Pacific, as well as both coast of the Americas. Seagrasses do not thrive in areas in low light intensity. If the water becomes too turbid (clouded with sediments), it can destroy the seagrasses and other dependent organisms. A massive die-off of seagrasses in Maryland's Chesapeake Bay estuary during the 1960s is largely due to diminished light associate with excessive sediments run-off from the land surrounding the land.

Seagrasses are hydrophytes, which means they generally live submerged beneath the water (Fig. 4.16). To survive in their subtidal habitat, these plants must be adapted to a saline environment with wave action and tidal currents. They must also be able to carry out pollination and seed dispersal under water.

The major stems of seagrasses, called rhizomes or long shoots, grow horizontally, usually just below the surface of the bottom sediments (Fig. 4.16). The rhizomes and roots of seagrasses help to stabilize the bottom and along with the leaves help trap large amounts of sediment. The leaves are flat and oval, ribbon-shaped, or cylindrical and flexible to better withstand water movement while remaining erect. Leaves either develop directly from the rhizomes or develop from small vertical side stems called short shoots. Roots develop from both the rhizomes and the base of short shoots. Unlike the fibrous roots of terrestrial grasses, roots of seagrasses are usually thicker and fleshlier. The flowers, normally inconspicuous, small and pale white, are usually located at the bases of the leaves. The stamens (male flower parts) and pistils (female flower parts) generally extend above the petals of the flower. As a rule, the pollen is released from the stamens in long, gelatinous strands that are carried by water currents to the pistils.

The internal structure of seagrasses is typical of a hydrophyte. One of the most characteristic features is the regular arrangement of air spaces in the soft tissues of the plant. This tissue plays an important role in flotation of the leaves and in gas exchange throughout the plant. The outer cell layer of the plant is called epidermis. Unlike most land plants, the epidermal cells of seagrasses contain maintaining the proper amount of salts in the cell's environment. Since water helps

Fig. 4.16 Seagrass community under the water

Fig. 4.17 Eel grass (*Zostera* sp.) is widely distributed in the intertidal and subtidal zone

to buoy the leaves of seagrasses, supporting tissues in these plants are much reduced and not very rigid.

Eelgrass (*Zostera marina*; Fig. 4.17) is widely distributed in the North Atlantic and Pacific Oceans and even extends into the Arctic Circle. It is generally a subtidal species, but may be found in the lower intertidal regions of shallow lagoons. It grows well on bottoms of sand, mud or a mixture of sand, gravel and mud.

Turtle grass (*Thalassia testudinum*) (Fig. 4.18) is the most common and abundant seagrass in the Caribbean Sea; it is found in calm water to depths of about 10 m (33 ft).

Fig. 4.18 Turtle grass: a common seagrass in the Caribbean coastal zone

The second most abundant and important Caribbean seagrass is **manatee grass** (*Syringodium filiforme*). Manatee grass grows in pure sands as well as mixed with turtle grass. It is usually found at depths of 10 m (33 ft) or less. Manatee grass tolerates lower salinities than other species and its reproduction is controlled by temperature.

Species of the genus *Phyllospadix*, commonly known as **surf grasses**, are unusual because they grow in intertidal areas where there is high surf. The various species often grow together in tight clusters with their rhizomes tightly packed and strongly attached to rocks or other solid surfaces. The roots have branching root hairs that form a dense mat that anchors the plants on the rocks and helps prevent *Phyllospadix* from being dislodged by the surf.

In India, six species of seagrass are reported from the Gulf of Kutch. *Thalassia hemprichii* occurs widely on the inward side of almost all the reef flats of the Gulf. The common seagrass found growing on the muddy substrate are *Halophila ovalis* (Fig. 4.19), *H. beccarii* and *Z. marina*.

Seagrass meadows occupy less than 0.2 % of the area of the world's oceans, but are estimated to bury 27.4 Tg C year^{-1}, roughly 10 % of the yearly estimated organic carbon in the oceans (Duarte 1999). Researches on storage potential of carbon by seagrass reveal that these producer communities may sequester carbon in soils composed of both autochthonous and allochthonous organic carbon. These soils are largely anaerobic, and as a result, the carbon in the soils can be preserved for millennia (Mateo et al. 1997, 2006; Orem et al. 1999). Fourqurean et al. (2012) compiled published and unpublished data on the organic carbon content of seagrass living biomass and organic carbon content and dry bulk density (DBD) of soils underlying seagrass meadows to deliver conservative, first-order estimates of the amount of organic carbon stored in these ecosystems. The database on organic carbon in seagrass meadows contained 3,640 observations from 946 distinct sampling locations across the world (Supplementary Information). The distribution of the data was geographically biased owing to an imbalance in research effort across regions (Orth et al. 2006), with most of the data from North America, Western Europe and Australia. Data were notably scarce from South America and Africa. Furthermore, given the spatial extent of seagrasses in the tropical Indo-Pacific, relatively few data points represented this region (Fig. 4.20).

According to Fourqurean et al. (2012), the amount of organic carbon stored in living seagrass biomass globally averaged 2.52 ± 0.48 MgC ha^{-1} (±95 % CI), two-thirds of which was buried in the

Fig. 4.19 *Halophila ovalis*

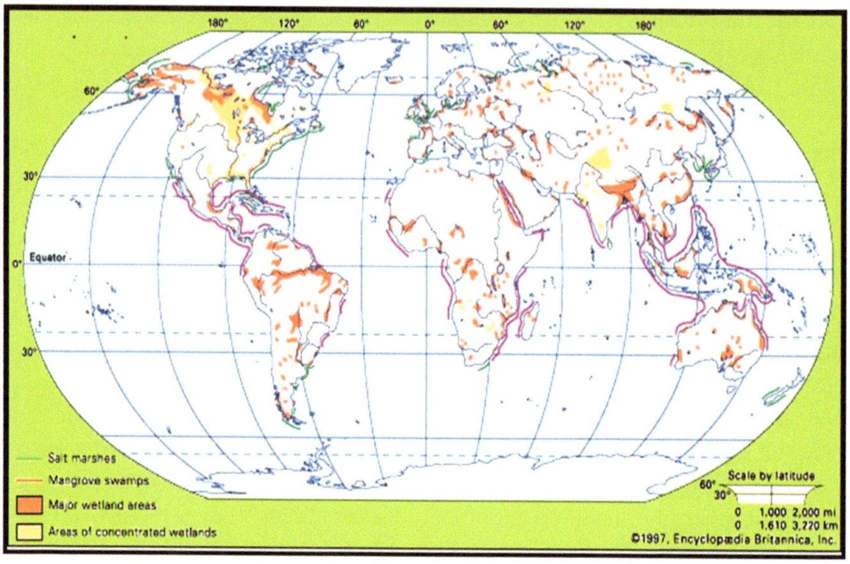

Fig. 4.20 Map showing seagrass bioregions

soil as rhizomes and roots. The largest pools of organic carbon stored in living seagrasses (with mean living biomass of 7.29 ± 1.52 MgC ha^{-1}) were found in Mediterranean meadows dominated by *Posidonia oceanica*, a large seagrass with extensive, long-lived rhizomes.

DBD of seagrass soils had a wide range. The median DBD of the entire data set was 0.92 g ml^{-1}, slightly lighter than the mean of 1.03 ± 0.02 g ml^{-1}. Organic carbon of seagrass soils varied widely, with a median measured 1.8 % of dry weight, and relatively infrequent high values, resulting in a global average of 2.5 %. Assuming that the median DBD and median organic carbon content of soils in the database represent the central tendency of the top metre of all seagrass soils worldwide, Fourqurean et al. (2012) estimated that the top metre of seagrass soils contains 165.6 Mg organic carbon ha^{-1}. It was also reported that soil organic carbon decreased with depth.

The factors affecting the capacity of storing carbon in seagrass are as follows:

1. Rate of primary production of seagrass (which is very high).
2. Capacity of seagrass to filter out particles from the water column and stored them in underlying soils.
3. Low degradation rate of organic carbon in oxygen-poor soil substratum.

Fourqurean et al. (2012) compile published and unpublished measurements of the organic carbon content of seagrass biomass and underlying soils in 946 distinct seagrass meadows across the globe. Using only data from sites for which full inventories exist, they estimated that, globally, seagrass ecosystems could store as much as 19.9 Pg organic carbon. They also made a very conservative approach incorporating depth-wise decrease of organic carbon in soil and estimated that the carbon pool in seagrass ecosystem lies between 4.2 and 8.4 Pg. They also estimated that loss of seagrass community due to dredging and filling activities, change in land-use pattern, degradation of water quality, etc., may result in the release of up to 299 Tg carbon per year to the atmosphere.

The present authors assessed the carbon in major species of seagrass sampled from the Gulf of Kutch region and documented the stored carbon within the range 30.9–33.8 % (Fig. 4.21).

Very scanty references are available on the mechanisms of carbon fixation in seagrasses and questions concerning the kinetics of the active forms of carbon utilized and possible photosynthetic pathways. Steemann Nielsen (1975)

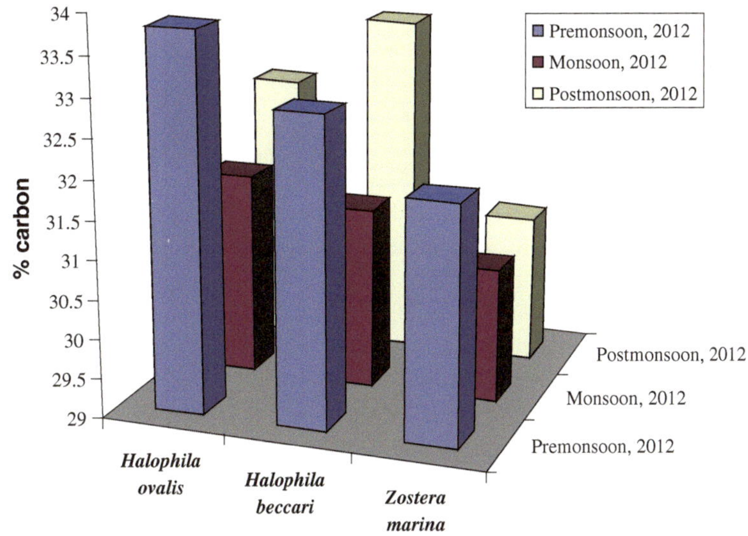

Fig. 4.21 Percentage of stored carbon in major seagrass species from Gulf of Kutch during 2012

postulated that the exogenous form of carbon assimilated by seagrasses is HCO_3^-. This may be attributed to extremely low concentration of carbon dioxide in seawater compared with HCO_3^- (about 200 times less) and the slow rate at which carbon dioxide could be resupplied to the plant from other forms of carbon. The complete dependency of few seagrass species on HCO_3^- was demonstrated by Beer et al. (1977). These species are *Halophila stipulacea*, *Thalassodendron ciliatum*, *Halodule uninervis* and *Syringodium isoetifolium*. It is of interest whether seagrass species belong to the C_3 or C_4 group of plants because many species are tropical and hence exposed to high temperatures, especially in shallow water. This is one of the conditions which might have contributed to the evolution of C_4 plants. While certain features such as the $\delta^{13}C$ values measured by Parker (1964), Smith and Epstein (1971), Doohan and Newcomb (1976) and Benedict and Scott (1976) suggest that various seagrass species belong to the C_4 type, other features such as the high rates of photorespiration (Hough 1976) and photosynthetic inhibition by high oxygen concentrations (Downton et al. 1976) are typically associated with the C_3 pathway. Results of a more appropriate test for the determination of C_3 or C_4 photosynthesis, i.e., short-time $\delta^{14}C$ experiments, have been reported previously only by Benedict and Scott (1976). They show that *Thalassia testudinurn* is a C_4 plant. In 1979, Beer and Waisel conducted short-time $\delta^{14}C$ pulse–chase experiments on two other seagrass species. Both species show a typical C_3 pattern with phosphoglyceric acid being the first major stable substance labelled. The label is subsequently transferred to phosphorylated sugars and phosphate esters and later to saccharose or aspartic acid. Although saccharose and aspartic acid were not separated in the scraping and counting of the chromatograms, two distinct spots could be identified on the X-ray film, the saccharose spot usually of higher density than the aspartic acid. In any case, neither malic nor aspartic acids were primary products of ^{14}C fixation showing that these seagrass species are not C_4 plants. Although ^{14}C pulse–chase experiments should ideally result in the same total activity of label in all treatments, the relative activity of fixed ^{14}C increased with increasing chase times. This is due to absorption of $H^{14}CO_3^-$ which continues to be metabolized after the plant has been transferred to a ~4C-free medium. Therefore, phosphoglyceric acid levels in 15- and 45-s chase periods are somewhat overestimated, while main labelled products are underestimated. The regain of spotted ~4C activity was about 60 %, including the activity of the whole plates. Some activity was lost to the running solvent in the tank, and about 10 % was lost as an evenly distributed smear over the whole plate. The two seagrass species investigated are unlike *T. testudinum* which was shown earlier to be a C_4 plant (Benedict and Scott 1976). They are also somewhat different from *H. stipulacea and Thalassodendron ciliatum* plants which, although mostly resembling C_3 plants, show a relatively higher labelling into malic and aspartic acids (Beer unpublished). Seagrasses do not exhibit the structural differentiation of photosynthetic tissue (Kranz anatomy) which is usually found in C_4 plants, and photosynthesis seems to be carried out mainly in the chloroplast-rich epidermis. C_4 photosynthesis as in *T. testudinurn* is therefore not associated with the Kranz anatomy (Benedict and Scott 1976). It is likely that variations in photosynthetic pathways found in seagrasses are linked to the levels of photosynthetic enzymes as well as HCO_3^- and/or carbon dioxide rather than to differences in anatomy. In some freshwater angiosperms, the ratio of phosphoenolpyruvate carboxylase to ribulosebiphosphate carboxylase differs as a response to seasonal changes of the environment (Bowes et al. 1978). If this is also true for seagrasses, one could expect to find variations in photosynthetic pathways both between and within species as a response to growth conditions. Also, nothing would hinder both pathways being more or less active simultaneously. The main exogenous form of carbon utilized by six seagrass species investigated so far is HCO_3^-. This agrees well with the high HCO_3^- compared with carbon dioxide concentrations found in sea water. The former concentration (about 2.5 mM) is saturating for photosynthesis in these plants. Whether HCO_3^- is fixed directly by phosphoenolpyruvate carboxylase or transformed to carbon dioxide and fixed via ribulosebiphosphate carboxylase probably depends on the levels of these and other enzymes as well as which form of carbon is available. ^{14}C pulse–chase experiments have shown that at least one seagrass species is a C_4 plant, while others are C_3 plants. Experiments based on other parameters sometimes show contradictory or intermediary photosynthetic pathways. The variations in photosynthetic pathways reflect an ability of seagrasses to adapt to various ecological conditions.

4.2.2 Salt Marsh Grass

Salt marsh communities are found on the shoreward side of mudflats in temperate and subarctic regions of the world. The dominant plant life in this community on many North American shores is cordgrass (*Spartina* sp.) that has moved to the shallow intertidal area. The cordgrass *Spartina alterniflora* dominates the marshlands of the Atlantic coast, while *Spartina foliosa* is dominant along the coast of California. On the Gulf Coast of Florida, the tallest and most marsh grass is *Juncus* sp., which is found in higher elevations (higher up on shore). The species found most seaward is tall cordgrass, *Spartina alterneriflora.* Tall cordgrass may grow as high as 3 m (10 ft) and is found growing in the low marsh, the lower part of the intertidal zone that is covered by tidal water much of the day. In the high marsh, the region closer to shore that is covered briefly by salt water each day is a thick carpet of short, fine grasses that usually do not grow taller than 60 cm (2 ft). These may include salt meadow hay (*Spartina patens*) and spike grass (*Distichlis spicata*), as well as *Juncus* sp.

The low marsh that is dominated by the tall cordgrass is typically flushed by the tide twice each day, whereas the high-marsh meadows are usually flooded by only the spring tides. In areas where the tall cordgrass grows thickly, leaves, stalks and debris are flushed by tidal currents so that the water is very clear. The beds of short cordgrass are not flushed by tidal currents, and dead leaves and debris accumulate on the marsh surface, forming a moist mat that is an ideal habitat for many species. The lower Gangetic region is dominated by the salt marsh grass *Porteresia coaractata* (which is locally called Dhani ghas).

Plants such as cordgrass (*Spartina*), rushes (*Juncus*) and pickle weed (*Salicornia* sp.) are adapted to survive in salt marshes. These plants generally grow in the middle to upper intertidal zones, where they are protected from the action of waves and tidal currents and where they are not completely submerged. Like seagrasses and mangroves, they are considered facultative halophytes, meaning that they can tolerate salty conditions as well as freshwater. Since they live in sediments with a high salt content, salt marsh plants tend to lose water to their environment by osmosis and display adaptations similar to those of desert plants to help them retain water. Both cordgrass and rushes have leaves covered by a thick cuticle to help retard water loss, and the stems and leaves contain well-developed vascular tissues for the efficient transport of water within the plant's tissues. Many of the plants that inhabit salt marshes, such as pickle weed, have thick, water-retaining, succulent leaves similar to those of plants that live in hot, arid terrestrial habitats.

Plants of the salt marsh have two types of roots one that acts to anchor the plant and another, short-lived type that functions in nutrient absorption. Stems develop from rhizomes and produce the leaves. Like seagrasses, salt marsh plants usually have shallow roots and rhizomes. This arrangement helps to stabilize coastal sediments and prevent shoreline erosion. The rhizomes of cordgrass also play an important role in recycling the nutrient phosphorous by transferring phosphates produce in the bottom sediments into the leaves and stems of the plant.

Salt marsh grass community with organic carbon-rich substrata sustains a wide spectrum of animal species (Fig. 4.22), which play important role in storing, sequestering and recycling blue carbon.

The estuaries of Indian subcontinent sustains an important salt marsh grass species namely *Porteresia coarctata*, which has a great role in stabilizing the islands. Their long network of roots binds the soil particles very intricately and retards the process of erosion. The growth, biomass and subsequent carbon storage capacity of the species are controlled by the ambient aquatic salinity. The species exhibits a luxuriant growth in relatively low-saline regions (salinity range within 5–20 psu) with subsequently higher carbon content, but at salinity around 30 psu, the growth is stunted. A time series analysis was conducted by the present authors on the stored carbon of the species sampled from the east and west coasts of India. Interestingly, the species sampled from the east coast exhibited higher biomass and carbon compared to those sampled from the west coast.

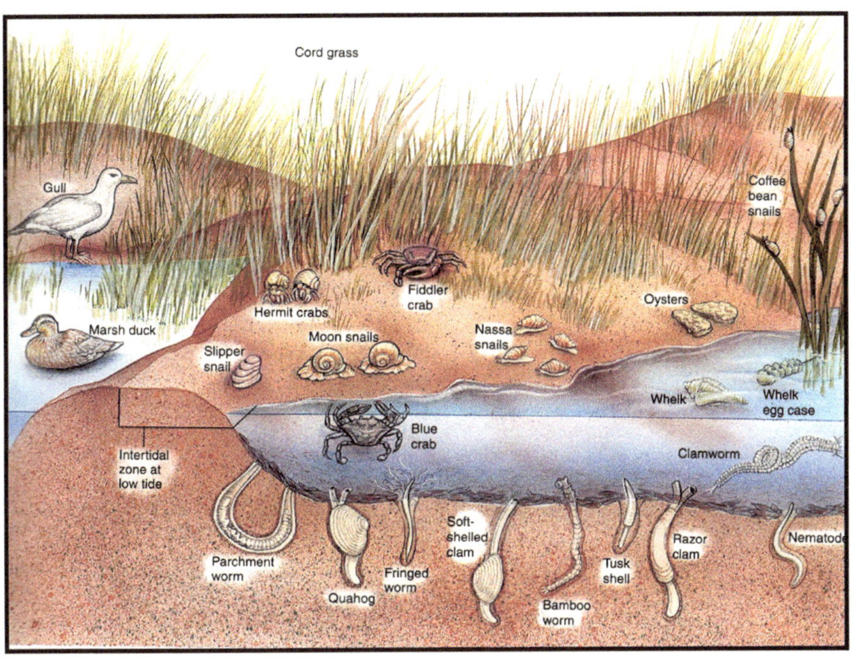

Fig. 4.22 Animals thriving in the salt marsh grass community

4.3 Seaweeds

The growth of seaweeds in the marine, estuarine or coastal waters depends on two important parameters. These are salinity of the surrounding and presence of light to drive the process of photosynthesis. Seaweed grows better where the suspended particulate matter (SPM) is low. In the lower Gangetic region, the growth of seaweed has been found to be more in the eastern sector of Indian Sundarbans compared to the western sector, where several industrial activities occur (Fig. 4.23). The intense industrial activities in the western part of the Indian Sundarbans release large amount of colloidal particles, oil and several waste materials of complex collector. This causes low penetration of solar radiation due to which the biomass of *Enteromorpha intestinalis, Ulva lactuca and Catenella repens* is relatively low compared to the eastern sector (Fig. 4.24).

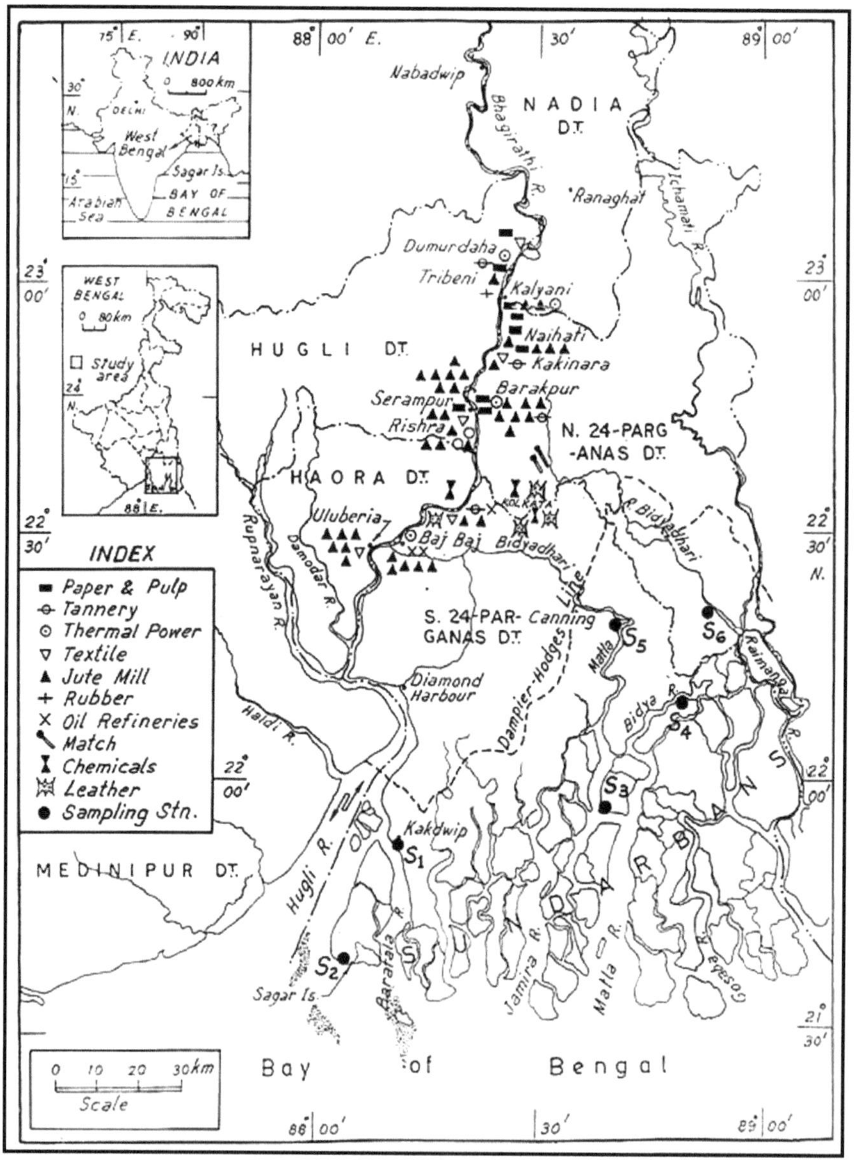

Fig. 4.23 Map showing the industrial units in the upstream of Gangetic delta region

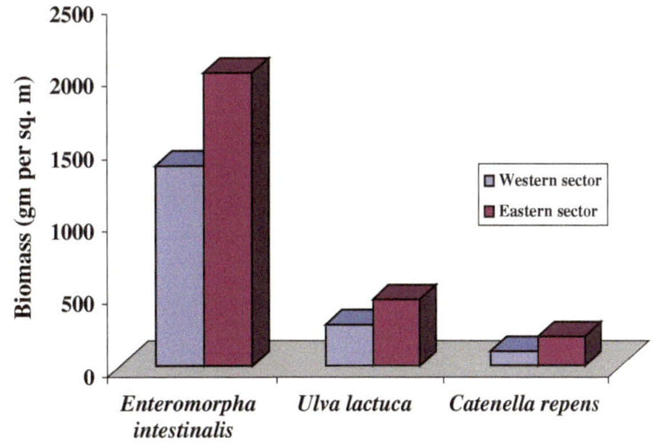

Fig. 4.24 Biomass of seaweeds (in gm m^{-2}) in western and eastern sectors of Indian Sundarbans

Seaweeds are noted for their capacity to utilize dissolved inorganic carbon (DIC) from the surrounding sea water (Smith and Bidwell 1989; Raven and Johnston 1991) for photosynthesis and growth, resulting in a decrease in the DIC concentration of sea water and a drop in the partial pressure of carbon dioxide (pCO_2) below atmospheric levels. The morphology and biomass of seaweeds are also regulated by physical conditions of the surrounding aquatic phase. Differences in morphology occur along wave exposure gradients in a variety of seaweeds. Narrower blades with several splits along their length typify Kelps from wave-exposed areas (*Laminaria hyperborea, L. digitala, Sacchoriza polyschides, Hedophullum sessile*), whereas plants with broader blades with fewer or no divisions occur in more sheltered conditions, even within the same species. Strong water movement imposes various stresses on the thallus: tension, shear forces, bending and twisting. Narrow, flat blades, especially if divided like streamers, provide a highly streamlined shape differences appear to be phenotypic growth responses. For instance, *Laminaria sacharina* can be induced to take on a more streamlined shape with a narrow blade and greater elongation rates simply by attaching weights to the extreme end of the blade. Similarly transplanted *Alaria* sp. has been shown to change from a broad-fronded sheltered form to an elongated type characteristic of exposed shores (Denny 1988; Norton 1991). Different seaweeds may cope with the same wave conditions in different ways, as shown by *Lessonia nigrissens* and *Durvillea antactica* which both occurs in the sublittoral of central and southern Chile. *Lessonia* has a strong, stiff stipe that bends with the flow. In contrast, the elastic stipe and stretchy blade of *Durvillea* allows it to align with the flow completely. Flexible stipes that can be bent parallel with water movements reduce the stress on the thallus and result in the plant being closer to the rock where movement is less. These structures respond best to the chaotic multidirectional water movements typical of exposed shores. In the North Atlantic, flexible elastic stipes are found in species from wave-exposed sites, such as *Alaria esculenta*, while *Laminaria digitata* has a more flexible stipe than the deeper and hence less wave-beaten *L. hyperborea.* Presumably, the stiff stipe of *L. hyperborea* is more important in maximizing access to light than in resisting wave action.

While avoiding dislodgement by wave movement is a major priority for rocky shore species, all seaweeds require some degree of water movement to break down the boundary layer around the thallus. Materials have to diffuse in and out of the plant through this layer of slowly moving fluid, so that the thicker the boundary layers, the slower the uptake of materials. In still water, the boundary layer can be several millimetres thick. Many species have a surface of spiny outgrowths (*Macrocystis pyrifera*), wavy margins, (*Laminaria saccharina*) or even holes or undulations in the blade (*Agarum cribosum, L. saccharina*) which are thought to enhance the turbulence in water flowing over the lamina and hence allow greater uptake of raw

Fig. 4.25 *Enteromorpha intestinalis*: the most widely available seaweed species in Indian Sundarbans with highest carbon content

materials for photosynthesis. There is also evidence that fucoids grow hair-like protrusions during periods of low nutrient concentration that these are involved in nutrient uptake (Hurd et al. 1993). The uptake of carbon by seaweed from ambient water is regulated by structures and their variations. A study undertaken in mangrove-dominated Indian Sundarbans exhibited significant variation in stored carbon collected from the same location (in terms of coordinates). The species-wise variation of stored carbon may be attributed to the morphological structure of the seaweed. Unlike *U. lactuca* and *C. repens*, the extremely coiled and spiral structure of *E. intestinalis* (Fig. 4.25) exposes more area of the species to ambient water, which enables relatively more capture of carbon through diffusion.

References

Adams JM, Faure H, Faure-Denard L, McGlade JM, Woodward FI (1990) Increases in terrestrial carbon storage from the last glacial maximum to the present. Nature 348:711–714

Ajtay GL, Ketner P, Duvigneaud P (1979) Terrestrial primary production and phytomass. Global Carbon Cycle 13:129–182

Alongi DM, Boto KG, Robertson AI (1992) Nitrogen and phosphorous cycle's. In: Robertson AI, Alongi DM (eds) Tropical mangrove ecosystems. American Geophysical Union, Washington, D.C, pp 251–292

Amarasinghe MD, Balasubramaniam S (1992) Structural properties of two types of mangrove stands on the northwestern coast of Sri Lanka. Hydrobiologia 247:17–27

Ball MC, Munns R (1992) Plant responses to salinity under elevated atmospheric conditions of CO_2. Aust J Bot 40:515–525

Ball MC, Cochrane MJ, Rawson HM (1997) Growth and water use of the mangroves *Rhizophora apiculata* and *R. stylosa* in response to salinity and humidity under ambient and elevated concentrations of atmospheric CO_2. Plant Cell Environ 20:1158–1166

Bazilevich NI (1974) Energy flow and biological regularities of the world ecosystems. Cavé

Bazilevich NI, Rodin LY, Rozov NN (1971) Geographical aspects of biological productivity. Sov Geogr Rev Transl 12:293–317; and Paper Vth Congress USSR Geographical Society, Leningrad, 1970

Beer S, Eschel A, Waisel Y (1977) Carbon metabolism in seagrasses. J Exp Bot 28:1180–1189

Benedict CR, Scott JR (1976) Photosynthetic carbon metabolism of a marine grass. Plant Physiol 57:876–880

Bohn HL (1976) Estimate of organic carbon in world soils. Soil Sci Soc Am J 40:468–470

Bohn HL (1978) Organic soil carbon and CO_2. Tellus 30:472–475

Bohn HL (1982a) Organic carbon in world soils. Soil Sci J Am 46:1118–1119

Bohn HL (1982b) Estimate of organic carbon in world soils: II. Soil Sci Soc Am J 46:1118–1119

Bolin B, Degens ET, Duvigneaud P, Kempe S (1979) The global biogeochemical carbon cycle. In: Bolin B, Degens ET, Kempe S, Kenter P (eds) The global carbon cycle, 1st edn. Wiley, Chichester, pp 1–56

Boto K, Wellington J (1984) Soil characteristics and nutrient status in a Northern Australian mangrove forest. Estuaries Coasts 7:61–69

Bouillon S, Rivera-Monroy VH, Twilley RR, Kairo JG (2009) Mangroves. In: Laffoley D, Grimsditch GD (eds) The management of natural coastal carbon sinks. IUCN

Bowes G, Holaday AS, Van TK, Hallen WT (1978) Photosynthetic and photorespiratory carbon metabolism in aquatic plants. In: Hall DO, Coombs J, Goodwin TW (eds) Photosynthesis. The Biochemical Society, London, pp 289–298

Briggs SV (1977) Estimates of biomass in a temperate mangrove community. Aust J Ecol 2:369–373. Sciences 68:338–354

Brown LR, Larsen J, Dorn JG, Moore F (2006) Time for Plan B: cutting carbon emissions 80 percent by 2020

Buringh P (1983) Organic carbon in soils of the world. Role Terr Veg Glob Carbon Cycle. Meas Remote Sens 23

Chave J, Andalo C, Brown S, Cairns MA, Chambers JQ, Eamus D, Fölster H, Fromard F, Higuchi N, Kira T, Lescure JP, Nelson BW, Ogawa H, Puig H, Riéra B, Yamakura T (2005) Tree allometry and improved estimation of carbon stocks and balance in tropical forests. Oecologia 145:87–99

Clough BF, Scott K (1989) Allometric relationship for estimating above ground biomass in six mangrove species. Forest Ecol Manage 27:117–127

Clough BF, Dixon P, Dalhaus O (1997) Allometric relationships for estimating biomass in multi-stemmed mangrove trees. Aust J Bot 45:1023–1031

Comley BWT, McGuinness KA (2005) Above- and below-ground biomass and allometry of four common northern Australian mangroves. Aust J Bot 53:431–436

Cox EF, Allen JA (1999) Stand structure and productivity of the introduced *Rhizophora mangle* in Hawaii'. Estuaries 22(2A):276–284

Dahdouh-Guebas F, Koedam N (2006) Empirical estimate of the reliability of the use of the point-centred quarter method (PCQM): solutions to ambiguous field situations and description of the PCQM + protocol. Forest Ecol Manage 228:1–18

Denny MW (1988) Biology and the mechanics of the wave-swept environment. Princeton University Press, Princeton

Dixon RK, Krankina ON (1993) Forest fires in Russia: carbon dioxide emissions to the atmosphere. Can J Forest Res 23:700–705

Donato DC, Kauffman JB, Murdiyarso D, Kurnianto S, Stidham M, Kanninen M (2011) Mangroves among the most carbon-rich forests in the tropics. Nat Geosci 4

Doohan ME, Newcomb EH (1976) Leaf ultrastructure and δ13C values of three seagrasses from the Great Barrier Reef. Aust J Plant Physiol 3:9–23

Downton WJS, Bishop DG, Larkum AWD, Osmond CB (1976) Oxygen inhibition of photosynthetic oxygen evolution in marine plants. Aust J Plant Physiol 3:73–79

Duarte CM (1999) Methods in comparative functional ecology. In: Pugnaire FI, Valladares F (eds) Handbook of functional plant ecology. Marcel Dekker, Inc., New York, pp 1–8

Ellison AM, Farnsworth EJ (1996) Anthropogenic disturbance of Caribbean mangrove ecosystems: past impacts, present trends, and future predictions. Biotropica 28:549–565

Eswaran H, Den Berg EV, Reich P (1993) Organic carbon in soils of the world. Soil Sci Soc Am J 57:192–194

FAO/Forest Trends (2007) Investing in the future: an assessment of private sector demand for engaging in markets & payments for ecosystem services, by S. Waage, with contributions from I. Mulder, K. ten Kate, S. Sherr, J.P. Roberts, A. Hawn, K. Hamilton, R. Bayon and N. Carroll. Forthcoming in PESAL (Payments for Environmental Services from Agricultural Landscapes) Papers series. Rome, FAO and Washington, DC, Forest Trends

Farnsworth EJ, Ellison AM, Gong WK (1996) Elevated CO_2 alters anatomy, physiology, growth, and reproduction of red mangrove (*Rhizophora mangle* L). Oecologia 108(4):599–609

Fourqurean JW, Duarte CM, Kennedy H, Marbà N, Holmer M, Mateo MA, Apostolaki ET, Kendrick GA, Krause-Jensen D, McGlathery KJ, Serrano O (2012) Seagrass ecosystems as a globally significant carbon stock. Nat Geosci 5:505–509

Fromard F, Puig H, Mougin E, Marty G, Betoulle JL, Cadamuro L (1998) Structure, above-ground biomass and dynamics of mangrove ecosystems: New data from French Guiana. Oecologia 115:39–53

Gorham E (1990) Biotic impoverishment in northern peatlands. In: Woodwell GM (ed) The earth in transition: patterns and processes of biotic impoverishment. Cambridge University, New York, NY, pp 65–98

Gorham E, Janssens JA (1992) Concepts of fen and bog reexamined in relation to bryophyte cover and the acidity of surface waters. Acta societalis botanicorum poloniae 61:7–20

Harmon ME, Ferrell WK, Franklin JF (1990) Effects on carbon storage conversion of old-growth forests to young forests. Science 247:699–702

Hough RA (1976) Light and dark respiration and release of organic carbon in marine macrophytes of the Great Barrier Reef region. Aust J Plant Physiol 3:63–68

Houghton J, Ding Y, Griggs D, Noguer M, Van der Linden P, Dai X, Maskell K, Johnson C (eds) (2001) Climate change. The scientific basis. Published for the Intergovernmental Panel on Climate Change. Cambridge University Press, Cambridge, New York, 881

Hurd CL, Galvin RS, Norton TA, Dring MJ (1993) Production of hyaline hairs by intertidal species of Fucus (Fucales) and their role in phosphate uptake. J Phycol 29:160–165

Hutchings PA, Saenger P (1985) Ecology of mangroves. University of Queensland Press, St. Lucie

Imbert D, Rollet B (1989) Phytmassaerienne et production primaire dans la mangrove du Grand Cul-de-sac Marine (Guadeloupe, *Antilles francaises*). Bull Ecol 20:27–39

IPCC-Working Group II (1990) Chapter 5: world ocean and coastal zones

Kira T, Shidei T (1967) Primary production and turnover of organic matter in different forest ecosystems of the Western Pacific. Jpn J Ecol 17:70–87

Komiyama A, Ogino K, Aksornkoae S, Sabhasri S (1987) Root biomass of a mangrove forest in southern Thailand. 1. Estimation by the trench method and the zonal structure of root biomass. J Trop Ecol 3:97–108

Komiyama A, Moriya H, Prawiroatmodjo S, Toma T, Ogino K (1988) Primary productivity of mangrove forest. In: Ogino K, Chihara M (eds) Biological system of mangroves. Ehime University, Matsuyama, pp 97–117

Komiyama A, Havanond S, Srisawatt W, Mochida Y, Fujimoto K, Ohnishi T, Ishihara S, Miyagi T (2000) Top/root biomass ratio of a secondary mangrove (*Ceriops tagal* (Perr.) CB Rob.) Forest. Forest Ecol Manage 139:127–134

Komiyama A, Poungparn S, Kato S (2005) Common allometric equations for estimating the tree weight of mangroves. J Trop Ecol 21:471–477

Kristensen E, Bouillon S, Dittmar T, Marchand C (2008) Organic matter dynamics in mangrove ecosystems. Aquat Bot. doi:10.1016/j.aquabot.2007.12.005, in press

Lee SY (1995) Mangrove outwelling: a review. Hydrobiologia 295:203–212

Lugo AE, Snedaker SC (1974) The ecology of mangroves. Annu Rev Ecol Syst 5:39–64

Mackey AP (1993) Biomass of the mangrove *Avicennia marina* (Forsk.) Vierh, near Brisbane, south eastern Queensland. Aust J Mar Freshw Res 44:721–725

Markov VD, Olunin AS, Ospennikova LA, Skobeeva EI, Khoroshev PI (1988) World peat resources. Moscow 'Nedra', pp 383 (in Russian)

Mateo MA, Romero J, Perez M, Littler MM, Littler DS (1997) Dynamics of millenary organic deposits resulting from the growth of the Mediterranean seagrass *Posidonia oceanic*. Estuar Coast Shelf Sci 44:103–110

Mateo I, Laborde I, Vicente VP (2006) Monitoring of tropical shallow water fish communities around the Eco Electrica liquefied natural gas import terminal and cogeneration plant in Guayanilla Bay, Puerto Rico. Proc Gulf Carib Fish Inst 57:633–652

Mitra A, Gangopadhyay A, Dube A, Andre CK, Schmidt AC, Banerjee K (2009) Observed changes in water mass properties in the Indian Sundarbans (Northwestern Bay of Bengal) during 1980–2007. Curr Sci 97 (10):1445–1452

Norton TA (1991) Conflicting constraints on the form of intertidal algae. Brit Phycol J 26:203–218

Odum WE, McIvor CC (1990) Mangroves. In: Ewel J, Myers R (eds) Ecosystems of Florida. University of Central Florida Press. USA, Chap 15, pp 517–548

Olson JS, Watts JA, Allison LJ (1983) Carbon in live vegetation of major world ecosystems. Oak ridge National Laboratory, Oak Ridge, ORNL-5862

Ong JE (1993) Mangroves—a carbon source and sink. Chemosphere 27:1097–1107

Ong JE, Gong WK, Wong CH (2004) Allometry and partitioning of the mangrove, *Rhizophora apiculata*. Forest Ecol Manage 188:395–408

Oohata S, Shinnozaki K (1979) A statical model of plant form—further analysis of the pipe model theory. Jpn J Ecol 29:323–335

Orem WH, Holmes CW, Kendall C, Lerch HE, Bates AL, Silva SR, Boylan A, Corum M, Marot M, Hedgeman C (1999) Geochemistry of Florida Bay sediments: nutrient history at five sites in eastern and central Florida Bay. J Coast Res 15:1055–1071

Orth RJ, Carruthers TJB, Dennison WC, Duarte CM, Fourqurean JW, Heck KL Jr, Hughes AR, Kendrick GA, Kenworthy WJ, Olyarnik S, Short FT, Waycott M, Williams SL (2006) A global crisis for seagrass ecosystems. Bioscience 56(2):987

Parker PL (1964) The biogeochemistry of the stable isotopes of carbon in a marine bay. Geochim Cosmochim Acta 28:1155–1164

Pool DJ, Snedaker SC, Lugo AE (1977) Structure of mangrove forests in Florida, Puerto Rico, Mexico and Costa Rica. Biotropica 9:195–212

Post WM, Emanuel WR, Zinke PJ, Stangenberger AG (1982) Soil carbon pools and world life zones. Nature 298:156–159

Putz FE, Chan HT (1986) Tree growth, dynamic and productivity in a mature mangrove forest in Malaysia. Forest Ecol Manage 17:211–230

Raven JA, Johnston AM (1991) Mechanisms of inorganic-carbon acquisition in marine phytoplankton and their implications for the use of other resources. Limnol Oceanogr 36:1701–1714

Ross MS, Ruiz PL, Telesnicki GJ, Meeder JF (2001) Estimating above-ground biomass and production in mangrove communities of Biscayne National Park, Florida (U.S.A.). Wetlands Ecol Manage 9:27–37

Saenger P (2002) Mangrove ecology, silviculture, and conservation. Kluwer Academic Publishers, Dordrecht

Saenger P, Snedaker SC (1993) Pantropical trends in mangrove above-ground biomass and annual litterfall. Oecologia 96(3):293–299

Schlesinger WH (1984) Soil organic matter: a source of atmospheric CO_2. In: Woodwell GM (ed) The role of terrestrial vegetation in the global carbon cycle: measurement by remote sensing. Wiley, New York, pp 111–127 (e-publishing)

Shinozaki K, Yoda K, Hozumi K, Kira T (1964) A quantitative analysis of plant forms the pipe model theory. I. Basic analysis. Jpn J Ecol 14:97–105

Sjors H (1980) Peat on earth: multiple uses or conservation? Ambio 9:303–308

Smith RG, Bidwell RGS (1989) Inorganic carbon uptake by photosynthetically active protoplasts of the red macroalga *Chondrus cnspus*. Mar Biol 102:1–4

Smith BN, Epstein S (1971) Two categories of 13C/12C ratios for higher plants. Plant Physiol 47:380–384

Snedaker SC, Araujo RJ (1998) Stomatal conductance and gas exchange in four species of Caribbean mangroves exposed to ambient and increased CO_2. Mar Freshw Res 49:325–327

Soares MLG, Schaeffer-Novelli YS (2005) Above-ground biomass of mangrove species. I. Analysis of models. Estuar Coast Shelf Sci 65:1–18

Steemann Nielsen E (1975) Marine photosynthesis. Elsevier, Amsterdam 142 pp

Steinke TD, Ward CJ, Rajh A (1995) Forest structure and biomass of mangrove in the Mgeni estuary, South Africa. Hydrobiologia 295:159–166

Tam NFY, Wong YS (1995) Spatial and temporal variations of heavy metal contamination in sediments of a mangrove swamp in Hong Kong. Mar Pollut Bull 31:254–261

Tamai S, Nakasuga T, Tabuchi R, Ogino K (1986) Standing biomass of mangrove forests in southern Thailand. J Jpn Forest Soc 68:384–388

Tomlinson PB (1986) The botany of mangroves. Cambridge University Press, Cambridge

Twilley RR (1995) Properties of mangrove ecosystems related to the energy signature of coastal environment. In: Hall C (ed) Maximum power. University of Colorado Press, Niwot, pp 43–62

Twilley RR, Lugo AE, Patterson-Zucca C (1986) Litter production and turnover in basin mangrove forests in southwest Florida. Ecology 67:670–683

Twilley RR, Chen RH, Hargis T (1992) Carbon sinks in mangroves and their implications to carbon budget of tropical coastal ecosystems. Water Air Soil Pollut 64:265–288

UNEP (1994) The convention on biological diversity: issues of relevance to Africa. Regional ministerial conference on the convention on biological diversity. October. UNEP/AMCEN/RCU 7/1 (A), 27 July

Whittaker RH, Likens GE (1975) The biosphere and man. In: Lieth H, Whittaker RH (eds) Primary Productivity of the biosphere. Springer, Berlin. Ecol Stud 14:305–328

Internet References

www.nature.com/naturegeoscience.com

http://www.recoftc.org/site/resources/Mangroves-more-Carbon-Rich-and-Important-for-Climate-Change.php

Phytoplankton Carbon Stock

5

...Oh ! phytoplankton green and tiny
Iron fertilization makes you more shiny
Absorb more carbon dioxide as you can
And save the Planet Earth from GHG emission...
The Authors

5.1 Marine and Estuarine Phytoplankton: Assessment and Distribution

Phytoplankton, the free-floating tiny microscopic floral particles, demonstrate a wide range of shapes and vary over several orders of magnitude in size, from submicron species such as the picoplanktonic prochlorophytes to diatoms measuring more than 1 mm in diameter. In mixed-pieces samples, high numbers of extremely small-sized species might actually contribute only a minor fraction of the overall biomass, whereas other, larger-sized species that are much less abundant in numbers might dominate the overall biomass. Thus, cell counts per se are inadequate as a measure of relative algal biomass. Several conventional biomass-related parameters, such as particulate organic carbon, ATP, or chl *a*, are known to vary significantly with environmental conditions, such as light and nutrient availability. Even the salinity of the aquatic system influences the biomass of phytoplankton species, although the effect is species specific. Thus, these parameters are not very perfect to differentiate between the contributions of different taxonomic groups and cannot be used to compare different species in a mixed assemblage or the same species under different environmental conditions. The most commonly used traditional biomass estimate for microalgae is cell biovolume, which is calculated from microscopically measured linear dimensions. Automated or semi-automated techniques for estimating algal biovolume using sophisticated laser or image analysis technologies are becoming increasingly available. However, these methods are not yet fully developed, and many researchers still use the conventional light-microscope-based biovolume estimates. Biovolume is also used for the conversion of cell counts into carbon units in studies of fluxes of organic matter in aquatic communities. Calculating phytoplankton carbon from biovolume rather than from particulate organic carbon eliminates the error due to detrital particulate matter contained in particulate organic carbon. Only a few attempts have been made to standardize the calculation of algal biovolume, and different researchers used different sets of equations to estimate the biovolume of phytoplankton. The matching of geometric shapes with phytoplankton structure requires careful attention, and the problem becomes acute for complex-shaped genera such as dinoflagellates, gomphonemoid diatoms and desmids. In this chapter, we aim to provide a set of geometric shapes and mathematical equations for calculating algal biovolume and stored carbon in dominant phytoplankton of tropical estuarine system.

5.1.1 Microscopic Method

Phytoplankton are free-floating microscopic unicells and colonies that thrive luxuriantly in the photic zone of the ocean, estuaries and other aquatic systems. They are the key players in maintaining the nutrient and energy flow through marine and estuarine food webs.

A. Mitra and S. Zaman, *Blue Carbon Reservoir of the Blue Planet*,
DOI 10.1007/978-81-322-2107-4_5, © Springer India 2015

The phytoplankton community encompasses both prokaryotic and eukaryotic species. Till the end of 1980s, living plankton flora of the world ocean amounted to 474–504 genera and 3,444–4,375 species including the insufficiently known or doubtful species (Sournia et al. 1991). To allow for undescribed species and to avoid a spurious impression of accuracy, Tett and Barton (1995) round up to a total (*N*) of 5,000 extant species of marine phytoplankton distinguishable by classical, mainly morphological, criteria. In the estuaries of Indian Sundarbans, about 102 species of phytoplankton were documented since 2004 (Mitra et al. 2004). In this chapter, we have identified and listed 137 species of diatoms, 26 species of dinoflagellates and 15 species of other algae (green and blue green algae) from the inshore waters of Bay of Bengal. Many of the species were recorded from the brackishwater ponds of Indian Sundarbans during 2010–2013 (**Annexure**). This documentation is the spin-off product of a pilot project entitled '… Study on carbon stock and response of estuarine phytoplankton to iron fertilization…' funded by DST, Govt. of India, during 2010–2011 to 2013–2014.

Assessment of phytoplankton standing stock and species composition is extremely important to understand the productivity of the aquatic system in the framework of climate change. Standing stock of phytoplankton can serve as the fingerprints of carbon-sequestering zone. For over a century, light microscopy was the basic tool to observe phytoplankton and the method is still in use with certain developments to determine the species composition of phytoplankton. The species composition and community structure of phytoplankton are important to understand the stress or favourable condition of the environment in which these free-floating organisms are dwelling. A brief description of the different species diversity indices are highlighted here with their respective mathematical expressions.

The nature and function of phytoplankton community are not determined by all the organisms or the species present in the community, but a relatively few species or species group generally exert the major controlling influence by virtue of their size, number, production efficiency or other activities. Within these, species groups that largely control the community energetics and that have profound influence on the survival strategy of other species are known as ecological dominants. The degree to which the dominance is concentrated in one, several or many species can be expressed through index of dominance that reflects the importance of a species or a group of species in the community as a whole. However, if the environmental parameters of the ecosystem or the habitat are highly favourable for the survival of all types of species, the value of index of dominance is eclipsed by the values of richness index, evenness index and Shannon–Weiner index, which express the uniform importance of almost all the species present in the community. A few useful indices for analysing the community structure of marine and estuarine phytoplankton are discussed here.

Index of Dominance

A single species or a group of species that modify the environment influences biologically controlled phytoplankton communities. These organisms are called dominants. It is not a very easy task to determine the criteria of dominance. The dominants in a community may be the most numerous, possess the highest biomass, occupy the most space, make the largest contribution to energy flow or mineral cycling or by some other means control or influence the rest of the community. Many ecologists have designated those phytoplankton species as dominant, which are numerically superior. The index of dominance (Simpson 1949) is calculated according to the following expression:

$$\text{Index of dominance}\,(C) = \sum \left(\frac{n_i}{N}\right)^2$$

where n = importance value for each species (number of individuals, biomass, production, etc.) and N = total of importance values.

The index of dominance is always lower where the dominancy is shared by a large number of species (Whittacker and Niering 1965), or the total populations of the community are uniformly distributed among different species that are usually witnessed in clean, pollution-free waters

(Osborne et al. 1976) or in habitats having very favourable physico-chemical variables. In fact, it is a general phenomenon that when the environmental stress in any particular habitat increases, the community in that system would be dominated by a fewer species (species that are more resistant to the existing set of physico-chemical and biological variables), which are often referred to as the opportunistic species. The magnitude of the index of dominance is usually high in stressful habitat.

Index of Similarity

Two phytoplankton communities (like the one in the Indian Ocean and the other in the Pacific Ocean), geographically wide apart from each, may also show similarity with respect to their species composition. The similarity arises due to the closeness of the two aquatic ecosystems with respect to environmental characteristics such as salinity, nutrient load and transparency.

The index of similarity between two sampling stations is calculated according to the expression:

$$\text{Index of similarity } (S) = \frac{2C}{A + B}$$

where A = number of species present at sampling station 'X', B = number of species present at sampling station 'Y' and C = number of species common to both the sampling stations.

Richness Index

The index reflects the suitability of a particular habitat for successful thriving and growth of different phytoplankton species. The value of the index usually decreases, when environment becomes unfavourable or stressful due to intrusion of some foreign matters that have adverse effect on the community, e.g., pollutants or any biological organisms that may be a parasite or predator to the existing group of species. Zooplankton often graze on phytoplankton making the environment highly stressful for phytoplankton. There are also reports on sharp decrease of phytoplankton due to oil pollution (Mitra 2000).

The mathematical expression of richness index is as follows:

$$\text{Richness index } (d) = \frac{S}{\sqrt{N}}$$

where S = number of species and N = total number of individuals of all the phytoplankton species.

Shannon–Weiner (1949) Species Diversity Index

It is the most commonly used index in ecological studies owing to the fact that it is dimensionless and independent of sample size and expresses the worth of each species. It is expressed as follows:

$$\text{Species diversity index } (\bar{H}) = p_i \log_e p_i$$

$$(\bar{H}) = -\sum_{i=1}^{S} \frac{n_i}{N} \log_e \frac{n_i}{N}$$

where p_i = importance probability for each species, n_i = importance value for each species and N = total of importance values.

The value of $\bar{H}$ has several ecological explanations. Higher diversity signifies longer food chains and more cases of symbiosis (mutualism, commensalisms, etc.) and greater probabilities for negative feedback control, which reduces oscillation and hence increases stability. The value of $\bar{H}$ is also a unique indicator of environmental stress. As the sensitive species gradually shift or are eliminated from a habitat with the increase of the magnitude of environmental stress, the species diversity index has been claimed as an effective statistics for predicting the change in environment (Wilhm and Dorris 1968; Cairns and Dickson 1971). Thus, decrease in the value of $\bar{H}$ reflects an increase in the magnitude of environment stress on the species/community, and the gradual restoration of the environment quality is indicated by an increment in the value of $\bar{H}$ through recruitment of new species.

5.1.2 Chemical Method

The standing stock of phytoplankton can also be assessed in the laboratory by chemical method

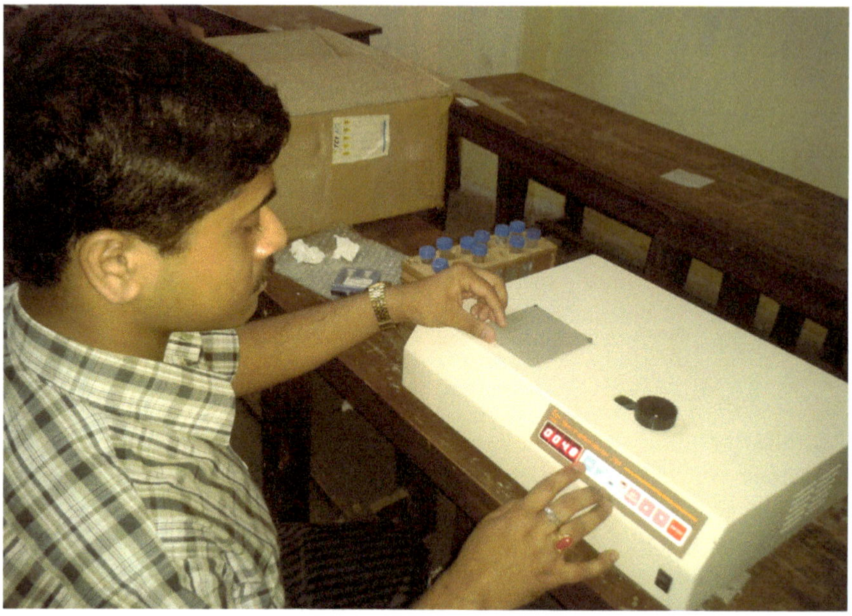

Fig. 5.1 Estimation of phytopigments through spectrophotometer

through analysis of phytopigments (Fig. 5.1). Chlorophyll is the key diagnostic marker of phytoplankton. The spectrophotometric method of chlorophyll estimation is widely used in researches related to phytoplankton distribution.

For chlorophyll estimation of marine and estuarine phytoplankton, 1 L of surface water, collected from the sampling stations during high tide condition, is filtered through a 0.45 μm Millipore membrane fitted with a vacuum pump. The residue along with the filter paper is dissolved in 90 % acetone and kept in a refrigerator for about 24 h in order to facilitate the complete extraction of the pigment. The solution is then centrifuged for about 20 min under 5,000 rpm, and the supernatant solution is considered for the determination of the chlorophyll pigment by recording the optical density at 750, 664, 647 and 630 nm with the help of spectrophotometer (Fig. 5.1). All the extinction values are corrected for a small turbidity blank by subtracting the 750 nm signal from all the optical densities, and finally, the phytoplankton chlorophyll is estimated as per the expression of Jeffrey and Humphrey (1975).

$$\text{Chl}\,a = 11.85\,\text{OD}_{664} - 1.54\,\text{OD}_{647} - 0.08\,\text{OD}_{630}$$

The values obtained from the equations were then multiplied by the volume of the extract (in ml) and divided by the volume of the water (in litre) filtered to express the chlorophyll content in mg m^{-3}. All the analyses are usually conducted in triplicate on the basis of collection of three samples from the same site in order to ensure the quality of the data.

In modern times, high-performance liquid chromatography (HPLC) (Fig. 5.2) is used to evaluate the pigment discrimination and quantification directly from sea water samples. This approach provides an idea about the distribution of phytoplankton in different regions of the ocean based on diagnostic carotenoids such as zeaxanthin (specific for cyanobacteria) and paraxanthine (specific for paraphycea).

5.1.3 Assessment Through Remote Sensing

Chlorophyll can also be identified by remote sensing. The critical Coastal Zone Colour Scanner (CZCS) has made it possible to establish the key features of phytoplankton distribution throughout the world ocean.

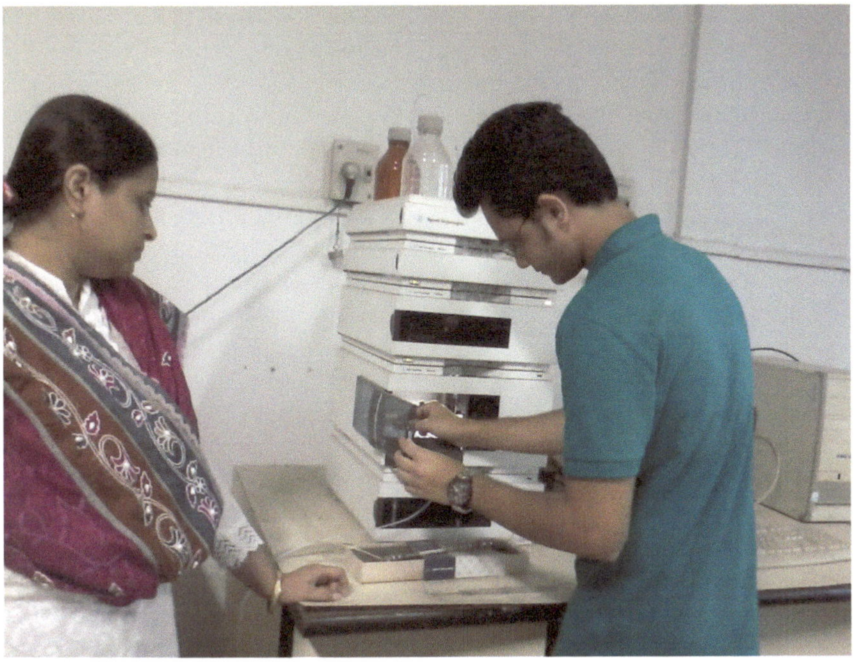

Fig. 5.2 Analysis of phytopigments through HPLC

Historically, satellite sensors have provided oceanographers with bulk phytoplankton pigment concentrations (e.g., chlorophyll *a*) over global scales with synoptic resolution (McClain 2009). CZCS was the beginning of the *era* of satellite remote sensing of ocean colour. It was a relatively simple sensor and was successful at the retrieval of chlorophyll *a* concentrations in open ocean waters. Models were then developed that utilized satellite-derived chlorophyll *a* data along with the information on photosynthetically available radiation and temperature at the sea surface to compute water column primary production by remote sensing. However, CZCS performed poorly in coastal waters. SeaWiFS and MODIS, the successor to the CZCS, had better radiometric precision and more wavebands and was designed to address the limitations of CZCS, in particular for applications in coastal waters. Satellites with even higher spectral resolution followed (e.g., MERIS). Though the newer generations of satellites were designed for improving performance in coastal waters, there has been increasing recognition of their value in the detection of phytoplankton functional types as well. These satellite missions have provided knowledge and understanding of a variety of events including frontal features and episodic blooms on a global basis. The scientific community has now moved on to predicting ecologically significant characteristics of the food web such as size structure as interpreted for phytoplankton size classes including picoplankton to microplankton. Ocean colour has been recognized as an essential climate variable by the global climate observation system (GCOS). As the global time series of ocean colour data grows (CZCS, MODIS and SeaWiFS), there is an increase in satellite products available to study decadal-scale variations in phytoplankton distribution and primary production. Understanding the spatial and temporal distribution of plant functional types (PFTs) has allowed the improvement and development of knowledge on biologically mediated fluxes of elements between the upper ocean and the ocean interior (Falkowski and Raven 1997). The performance of biogeochemical models in the ocean has improved substantially because of incorporating PFTs into ecosystem models. The spatial variability and concentration of various PFTs are critical to improving primary productivity estimates and understanding the feedbacks of climate change. The ability to observe PFTs on a global scale that relate to key biogeochemical processes such as nitrogen fixation, silicification and calcification is valuable to studies of marine elemental cycles. Despite the paucity of data on functional groups, our understanding of ecosystem linkages is improving as we accrue larger amounts of data. Major divisions of phytoplankton taxonomic groups such as diatoms, coccolithophores, dinoflagellates, chlorophytes and cyanobacteria are often separated into distinct functional groups, as these taxonomic groups have unique biogeochemical signatures. Aiken et al. (2009) showed how different size classes of phytoplankton are distributed globally, and size is often related to function (Nair et al. 2008). Nonetheless, there are limitations to discriminating phytoplankton taxonomic composition or even functional groups by remote sensing, due to our inability to discriminate various phytoplankton types optically. Remote sensing reflectance Rrs (λ) is influenced by absorption and backscattering by sea water, phytoplankton, coloured dissolved organic matter, detrital matter and other suspended materials (Garver and Siegel 1997). Remote sensing of particular types of phytoplankton is only possible if optical characteristics are identified for those types that can be used to distinguish them from all other types of materials in the water.

New technological developments and improved scientific knowledge have allowed for the development of several approaches for detecting phytoplankton biomass and some functional groups of phytoplankton including coccolithophores (Balch et al. 1991, 1996) and *Trichodesmium* (Subramanian and Carpenter 1994; Subramanian et al. 1999a, b; Hu et al. 2010). More recently, algorithms have been developed to distinguish additional phytoplankton groups and size classes (Sathyendranath et al. 2004; Alvain et al. 2005; Moisan et al. 2011a, b). HPLC is the in situ method of choice to serve as ground truth for satellite products due to its accuracy and rapid processing. A relatively new analysis tool called CHEMical TAXonomy (CHEMTAX) has been developed that is designed to yield information on the phytoplankton composition using HPLC data (Mackey et al. 1996), but it has to be recognized that HPLC models have their limitations (Latasa 2007).

Algorithm development for remote sensing is focused on estimating the quantities of various optically active constituents in the ocean. Various pigments present in phytoplankton play a role in determining the total absorption coefficient of phytoplankton, which is a key determinant of the spectral variability in remote sensing reflectance. Many researchers utilize a reconstruction model to demonstrate the spectral variability of photoprotective and photosynthetic pigments (Sathyendranath et al. 1987; Bidigare et al. 1990). There are differences in the weight-specific absorption spectra of various in vitro (extracted) phytoplankton pigments (Fig. 5.3).

The comparison between maximal peaks of mass-specific absorption spectra and centre wavelengths on ocean colour satellite platforms proves that coupling taxonomic and optical properties is quite challenging (Table 5.1). Researches have been undertaken to upgrade sensors with increased temporal and spectral resolution that will provide more spectral information for the development of algorithms. Scientists are also putting their efforts to develop hyperspectral sensors that would cover pigment-specific peaks at relatively high spatial and temporal resolution, which would allow retrieval of photosynthetic and photoprotective pigments.

The multispectral radiance measurements from the satellite sensor can be corrected for the influence of the atmosphere to yield remote sensing reflectance.

The suspended solids in the aquatic phase often influence the spectral signatures of phytopigment. Hence, development of site-specific algorithm has been suggested by researchers. On this issue, a study was conducted in the Indian Sundarbans estuaries during 2005 to assess the influence of suspended solid on spectral signatures of phytoplankton.

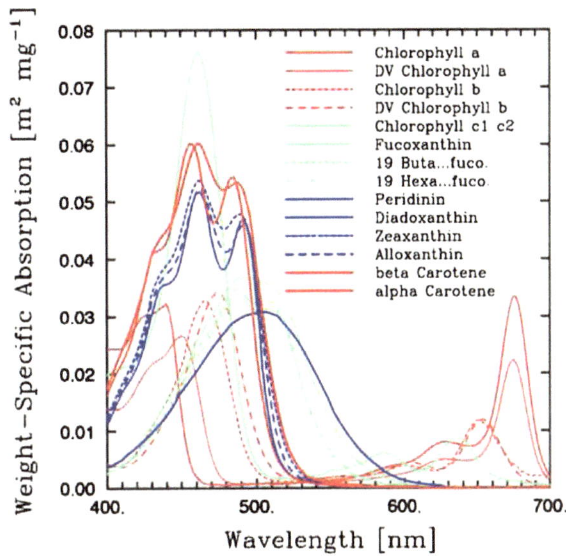

Fig. 5.3 Weight-specific absorption spectra of various in vitro (extracted) phytoplankton pigments

Table 5.1 Past and present ocean colour sensors and their respective centre wavelengths (nm)

SeaWiFS	MODIS	MERIS	OCM2	OCTS	CZCS
412	412	413	415	412	443
443	443	443	442	443	520
490	469	490	491	490	550
510	488	510	512	516	670
555	531	560	557	565	
670	547	620	620	667	
	555	665			
	645	681			
	667	709			
	678				

5.2 Assessment of Phytoplankton Carbon

The twenty-first century is the *era* of human population explosion, rapid urbanization and intense industrialization, leading to change in land use pattern and shrinkage of aquatic system. In many parts of the world, man has changed the original characters of the aquatic ecosystem for short-term gain; for example, in India, Bangladesh, Thailand and the Philippines, man has artificially channelized the sea water in the agricultural field or mangrove system for culturing shrimp that require brackishwater for growth. The mangroves, being an important burial site of carbon, are depleting due to such human activities. Anthropogenic activities that influence the climate of the planet Earth are mainly those that affect the atmosphere. This encompasses increased exposures to ultraviolet radiation (UVR; 280–400 nm) and higher carbon dioxide levels. The effects of these changes are propagated to aquatic systems through the surface layer (Sabine et al. 2004), where most of the biological processes that sustain life and biogeochemical processes take place. Phytoplankton play a key role in determining the effects of environmental change on the surface layer since they are responsible for dissolved inorganic carbon (DIC) sequestration through photosynthetic processes. Carbon dioxide required for photosynthesis diffuses through the cell membrane of most phytoplankton species, while both carbon dioxide and HCO_3^- can be incorporated through adenosine triphosphate (ATP)-dependent mechanisms called carbon dioxide-concentrating mechanisms (CCMs) (Raven 1997). It is expected that long-term elevation of carbon dioxide would downregulate the activity of CCMs, increasing the energy available for other cellular processes and resulting in increased carbon fixation and growth rates (Raven 1991). However, the effect of elevated carbon dioxide on growth and photosynthesis differs among studies, and more work is still needed to understand phytoplankton response to a rise in carbon dioxide concentrations (Riebesell et al. 1993; Hein and Sand-Jensen 1997; Tortell et al. 2000). A proper evaluation of future carbon dioxide concentrations depends not only on estimates of the carbon emitted from anthropogenic activities and stored in the atmosphere, but also on our knowledge of the physical and biological processes affecting ocean–atmosphere feedback. Recent results have suggested that variation among experiments can be related to both interspecific differences in the mechanisms responsible for carbon incorporation and the dependence of DIC uptake on other factors, such as nutrients and light availability (Leonardos and Richard 2005). The salinity level of the aquatic phase also regulates the phytoplankton volume in marine and estuarine environment (Mitra et al. 2012; Mitra 2013), which subsequently influences the carbon content in the phytoplankton.

The Tasio project's research (Cermeño et al. 2008) analysed phytoplankton community composition across four latitudinal transects in the Atlantic Ocean to look at the processes controlling the biogeographical distributions of phytoplankton and their role in the carbon cycle. The project team members applied a diagnostic analysis to a coupled atmosphere–ocean general circulation model to make projections about the response of diatoms and coccolithophorids in the next century and their potential impact on the sequestration of atmospheric carbon. They stated from the database that the distribution of these two groups of phytoplankton is controlled by the rate of nutrient input to the system. If nutrients enter the upper oceans (through continental runoff), diatoms dominate in the system, but if nutrients are supplied slowly, the coccolithophorids are selected. Tasio's results thus reflect the view that the relative distribution of diatoms and coccolithophorids is strongly dependent on the mechanisms that supply nutrients into the upper mixed layer of the ocean. The project suggests that this mechanistic connection is the major factor responsible for the succession and domain shifts of these two phytoplankton functional groups in the ocean.

Laboratory analysis also confirmed that the distribution of phytoplankton is correlated with the injection of nutrients into the upper ocean. In nature, high latitude, temperate and upwelling systems receive substantial amounts of nutrients from deep-waters via wind-driven vertical mixing, leading to the dominancy of diatoms. In contrast, phytoplankton inhabiting low-latitude environments largely rely on slow nutrient diffusion from deep-waters to sustain their standing stocks, and coccolithophorids dominate. The discussion thus throws light on the standing stock and taxonomic diversity of phytoplankton that mainly govern the magnitude and direction of carbon dioxide exchange in the ocean–atmosphere matrix. The role of salinity is also vital as this variable not only influences the phytoplankton diversity, but also controls the cell volume and subsequently the carbon storage capacity (Mitra et al. 2012).

Carbon fixation by photoautotrophic algae has the potential to diminish the release of carbon dioxide into the atmosphere and help alleviate the trend towards global warming. Primary producers of coastal and marine ecosystems such as phytoplankton, seaweed and seagrass are excellent carbon-sequestering agents than their terrestrial counterparts (Zou 2005). More than 36.5 Gt of carbon dioxide is captured each year by planktonic algae through photosynthesis in the oceans (Gonzalez et al. 2008). Zooplankton dynamics are a major controlling factor in the sedimentation of particulate carbon in open oceans (Bishop and Wood 2009). Of the captured carbon dioxide, an estimated 0.5 GtC year^{-1} is stored at the seabed (Seiter et al. 2005).

In this chapter, we attempt to project a quantitative estimate of carbon-sequestering capacity of the phytoplankton that are distributed along the Indian Sundarbans estuaries based on our observation on shapes and subsequent volumes of various phytoplankton species during 2011. Some common shapes of phytoplankton are highlighted in Fig. 5.4 in the form table. The readers may consult Baltic Sea Environment Proceedings No. 106 (www.helcom.fi/stc/files/Publications/Proceedings/bsep106.pdf) for further study and research in this vertical.

Carbon uptake by phytoplankton and its export as organic matter to the ocean interior (a mechanism known as the biological pump) lowers the partial pressure of carbon dioxide in the upper waters and facilitates the diffusive drawdown of atmospheric carbon dioxide. However, precipitation of calcium carbonate by coccolithophores increases the partial pressure of carbon dioxide and promotes outgassing from the ocean to the atmosphere (known as the alkalinity pump). While these carbon fluxes have been, over the past 100 million years, modulated by the abundance of diatoms and coccolithophorids, resulting in biological feedback on atmospheric carbon dioxide and Earth's climate, the processes, which determine the distribution of these phytoplankton groups, remain poorly understood. Considering this gap area, the present chapter deals with the storage of carbon by the phytoplankton cells, which is a direct function of their volume. Very few studies have been conducted in the tropical ocean and estuaries on the phytoplankton carbon storage capacity and the inter-relationships of the same with abiotic parameters. In this chapter, we have highlighted a case study of variation of carbon storage by phytoplankton in different salinity, which is a major seasonal feature of tropical waters. The highest volume of phytoplankton and carbon content was observed in the monsoon season in the estuaries of the lower Gangetic region. We observed a completely contrasting picture in the pre-monsoon season when the salinity is highest in the estuarine water due to minimum precipitation and maximum evaporation rate. Under this situation, the volume of phytoplankton and subsequent stored carbon was lowest (Tables 5.2 and 5.3).

The regulatory role of salinity on stored carbon in phytoplankton is highlighted in Table 5.4. Basically, we infer that the highest carbon in the phytoplankton during monsoon is due to their maximum volume acquired during this season as a result of high dilution factor of the ambient aquatic phase. The direct relationships between cell volume and cell carbon in different types of phytoplankton are illustrated clearly in Figs. 5.5, 5.6, 5.7, 5.8, 5.9 and 5.10.

It is extremely important to understand the enzymatic role in storing carbon dioxide by phytoplankton. Carbonic anhydrase converts inorganic carbon from the form it takes in sea water (bicarbonate) to carbon dioxide that is used in phytoplankton photosynthesis. The carbonic anhydrase enzyme usually contains zinc, which is scarce in surface waters of the open ocean. Because carbon acquisition is essential to cell growth, the amount of zinc may limit the rate of primary production and carbon dioxide fixation. Thus, carbon cannot be acquired if organisms lack zinc, or a means of compensating for its absence.

Lane and Morel (2000a) have identified and characterized a diatom carbonic anhydrase enzyme that can substitute cobalt for zinc without a loss of enzyme activity. In these experiments, zinc was limited in the diatom growth medium and cobalt was made available. Carbonic anhydrase activity was still observed in the *Thalassiosira weissflogii* culture in the absence of zinc. Protein purification yielded a cobalt-containing carbonic anhydrase enzyme, which cochromatographed with the Zn-containing enzyme on an electrophoresis gel, indicating that the new enzyme is just like the old one, except for the metal. Extended X-ray absorption fine structure (EXAFS) studies allowed comparison of the coordination environment of the two enzymes and provided further evidence that the zinc- and cobalt-containing enzymes are otherwise identical.

Lane and Morel (2000b) also identified another carbonic anhydrase enzyme containing cadmium in the same diatom *Thalassiosira*

Common shape of phytoplankton	Figure	Formula
Sphere	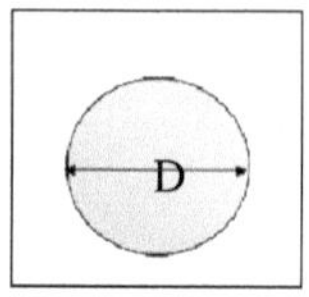	$V = \pi/6 * d^3$ Where, V = Volume d = Diameter
Cylinder	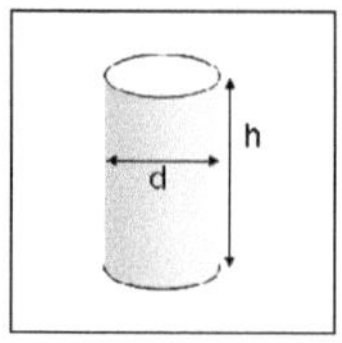	$V = \pi/4 * d^2 * h$ Where, V = Volume d = Diameter h = Height
Trapezoid	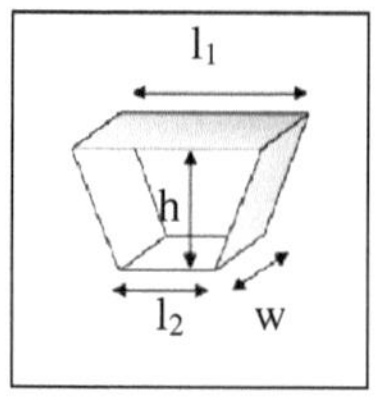	$V = 1/2*h*w*(l_1+l_2)$ Where, V = Volume h = Height w = Wide l_1 = Length l_2 = Length
Rotational Ellipsoid	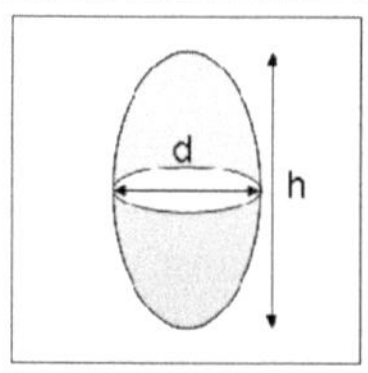	$V = \pi/6 * d^2 * h$ Where, V = Volume d = Diameter h = Height
Cone	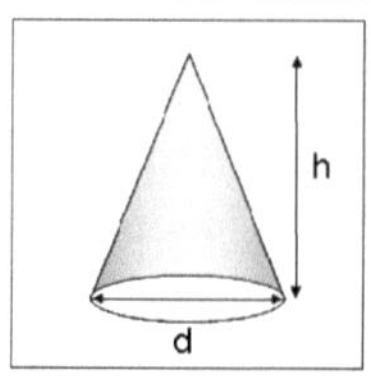	$V = \pi/12 * d^2 * h$ Where, V = Volume d= Diameter h = Height
Parallelepiped	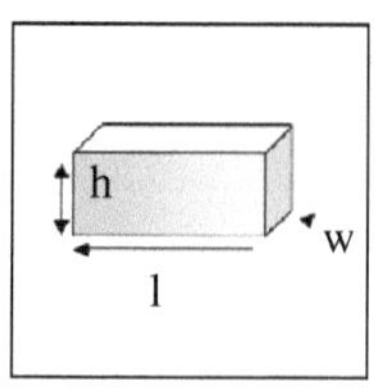	$V = l * w * h$ Where, V = Volume l = Length w = Wide h = Height

Fig. 5.4 Few common shapes of phytoplankton along with their volume expression

weissflogii. This carbonic anhydrase enzyme, which is different from the zinc/cobalt version, is the first reported case of an enzyme that uses cadmium. This discovery also helps to explain another mystery. It has been known for twenty years that cadmium, like other trace metals, tends to be present in ocean exactly in proportion to other essential nutrients such as phosphate, nitrate and silicate. This would be easy to understand whether cadmium had a biological role and will

Table 5.2 Seasonal variation of cell volume (in μm^3) of phytoplankton in Indian Sundarbans

S. no.	Species	Western sector			Central sector		
		Pre-monsoon	Monsoon	Postmonsoon	Pre-monsoon	Monsoon	Postmonsoon
1	*Coscinodiscus excentricus*	10637.39	14968.71	12477.54	8952.38	12632.06	10132.49
2	*Coscinodiscus jonesianus*	12132.22	13996.13	12336.21	8797.29	10700.41	9133.41
3	*Coscinodiscus lineatus*	4196.70	4778.10	4317.48	2250.07	2762.49	2384.49
4	*Coscinodiscus radiates*	44451.19	45918.19	45240.19	31950.14	35252.09	33519.31
5	*Coscinodiscus gigas*	4266.90	4942.07	4387.07	4158.55	4699.30	4316.18
6	*Coscinodiscus oculus-iridis*	3839.30	4551.30	3978.08	3404.25	4046.49	3522.94
7	*Planktoniella sol*	264994.76	271405.63	269383.6	218202.63	227153.21	225131.2
8	*Cyclotella striata*	25031.36	26984.78	25962.78	21894.74	24735.60	23381.79
9	*Thalassiosira subtilis*	770.75	1327.55	966.22	651.18	1209.82	867.82
10	*Ceratium tripos*	357.90	580.75	442.07	324.78	533.65	413.65
11	*Skeletonema costatum*	1503.43	1738.43	1636.43	1255.07	1555.88	1420.80
12	*Paralia sulcata*	19855.18	21165.56	20709.56	19240.80	20768.98	20300.37
13	*Rhizosolenia crassispina*	3764.91	4726.57	4141.13	3757.13	4301.63	3603.63
14	*Rhizosolenia setigera*	4341.69	5465.69	4760.69	4171.14	5334.15	4724.94
15	*Rhizosolenia alata*	3726.28	4370.98	4098.28	3764.90	4590.08	4152.08
16	*Ceratium teres*	1186.60	1711.99	1395.95	1005.77	1474.05	1185.05
17	*Ceratium trichoceros*	994.36	1280.98	1160.36	1009.27	1284.20	1170.16
18	*Bacteriastrum delicatulum*	39595.75	41915.00	40684.01	36521.99	39088.83	37792.22
19	*Bacteriastrum varians*	38028.52	42328.52	41028.52	37024.92	42018.74	40750.11
20	*Bacteriastrum comosum*	36959.33	40688.09	39759.33	35577.78	39671.08	38358.4
21	*Chaetoceros didymus*	4287.00	5705.20	4652.00	2241.10	3861.23	2259.56
22	*Chaetoceros peruvianus*	3844.48	4727.96	4161.48	1901.92	2565.09	1832.76
23	*Chaetoceros compressus*	3170.71	4215.71	3467.71	1649.00	2467.42	1634.5
24	*Ditylum sol*	1678.78	2048.78	1863.78	1644.05	2018.14	1838.47
25	*Triceratium favus*	2552.06	3557.06	2856.06	2151.66	2704.98	2338.18
26	*Triceratium reticulatum*	3704.79	4414.38	3937.79	3188.87	3821.97	3478.13
27	*Biddulphia sinensis*	16700.84	19441.51	17395.84	15691.40	17228.50	16495.92

(continued)

Table 5.2 (continued)

S. no.	Species	Western sector			Central sector		
		Pre-monsoon	Monsoon	Postmonsoon	Pre-monsoon	Monsoon	Postmonsoon
28	*Biddulphia mobiliensis*	27305.85	29792.85	28805.85	26456.24	29066.02	28315.68
29	*Hemidiscus hardmanninus*	86.89	264.89	112.89	85.80	242.27	108.37
30	*Climacosphenia elongate*	8439.61	9428.61	8975.61	8316.51	9318.79	8852.51
31	*Fragilaria oceanica*	33704.85	36691.85	35704.85	33211.39	36309.34	35155.48
32	*Rhaphoneis amphiceros*	3618.27	4539.27	3783.27	3213.38	3901.00	3290.73
33	*Thalassionema nitzschioides*	357.36	598.70	442.36	212.06	404.23	283.30
34	*Thalassiothrix longissima*	2079.97	2613.97	2279.97	1981.33	2507.89	2152.19
35	*Thalassiothrix frauenfeldii*	356.87	521.10	425.87	240.10	409.46	299.45
36	*Asterionella japonica*	98.89	209.70	141.22	85.39	197.536	127.60
37	*Ceratium extensum*	159.20	283.72	193.72	154.16	266.12	185.46
38	*Gyrosigma balticum*	2854.95	3230.95	2932.95	2548.63	2969.27	2654.14
39	*Pleurosigma normanii*	1396.71	1750.71	1520.71	1216.92	1558.98	1327.03
40	*Pleurosigma elongatum*	49151.35	55262.35	53251.35	41648.81	48377.66	45702.66
41	*Diploneis smithii*	2404.30	2924.06	2674.95	2228.71	2731.72	2480.27
42	*Cymbella marina*	391.05	630.09	506.72	386.43	613.50	492.50
43	*Nitzschia sigma*	404.62	694.05	537.62	318.33	629.39	449.50
44	*Nitzschia closterium*	425.50	718.91	581.50	439.20	732.92	587.80
45	*Ceratium furca*	442.74	724.22	586.74	475.34	745.29	613.20
46	*Trichodesmium erythraea*	3667.96	4232.77	3927.96	4479.54	5170.93	4868.80
47	*Chlorella marina*	568.54	742.54	646.54	585.46	758.90	661.50

Table 5.3 Mean seasonal variation of phytoplankton cell carbon content (in pg) in estuarine waters of Indian Sundarbans

S. no.	Species	Cell carbon (in picogram)					
		Western sector			Central sector		
		Pre-monsoon	Monsoon	Postmonsoon	Pre-monsoon	Monsoon	Postmonsoon
1	*Coscinodiscus excentricus*	531.08	700.59	604.44	461.76	610.50	510.54
2	*Coscinodiscus jonesianus*	590.84	663.45	598.88	455.26	533.63	469.32
3	*Coscinodiscus lineatus*	249.79	277.50	255.60	150.67	177.95	157.93
4	*Coscinodiscus radiates*	1693.66	1738.85	1717.99	1295.75	1403.33	1347.12
5	*Coscinodiscus gigas*	253.17	285.20	258.94	247.94	273.79	255.54
6	*Coscinodiscus oculus-iridis*	232.39	266.77	239.18	210.79	242.51	216.74
7	*Planktoniella sol*	7205.07	7346.10	7301.69	6154.72	6358.68	6312.74
8	*Cyclotella striata*	1063.07	1129.87	1095.04	953.69	1052.88	1005.89
9	*Thalassiosira subtilis*	63.19	98.22	75.90	55.12	91.09	69.58
10	*Ceratium tripos*	93.83	139.48	111.55	86.66	130.15	105.64
11	*Skeletonema costatum*	108.64	122.22	116.38	93.85	111.70	103.78
12	*Paralia sulcata*	880.98	927.85	911.60	858.81	913.73	896.97
13	*Rhizosolenia crassispina*	228.35	275.07	247.10	228.73	254.84	220.75
14	*Rhizosolenia setigera*	256.76	309.47	276.68	248.55	303.42	274.99
15	*Rhizosolenia alata*	226.83	258.17	245.03	228.73	268.61	247.63
16	*Ceratium teres*	250.42	338.10	286.06	218.70	299.10	250.15
17	*Ceratium trichoceros*	216.67	266.61	245.88	219.33	267.17	247.57
18	*Bacteriastrum delicatulum*	1542.00	1614.86	1576.29	1444.19	1525.97	1484.79
19	*Bacteriastrum varians*	1492.32	1627.77	1587.10	1460.29	1618.09	1578.36
20	*Bacteriastrum comosum*	1458.19	1576.41	1547.17	1413.83	1544.38	1502.80
21	*Chaetoceros didymus*	254.14	320.43	271.55	150.18	233.47	151.18
22	*Chaetoceros peruvianus*	232.65	275.14	248.08	131.47	167.56	127.58
23	*Chaetoceros compressus*	198.99	250.70	213.98	117.10	162.37	116.26
24	*Ditylum sol*	118.81	139.64	129.33	116.82	137.95	127.89
25	*Triceratium favus*	166.87	218.44	182.82	145.30	174.94	155.44
26	*Triceratium reticulatum*	225.77	260.24	237.22	199.91	231.54	214.49

(continued)

Table 5.3 (continued)

S. no.	Species	Cell carbon (in picogram)					
		Western sector			Central sector		
		Pre-monsoon	Monsoon	Postmonsoon	Pre-monsoon	Monsoon	Postmonsoon
27	*Biddulphia sinensis*	765.66	866.06	791.39	727.90	785.22	758.03
28	*Biddulphia mobiliensis*	1140.76	1224.33	1191.33	1111.89	1200.05	1174.86
29	*Hemidiscus hardmanninus*	10.76	27.58	13.30	10.65	24.72	12.87
30	*Climacosphenia elongate*	440.19	481.58	462.73	434.98	477.03	457.58
31	*Fragilaria oceanica*	1353.16	1449.63	1417.93	1337.08	1437.37	1400.20
32	*Rhaphoneis amphiceros*	221.48	266.20	229.64	201.16	235.42	205.08
33	*Thalassionema nitzschioides*	33.88	51.49	40.28	22.19	37.44	28.06
34	*Thalassiothrix longissima*	141.36	170.15	152.29	135.90	164.52	145.33
35	*Thalassiothrix frauenfeldii*	33.84	46.00	39.06	24.54	37.83	29.36
36	*Asterionella japonica*	11.95	21.99	15.96	10.61	20.95	14.70
37	*Ceratium extensum*	48.33	77.58	56.76	47.07	73.61	54.77
38	*Gyrosigma balticum*	182.76	202.05	186.79	166.69	188.67	172.26
39	*Pleurosigma normanii*	102.35	122.92	109.66	91.53	111.89	98.19
40	*Pleurosigma elongatum*	1837.50	2020.70	1960.86	1606.54	1814.00	1732.29
41	*Diploneis smithii*	158.99	186.34	173.36	149.50	176.34	163.05
42	*Cymbella marina*	36.45	53.67	44.97	36.09	52.52	43.95
43	*Nitzschia sigma*	37.47	58.04	47.19	30.85	53.62	40.80
44	*Nitzschia closterium*	39.03	59.72	50.29	40.05	60.67	50.73
45	*Ceratium furca*	111.69	167.13	140.66	118.38	171.01	145.83
46	*Trichodesmium erythraea*	262.39	292.57	276.42	305.45	340.65	325.42
47	*Chlorella marina*	63.62	77.94	70.15	65.06	72.40	71.38

Table 5.4 Inter-relationship between phytoplankton cell carbon and salinity

Species	Correlation coefficient (*r*) between phytoplankton cell carbon and salinity			*p* value		
	A	*B*	*C*	*A*	*B*	*C*
Coscinodiscus excentricus	−0.9673	−0.9088	−0.6294	<0.01	<0.01	<0.01
Coscinodiscus jonesianus	−0.5355	−0.5821	−0.5055	<0.01	<0.01	<0.01
Coscinodiscus lineatus	−0.6674	−0.6504	−0.6250	<0.01	<0.01	<0.01
Coscinodiscus radiatus	−0.6718	−0.5723	−0.6274	<0.01	<0.01	<0.01
Coscinodiscus gigas	0.0654	−0.1021	0.0790	IS	IS	IS
Coscinodiscus oculus-iridis	−0.7708	−0.7077	−0.7550	<0.01	<0.01	<0.01
Planktoniella sol	−0.8030	−0.8463	−0.7965	<0.01	<0.01	<0.01
Cyclotella striata	−0.3647	−0.3334	−0.3611	IS	IS	IS
Thalassiosira subtilis	−0.2364	−0.2766	−0.2103	IS	IS	IS
Ceratium tripos	−0.1692	−0.1171	−0.1924	IS	IS	IS
Skeletonema costatum	−0.1591	−0.0072	−0.0927	IS	IS	IS
Paralia sulcata	−0.3471	−0.3547	−0.3450	IS	IS	IS
Rhizosolenia crassispina	−0.0707	−0.4175	−0.3827	IS	IS	IS
Rhizosolenia setigera	−0.4142	−0.3765	−0.3360	IS	IS	IS
Rhizosolenia alata	−0.3186	0.0466	−0.2770	IS	IS	IS
Ceratium teres	−0.1247	−0.2556	−0.1973	IS	IS	IS
Ceratium trichoceros	−0.3453	−0.1612	−0.2761	IS	IS	IS
Bacteriastrum delicatulum	−0.2859	−0.3076	−0.2709	IS	IS	IS
Bacteriastrum varians	−0.1518	0.0903	0.1061	IS	IS	IS
Bacteriastrum comosum	−0.2296	−0.2015	−0.2596	IS	IS	IS
Chaetoceros didymus	−0.3033	−0.3837	−0.3691	IS	IS	IS
Chaetoceros peruvianus	−0.3026	−0.2977	−0.3872	IS	IS	IS
Chaetoceros compressus	−0.2859	−0.1558	−0.3836	IS	IS	IS
Ditylum sol	−0.0872	−0.0748	−0.0623	IS	IS	IS
Triceratium favus	−0.3937	−0.0605	−0.3720	IS	IS	IS
Triceratium reticulatum	−0.4089	−0.1785	−0.3317	IS	IS	IS
Biddulphia sinensis	−0.3512	−0.0733	−0.3595	IS	IS	IS
Biddulphia mobiliensis	−0.1966	−0.1630	−0.0486	IS	IS	IS
Hemidiscus hardmanninus	−0.3699	0.3786	−0.3962	IS	IS	IS
Climacosphenia elongate	−0.3814	−0.3488	−0.3487	IS	IS	IS
Fragilaria oceanica	0.6087	0.6917	0.6494	<0.01	<0.01	<0.01
Rhaphoneis amphiceros	−0.3125	0.0127	−0.4025	IS	IS	IS
Thalassionema nitzschioides	−0.3351	−0.1317	−0.3896	IS	IS	IS
Thalassiothrix longissima	0.1337	−0.2601	−0.3458	IS	IS	IS
Thalassiothrix frauenfeldii	−0.3426	−0.2797	−0.3882	IS	IS	IS
Asterionella japonica	−0.2804	−0.3653	−0.3392	IS	IS	IS
Ceratium extensum	0.0439	0.3595	−0.3795	IS	IS	IS
Gyrosigma balticum	−0.3085	−0.3851	−0.3732	IS	IS	IS
Pleurosigma normanii	−0.2107	−0.2793	−0.2186	IS	IS	IS
Pleurosigma elongatum	−0.2648	−0.2591	−0.2329	IS	IS	IS
Diploneis smithii	−0.0587	−0.3192	−0.2034	IS	IS	IS

(continued)

Table 5.4 (continued)

Species	Correlation coefficient (*r*) between phytoplankton cell carbon and salinity			*p* value		
	A	*B*	*C*	*A*	*B*	*C*
Cymbella marina	0.0415	−0.3165	−0.3705	IS	IS	IS
Nitzschia sigma	−0.3280	−0.3674	−0.3703	IS	IS	IS
Nitzschia closterium	−0.3477	−0.3463	−0.3847	IS	IS	IS
Ceratium furca	−0.2220	−0.3118	−0.3300	IS	IS	IS
Trichodesmium erythraea	0.3508	0.4193	0.3436	IS	IS	IS
Chlorella marina	−0.7043	−0.7633	−0.7201	<0.01	<0.01	<0.01

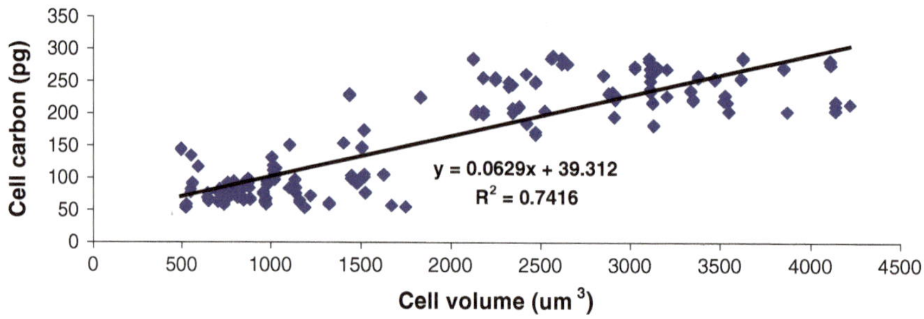

Fig. 5.5 Inter-relationship between cell carbon content (pg) and cell volume (μm^3) for cyanobacteria and green algae in central sector

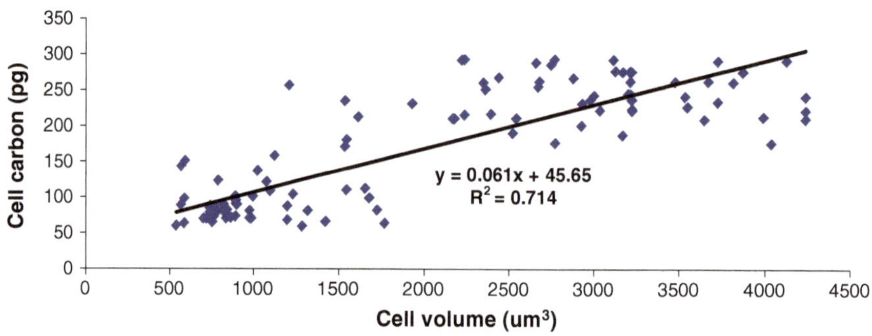

Fig. 5.6 Inter-relationship between cell carbon content (pg) and cell volume (μm^3) for cyanobacteria and green algae in western sector

also add a new dimension to cadmium, which, otherwise, is considered as toxic metal. The research can also throw light on the direct proportionality of metals and nutrients in oceans. The essential metals are scarce where nutrients are scarce because the metals are taken in and used by organisms to help them utilize the available nutrients. Again, the metals are abundant where nutrients are abundant because organisms decay, liberating both nutrients and metals.

Cullen et al. (1999) also observed, in laboratory cultures of *T. weissflogii*, an inverse correlation between the concentration of the cadmium-containing carbonic anhydrase and dissolved carbon dioxide and zinc. Taken together, these results—one from the laboratory and one from the

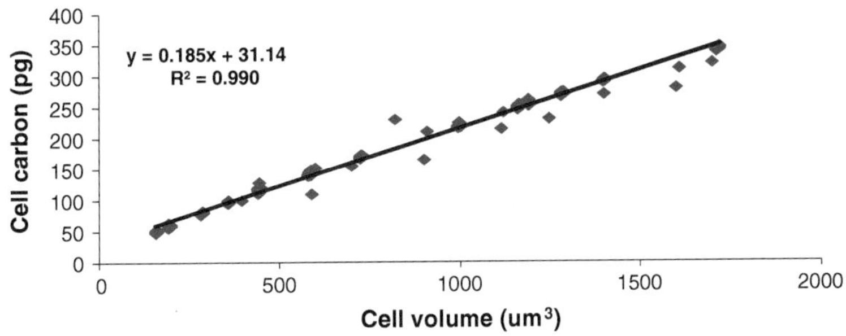

Fig. 5.7 Inter-relationship between cell carbon content (pg) and cell volume (μm^3) for dinoflagellates in western sector

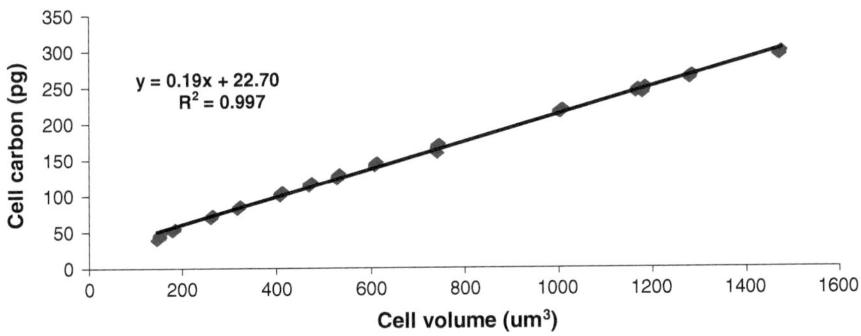

Fig. 5.8 Inter-relationship between cell carbon content (pg) and cell volume (μm^3) for dinoflagellates in central sector

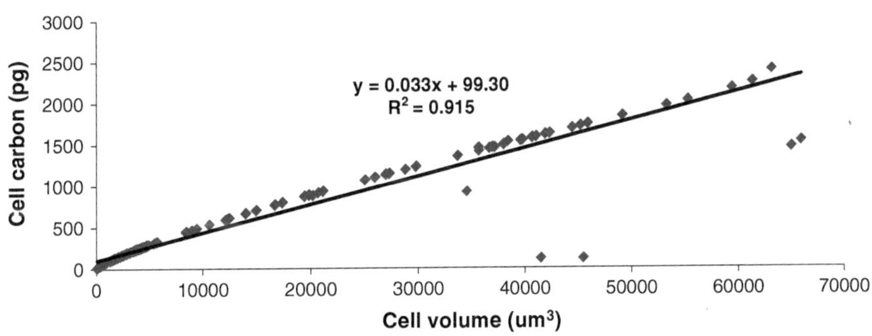

Fig. 5.9 Inter-relationship between cell carbon content (pg) and cell volume (μm^3) for diatoms in western sector

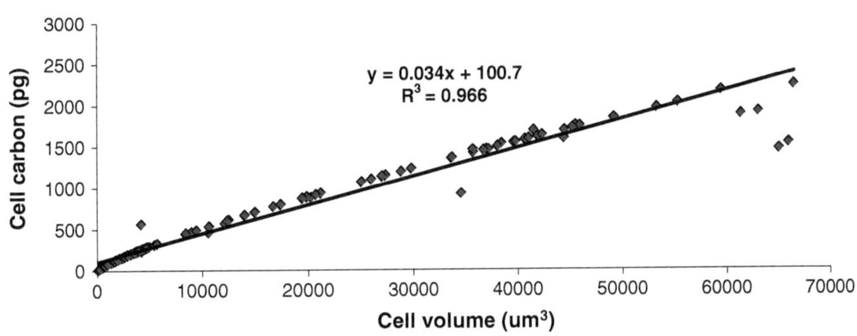

Fig. 5.10 Inter-relationship between cell carbon content (pg) and cell volume (μm^3) for diatoms in central sector

field—strongly support the hypothesis that phytoplankton use cadmium to make carbonic anhydrase and that this is the reason cadmium is present in sea water in direct proportion to nutrients such as phosphorous. The authors measured these effects (the correlations between cadmium concentration and the concentrations of carbon dioxide and zinc) as a function of phytoplankton size. They found that the correlation exists in every size class, providing strong evidence that the cadmium-containing carbonic anhydrase occurs in many species of phytoplankton, not just in *T. weissflogii* where it was first observed.

5.3 Artificial Enhancement of Phytoplankton: Bioremediation or Biodeterioration?

Bioremediation is the use of living organisms (primarily micro-organisms) for the removal of pollutants or any harmful substances from the biosphere. It relies on biological processes to minimize an unwanted environmental impact of the pollutants. The micro-organisms, in particular, have the abilities to degrade, detoxify and even accumulate the harmful organic as well as inorganic compounds. Besides them, higher plants have also been reported to remove such pollutants, primarily through their ability to accumulate these in their tissues. The storage and sequestration of carbon by autotrophs also come under the domain of bioremediation, as it is one of the most eco-friendly approaches to reduce carbon dioxide level of the atmosphere/ambient media. The phytoplankton present in the marine and estuarine environments absorb carbon dioxide for photosynthesis.

Marine phytoplankton have the potential to significantly buffer future increases in atmospheric carbon dioxide levels. However, in order for carbon dioxide fertilization to have an effect on carbon sequestration to the deep ocean, the increase in dissolved carbon dioxide must stimulate primary productivity; that is, marine phototrophs must be carbon dioxide limited (Riebesell et al. 1993). Estimation of the extent of bicarbonate (HCO_3^-) uptake in the oceans is therefore required to determine whether the anthropogenic carbon sources will enhance carbon flux to the deep ocean. A short term carbon-dioxide disequilibrium experiment was conducted during the Southern Ocean Iron Experiment (SOF_{ex}) with ^{14}C isotope and a significant uptake of HCO_3^- was observed by Southern Ocean phytoplankton. Since the majority of DIC in the ocean is in the form of bicarbonate, the biological pump may therefore be insensitive to anthropogenic carbon dioxide. Approximately half of the DIC uptake observed was attributable to direct HCO_3^- uptake, the other half being direct carbon dioxide uptake mediated either by passive diffusion or by active uptake mechanisms. The increase in growth rates and decrease in carbon dioxide concentration associated with the iron fertilization did not trigger any noticeable changes in the mode of DIC acquisition, indicating that under most environmental conditions, the carbon dioxide-concentrating mechanism (CCM) is constitutive. A low-carbon dioxide treatment induced an increase in uptake of carbon dioxide, which was attributed to increased extracellular carbonic anhydrase activity, at the expense of direct HCO_3^- transport across the plasmalemma. Isotopic disequilibrium experimental results are consistent with Southern Ocean carbon stable isotope fractionation data from this and other studies. Although iron fertilization has been shown to significantly enhance phytoplankton growth and may potentially increase carbon flux to the deep ocean, an important source of the inorganic carbon taken up by phytoplankton in this study was HCO_3^- whose concentration is negligibly affected by the anthropogenic rise in carbon dioxide.

Bioremediation has emerged during recent past as most ideal alternative, environment-friendly and ecologically sound technology for removing pollutants from the environment, restoring contaminated sites and preventing further pollution. This technology has the potential to be more socially acceptable when compared to physical or chemical processes and is, therefore, expanding the range of organisms to be used for pollution clean-up. Bioremediation in fact forms a vital component of the so-called green technology of maintaining the nature's overall ecological balance, an issue of top priority of environmental awareness and public policy. Bioremediation is the alliance of a biological system and an engineering system for the remediation of undesirable chemical pollution. The objective of bioremediation is to exploit naturally occurring biodegradative processes for clean-up of the polluted environments.

One of the prime objectives of this chapter is to analyse in detail the storage capacity of carbon by phytoplankton due to iron fertilization in oceans and brackishwater bodies and whether the process can be considered under the bioremediation vertical.

The twenty-first century is the *era* of advanced science and technology and its applications in various spheres of life. In this age of satellites, it is not very difficult to answer the basic question that whether adding iron to the ocean can trigger the phytoplankton bloom. When storms over land transfer iron-rich dust into the sea, satellite images show marbled swaths of green phytoplankton spinning across waters. Those regions were previously blue and barren. Satellites also exhibit plankton blooms near the Galápagos and other islands where iron-rich deep-waters naturally well up to surface. Even phytoplankton blooms spurred by experimental additions of iron to the ocean can be detected by satellites. Marine scientists who conducted the experiments on iron fertilization in ocean also reported an almost instantaneous change in the colour of the ocean water.

Twelve experiments so far carried out by researchers on the regulatory role of adding iron on marine and estuarine phytoplankton have not critically addressed the questions of how much carbon dioxide taken up by a bloom is drawn out of the air and transferred into the deep sea and how long it remains sequestered there. Four experiments took place in the northwest Pacific, two experiments were in the equatorial Pacific, and six experiments were conducted in the Southern Ocean (Fig. 5.11). All these twelve experiments reported up to 15-fold increases in the chlorophyll content of surface waters.

Iron is basically an essential trace element for phytoplankton as it is necessary for a number of cellular functions including the synthesis of chlorophyll. The recent research on iron fertilization put forward the concept that phytoplankton growth can be stimulated by adding iron to high-nitrate low-chlorophyll (HNLC) waters. Many scientists also forwarded the view that the efficiency of this biological pump can be enhanced by fertilizing the oceans with iron. Increased efficiency means sucking up more carbon dioxide from the atmosphere into the oceans. This can be an effective road map to mitigate climate change due to rise of carbon dioxide in the atmosphere. However, such

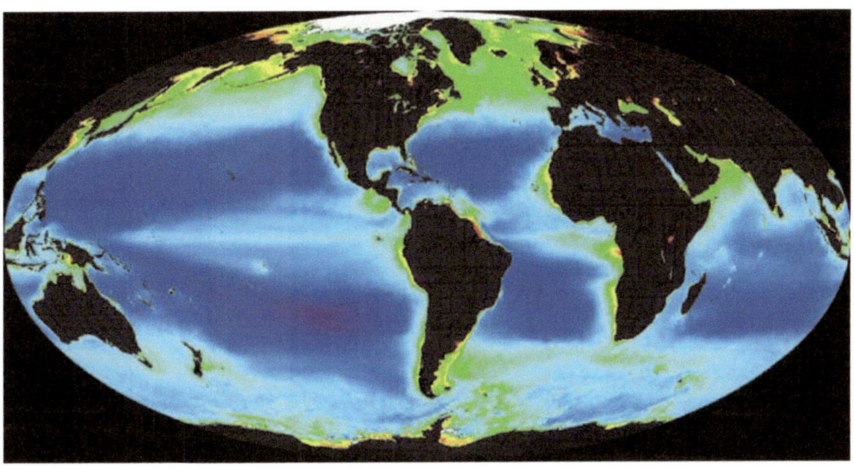

Fig. 5.11 Locations of twelve small-scale experiments over the past decade in several oceans of the world

researches are based on an incomplete understanding and highly simplified interpretation of current scientific knowledge. The major drawbacks of the concept of reducing carbon dioxide level in the atmosphere by triggering the growth of phytoplankton of the ocean through iron fertilization are listed here in points:

1. Ecological perturbations in the marine and estuarine environments due to iron fertilization have not been given proper importance.
2. The geophysical changes of the ecosystem that may result due to large-scale iron fertilization have not been accounted in this research programme.
3. In order to remove significant amount of atmospheric carbon dioxide, a huge area of the ocean needs to be fertilized which may not be particularly feasible.
4. Commercial iron fertilization may lead to harmful algal bloom (HAB), which has adverse effect on other marine lives.
5. Iron addition and subsequent phytoplankton bloom associated with increased particulate organic export and remineralization could reduce dissolved oxygen (DO) level in subsurface water. This has the high probability to pose a negative impact on the marine organisms.
6. As the bloom will progress due to iron fertilization, the nutrient load of the ambient water will sharply decrease. This could result in a reduction of these nutrients downcurrent from an iron-fertilized area. Finally, lack of nutrients may cause a detrimental effect on phytoplankton, resulting in a reduction in overall biological productivity. Predictive mathematical models have suggested that commercial-scale iron fertilization of the oceans could have a significant adverse impact on marine fisheries.
7. There is also the concern that the changes in water chemistry caused by iron fertilization could lead to higher emission of nitrous oxide due to decomposition of organic matter in deep-water. Some researchers have contradicted the concern with the idea of fertilizing low-nutrient regions of the oceans instead of high-nutrient ones. This scientific school believes that excess phytoplankton density (bloom condition) could create a larger aerosol effect leading to block incoming solar radiation that may cause global cooling.
8. It is difficult to track phytoplankton blooms in the vast ocean with dynamic circulation pattern. The added iron rapidly dilutes, sinks and reacts

Fig. 5.12 Phytoplankton bloom in the experimental pond after adding ferrous sulphate

Table 5.5 Monthly variation of body weight (gm) in *Macrobrachium rosenbergii* in the selected ponds

Month	Control	Pond 1 (0.5 ppm)	Pond 2 (1.5 ppm)	Pond 3 (2.5 ppm)
1	3.21	3.18	5.42	3.30
2	8.00	5.29	16.60	10.20
3	16.89	16.57	21.98	17.12
4	20.49	23.16	33.83	26.98
5	32.94	34.72	46.55	38.53
6	41.03	45.94	58.22	51.32
7	52.55	57.04	69.92	62.95
8	61.76	66.17	80.48	72.04

Figures in bracket indicate the applied doses of ferrous sulphate in the pond water

with sea water, becoming virtually undetectable after a few days. Researchers therefore have to add minute amounts of an inert tracer like sulphur hexafluoride (SF_6), which itself is a potent greenhouse gas. One kilogram of SF_6 added in an experiment is equivalent to releasing 7 tonnes of carbon dioxide.

Considering the drawbacks of iron fertilization in open ocean, a case study was undertaken to find the potentiality of aquacultural ponds in holding and sequestering carbon. The merits of the programme are as follows:

1. The question of huge input of iron does not arise because the water body has a limited surface area or volume.
2. Standardization of iron fertilization can be easily done because of the closed nature of the system due to which the wave action, tidal phenomenon and other parameters can be controlled.

Fig. 5.13 Freshwater prawn production from iron-fertilized pond

Table 5.6 Mean cell volume (μm^3) in diatoms in the selected ponds

S. no.	Species	Pond 1 (0.5 ppm)	Pond 2 (1.5 ppm)	Pond 3 (2.5 ppm)
1	*Coscinodiscus excentricus*	9650.44	10611.320	8546.3
2	*Coscinodiscus jonesianus*	11222.25	13453.88	12652.66
3	*Coscinodiscus lineatus*	4195.77	5004.32	4989.1
4	*Coscinodiscus radiates*	4342.19	5163.08	4764.55
5	*Coscinodiscus gigas*	4150.9	6124.27	5285.44
6	*Coscinodiscus oculus-iridis*	5685.32	6199.85	5823.44
7	*Coscinodiscus concinnus*	2896.33	4001.29	3842.55
8	*Coscinodiscus perforates*	3842.66	4734.55	4181.21
9	*Coscinodiscus asteromphalus*	4576.9	5678.34	4980.23
10	*Coscinodiscus thorii*	3998.34	4562.09	4127.9
11	*Coscinodiscus granii*	3480	6789.12	5100.04
12	*Cyclotella* sp.	22156.32	30355.12	27432.77
13	*Cyclotella striata*	25985.34	28856.12	26149.08
14	*Cyclotella stylorum*	17156.44	19222.41	18435.66
15	*Thalassiosira subtilis*	944.56	1210.33	981.75
16	*Thalassiosira* sp.	856.92	1432.75	901
17	*Thalassiosira decipiens*	913.72	1217.15	1010.35
18	*Thalassiosira punctigera*	757.88	913.56	842.5
19	*Thalassiosira hyaline*	681.29	744.45	701.28
20	*Thalassiosira eccentric*	952.21	1113.66	1054.32
21	*Skeletonema costatum*	1385.43	1812.55	1659.71
22	*Paralia sulcata*	18840.55	19292.39	18956.1
23	*Planktoniella sol*	211886.32	256931.85	230121.67
24	*Planktoniella blanda*	105115.19	121562.3	119358.44
25	*Rhizosolenia setigera*	5348.69	5851.44	5630.32
26	*Rhizosolenia alata*	3751.88	3929.49	3862.45
27	*Rhizosolenia hebetate*	4123.66	5028.1	4888.19
28	*Rhizosolenia styliformis*	5034.66	6130.29	5841
29	*Rhizosolenia robusta*	7899.75	9256.33	8141.89
30	*Rhizosolenia stolterfothii*	6215.77	7100.56	6853.49
31	*Rhizosolenia cylindrus*	5298.55	5946.2	5755.18
32	*Rhizosolenia shrubsolei*	4385.66	4611.73	4578.54
33	*Rhizosolenia imbricata*	6215	7000.35	6859.82
34	*Lauderia annulata*	29434.34	31165.49	30458.38
35	*Bacteriastrum* sp.	32519.71	40605.22	31672.67
36	*Bacteriastrum delicatulum*	38895.72	41468.21	39224.47
37	*Bacteriastrum varians*	28028.52	32455	29611.09
38	*Bacteriastrum comosum*	46959.33	49110.52	47564.6
39	*Bacteriastrum hyalinum*	47852.1	51677.35	49810.55
40	*Chaetoceros didymus*	3156.32	4029.99	3898.44
41	*Chaetoceros curvisetus*	2985.77	5100.62	4781.9
42	*Chaetoceros diversus*	3185.22	4629.77	4244.51
43	*Chaetoceros messanensis*	2621.51	4050.66	2985.42
44	*Chaetoceros peruvianus*	4652.85	5087.19	4985.66

(continued)

Table 5.6 (continued)

S. no.	Species	Pond 1 (0.5 ppm)	Pond 2 (1.5 ppm)	Pond 3 (2.5 ppm)
45	*Chaetoceros eibenii*	4985.21	5324.4	5182.75
46	*Chaetoceros lorenzianus*	2156.42	2833.57	2600.88
47	*Chaetoceros compressus*	3270.77	4135.4	3928.95
48	*Chaetoceros decipiens*	6176.85	7035.12	6839.47
49	*Chaetoceros atlanticus*	5949.44	6381.29	6125.75
50	*Chaetoceros subtilis*	5185.37	5721.06	5443.11
51	*Chaetoceros convolutes*	4384.2	4711.36	4644.52
52	*Chaetoceros holsaticum*	3961	4255.39	4017.5
53	*Chaetoceros gracile*	2622.74	2981.8	2844.37
54	*Chaetoceros cinctum*	2432.54	2681.09	2572.5
55	*Chaetoceros affinis*	2834.49	3142.66	2988.7
56	*Chaetoceros danicus*	5138	5764.93	5438.75
57	*Chaetoceros constrictus*	2964.42	3100.97	2979.18
58	*Ditylum sol*	1532.44	1987.09	1764.21
59	*Ditylum brightwellii*	2064.85	2555.2	2386.13
60	*Lithodesmium undulatum*	8179.44	9032.3	8539.75
61	*Triceratium favus*	2553.04	2986.82	2782.7
62	*Triceratium reticulatum*	3176.54	5480.6	4638.49
63	*Triceratium* sp.	4017.55	5111.67	4752.22
64	*Biddulphia sinensis*	16712.52	19000.94	18316.41
65	*Biddulphia mobiliensis*	26315.5	29404	28666.12
66	*Biddulphia regia*	22316.72	24804.55	23119.22
67	*Eucampia zodiacus*	1851.94	2032.99	1964.76
68	*Hemidiscus cuneiformis*	194.32	216.88	118.77
69	*Climacosphenia elongate*	3034.46	4112	3976.14
70	*Fragilaria oceanica*	33104.55	39321.66	36428.45
71	*Rhaphoneis amphiceros*	5354.77	6082.21	5944.6
72	*Thalassionema nitzschioides*	351.44	504.92	419.05
73	*Thalassionema* sp.	1026.44	1985.23	1751.49
74	*Thalassiothrix longissima*	2164.54	2344.99	2281.02
75	*Thalassiothrix frauenfeldii*	356.88	452.19	391.85
76	*Thalassiothrix* sp.	656.85	829.25	704.17
77	*Thalassiothrix nitzschioides*	585.29	609.42	593.17
78	*Asterionella japonica*	117.65	201.32	198.44
79	*Asterionellopsis glacialis*	446.12	511.77	498.54
80	*Diatoma vulgare*	1256.11	1572.32	1388.96
81	*Diatoma* sp.	948.45	1115.69	982.2
82	*Cocconeis* sp.	1682	1954.33	1856.21
83	*Cocconeis scutellum*	2014.22	2977.54	2866.82
84	*Gyrosigma balticum*	2951.27	2989.16	2972.44
85	*Gyrosigma* sp.	6144	6988.76	6712.45
86	*Pleurosigma normanii*	1285.42	1611.21	1404.55
87	*Pleurosigma elongatum*	41054.68	45122.49	43601.28
88	*Pleurosigma directum*	40090.8	47146.11	46222.48

(continued)

Table 5.6 (continued)

S. no.	Species	Pond 1 (0.5 ppm)	Pond 2 (1.5 ppm)	Pond 3 (2.5 ppm)
89	*Diploneis smithii*	1985.42	2314.08	2166.37
90	*Navicula longa*	5438.66	6015.29	5888.7
91	*Navicula rhombic*	3856.88	4023.41	3985.72
92	*Navicula pennata*	3744.82	4012.53	3985.76
93	*Navicula* sp.	3843.78	4197.95	3909.18
94	*Navicula cancellata*	8648.21	9032.16	8111
95	*Cymbella marina*	403.52	614.88	599.21
96	*Nitzschia sigma*	400.19	482.54	450.06
97	*Nitzschia longissima*	513.85	667.2	604.29
98	*Nitzschia closterium*	719.44	608.88	765.08
99	*Nitzschia striata*	815.42	900.33	866.8
100	*Nitzschia seriata*	652.49	719.5	678.32
101	*Pseudo-nitzschia pungens*	1035.85	1467.09	1281.52
102	*Pseudo-nitzschia australis*	2364.54	3091.38	2844.55
103	*Pinnularia alpine*	612.66	898.11	702.55
104	*Pinnularia* sp.	385.65	411	391.66
105	*Pinnularia Trevelyan*	1796.22	2814.44	2190
106	*Bacillaria paradoxa*	4115.75	5090.88	4066.02
107	*Bacillaria paxillifer*	1398.86	2031.32	1988.09
108	*Triceratium spinosum*	398.42	566.49	493.1
109	*Phaeocystis* sp.	2039.44	3146.82	2981.99
110	*Amphora hyaline*	4136.44	5032.44	4984.66
111	*Amphipleura* sp.	392.39	501	400.89
112	*Amphiprora constrica*	1603.98	1838.21	1792.55
113	*Halosphaera viridis*	298.33	401.67	306
114	*Halosphaera* sp.	2014.88	3140.32	2968.55
115	*Corethron crinophyllum*	542.99	601.66	593.55
116	*Guinardia flaccida*	711.62	802.55	796.22
117	*Hemiaulus sinensis*	911.31	1386.66	1022.45
118	*Melosira nummuloides*	2349.66	3011.79	2809.66
119	*Melosira granulate*	9149.2	11667.67	10980.11
120	*Melosira variance*	8067.2	9114.53	8714.77
121	*Melosira* sp.	704.28	906.39	821.44
122	*Bellerochea malleus*	154.66	201.3	180.4
123	*Cerataulina pelagic*	309.44	511.77	408
124	*Leptocylindrus danicus*	398.86	411.38	405.55
125	*Leptocylindrus minimus*	1298.49	1633.08	1405.33
126	*Asteromphalus* sp.	504.55	599.77	580
127	*Hyalodiscus* sp.	398.22	1032.05	496.89
128	*Actinocyclus octanarius*	995.29	1386.21	1144.5
129	*Actinoptychus* sp.	1561.99	2115	1980.32
130	*Eunotia* sp.	598.66	708.74	611.88
131	*Tabellaria* sp.	6098.22	9133.14	7196.77
132	*Grammatophora marina*	4968.29	6014.33	5138.09

(continued)

Table 5.6 (continued)

S. no.	Species	Pond 1 (0.5 ppm)	Pond 2 (1.5 ppm)	Pond 3 (2.5 ppm)
133	*Closterium* sp.	898.9	1416.44	1284.55
134	*Striatella* sp.	2961.2	3640.44	3239.39
135	*Striatella unipunctata*	3460.22	4988.44	3966.8
136	*Diplosalis* sp.	809.16	949.44	813.66
137	*Licmophora ehrenbergii*	541.88	709.44	608.8

Figures in bracket indicate the applied doses of ferrous sulphate in the pond water

3. Ecological perturbations will be negligible in aquaculture ponds because unlike open oceans, the species richness is not too high in this closed system. Only the target species is cultured in the pond, and essentially, the food chain consists of phytoplankton, zooplankton and target (cultured) species.
4. The aquaculture ponds are closed systems with depth ranging from 1 to 5 m. The substratum is basically soil. The question of

Table 5.7 Mean cell volume (μm^3) in dinoflagellates in the selected ponds

S. no.	Species	Pond 1 (0.5 ppm)	Pond 2 (1.5 ppm)	Pond 3 (2.5 ppm)
1	*Prorocentrum gracile*	856.39	954.22	901.08
2	*Prorocentrum micans*	1024.85	1391.66	1154.45
3	*Prorocentrum concavum*	1381.43	1600.39	1417.55
4	*Ceratium furca*	1185.29	1666.54	1498.5
5	*Ceratium fusus*	1084.91	1346.75	1111.09
6	*Ceratium teres*	946.55	1294.16	1088.57
7	*Ceratium minutum*	549.76	609.22	585.4
8	*Ceratium tripos (Ceratium tripos)*	1876.44	2085.15	1972.09
9	*Ceratium trichoceros*	498.54	606.98	514.11
10	*Ceratium horridum*	1029.97	1413.29	1208.75
11	*Ceratium inflatum*	986.79	1152.47	1066.48
12	*Ceratocorys horrida*	459.66	782.21	695.45
13	*Protoperidinium* sp.	1387.66	1941.08	1694.25
14	*Protoperidinium crassipes*	2025.55	2411.6	2282.29
15	*Protoperidinium depressum*	1376.49	1718.62	1455.48
16	*Protoperidinium ovatum*	822.17	1049.62	954.45
17	*Protoperidinium pellucidum*	765.2	840.13	798.29
18	*Protoperidinium conicum*	2011.17	2642.54	2388.39
19	*Pyrocystis fusiformis*	985.77	2146.32	1911
20	*Pyrocystis* sp.	356.25	458.59	391.77
21	*Dinophysis caudata*	6154.05	6289.39	6199.1
22	*Dinophysis acuta*	2404.13	2817.19	2643.45
23	*Dinophysis norvegica*	1698.49	1934.6	1855.11
24	*Dinophysis* sp.	1317.44	1788.2	1513.55
25	*Alexandrium* sp.	6928.38	9144.72	7000.15
26	*Preperidinium meunieri*	11141.62	13984.44	12199.05

Figures in bracket indicate the applied doses of ferrous sulphate in the pond water

Table 5.8 Mean cell volume (μm^3) in other algae in the selected ponds

S. no.	Species	Pond 1 (0.5 ppm)	Pond 2 (1.5 ppm)	Pond 3 (2.5 ppm)
1	*Trichodesmium* sp.	3552.14	4005.29	3914.66
2	*Trichodesmium thiebautii*	2849.42	2911.68	2899.95
3	*Oscillatoria* sp.	698.44	803.28	714.66
4	*Oscillatoria limosa*	699.49	734.22	710.05
5	*Gloeocapsa* sp.	153	178.69	165.25
6	*Stigonema* sp.	749.75	813.66	795.28
7	*Cylindrospermopsis raciborskii*	835.66	892.44	854.1
8	*Anabaena* sp.	354.29	496.38	398.72
9	*Dictyocha* sp.	103.28	211.44	196.71
10	*Ditylum sol*	1982.44	2004.91	1995.76
11	*Chlorella salina*	568.44	601	575.75
12	*Amphora hyalina*	2111.7	2802.55	2644.12
13	*Netrium* sp.	456.55	609.82	517.44
14	*Draparnaldia* sp.	1468.22	1802.1	1717.45
15	*Gloeocapsa* sp.	213.11	288.42	265.07

Figures in bracket indicate the applied doses of ferrous sulphate in the pond water

massive geophysical changes does not arise when iron fertilization is done.

5. In aquaculture pond, the algal bloom can be converted into animal biomass as the cultured species feed and phytoplankton and the question of DO fall does not arise.
6. The HAB species can be detected and managed as aquaculturists monitor the water quality and check the pond health almost on an hourly basis.
7. Hydrogen sulphide gas produced by microbial activity in the aquaculture ponds can be precipitated through the application of ferrous sulphate, which may otherwise be toxic for the cultured species.

Different doses of iron were mixed in rainwater-harvested ponds selected in deltaic Sundarbans to trigger up the bloom of phytoplankton (Fig. 5.12).

The excess phytoplankton was utilized as the natural food of cultured species (*Macrobrachium rosenbergii*). The best growth (in terms of biomass) was observed in the pond where 1.5 ppm ferrous sulphate was applied (Table 5.5 and Fig. 5.13). The protein content was also highest in the pond treated with 1.5 ppm ferrous sulphate, which is an effect of growth of more phytoplankton in the pond treated with 1.5 ppm ferrous sulphate compared to the other 3 ponds (Tables 5.6, 5.7, 5.8, 5.9 and 5.10).

Table 5.9 Mean cell carbon (pg) in diatoms in the selected ponds

S. no.	Species	Pond 1 (0.5 ppm)	Pond 2 (1.5 ppm)	Pond 3 (2.5 ppm)
1	*Coscinodiscus excentricus*	490.75	530.02	432.34
2	*Coscinodiscus jonesianus*	554.64	642.52	594.33
3	*Coscinodiscus lineatus*	249.74	288.11	279.42
4	*Coscinodiscus radiatus*	256.79	295.50	269.17
5	*Coscinodiscus gigas*	247.57	339.39	292.80
6	*Coscinodiscus oculus-iridis*	319.52	342.78	316.75
7	*Coscinodiscus concinnus*	184.91	240.31	226.09
8	*Coscinodiscus perforatus*	232.56	275.45	242.12
9	*Coscinodiscus asteromphalus*	267.99	319.20	279.01
10	*Coscinodiscus thorii*	240.17	267.28	239.62
11	*Coscinodiscus granii*	214.59	368.97	284.45
12	*Cyclotella* sp.	962.92	1243.03	1113.26
13	*Cyclotella striata*	1095.81	1193.01	1070.82
14	*Cyclotella stylorum*	782.55	858.14	806.51
15	*Thalassiosira subtilis*	74.53	91.12	74.76
16	*Thalassiosira* sp.	68.87	104.48	69.73
17	*Thalassiosira decipiens*	72.55	91.54	76.52
18	*Thalassiosira punctigera*	62.34	72.54	66.04
19	*Thalassiosira hyaline*	57.18	61.44	56.91
20	*Thalassiosira eccentric*	75.01	85.17	79.21
21	*Skeletonema costatum*	101.68	126.43	114.45
22	*Paralia sulcata*	844.29	860.68	824.92
23	*Planktoniella sol*	6009.83	7026.76	6247.49
24	*Planktoniella blanda*	3403.79	3829.70	3668.48
25	*Rhizosolenia setigera*	304.09	327.07	308.21
26	*Rhizosolenia alata*	228.09	236.81	227.04
27	*Rhizosolenia hebetata*	246.26	289.22	274.83
28	*Rhizosolenia styliformis*	289.53	339.66	317.53
29	*Rhizosolenia robusta*	417.21	474.43	415.68
30	*Rhizosolenia stolterfothii*	343.49	382.64	361.48
31	*Rhizosolenia cylindrus*	301.78	331.36	313.74
32	*Rhizosolenia shrubsolei*	258.87	269.64	260.62
33	*Rhizosolenia imbricata:*	343.46	378.25	361.75
34	*Lauderia annulata*	1212.36	1269.88	1211.84
35	*Bacteriastrum* sp.	1314.45	1573.81	1250.87
36	*Bacteriastrum delicatulum*	1519.86	1600.88	1487.76
37	*Bacteriastrum varians*	1165.19	1312.33	1184.42
38	*Bacteriastrum comosum*	1770.76	1836.26	1739.54
39	*Bacteriastrum hyalinum*	1798.01	1913.72	1805.87
40	*Chaetoceros didymus*	198.26	241.71	228.75
41	*Chaetoceros curvisetus*	189.52	292.60	269.97
42	*Chaetoceros diversus*	199.73	270.50	245.09
43	*Chaetoceros messanensis*	170.54	242.71	184.24
44	*Chaetoceros peruvianus*	271.59	291.98	279.26
45	*Chaetoceros eibenii*	287.22	302.97	288.18

(continued)

Table 5.9 (continued)

S. no.	Species	Pond 1 (0.5 ppm)	Pond 2 (1.5 ppm)	Pond 3 (2.5 ppm)
46	*Chaetoceros lorenzianus*	145.56	181.65	164.75
47	*Chaetoceros compressus*	204.07	246.82	230.21
48	*Chaetoceros decipiens*	341.75	379.78	360.88
49	*Chaetoceros atlanticus*	331.51	350.89	330.02
50	*Chaetoceros subtilis*	296.54	321.15	299.87
51	*Chaetoceros convolutus*	258.80	274.36	263.66
52	*Chaetoceros holsaticum*	238.35	252.62	234.40
53	*Chaetoceros gracile*	170.61	189.32	177.15
54	*Chaetoceros cinctum*	160.50	173.68	163.29
55	*Chaetoceros affinis*	181.70	197.56	184.41
56	*Chaetoceros danicus*	294.34	323.14	299.67
57	*Chaetoceros constrictus*	188.42	195.43	183.93
58	*Ditylum sol*	110.34	136.22	120.26
59	*Ditylum brightwellii*	140.53	167.04	153.63
60	*Lithodesmium undulatum*	429.15	465.10	432.08
61	*Triceratium favus*	166.92	189.58	174.03
62	*Triceratium reticulatum*	199.29	310.16	263.38
63	*Triceratium* sp.	241.10	293.11	268.61
64	*Biddulphia sinensis*	766.09	850.12	802.27
65	*Biddulphia mobiliensis*	1107.09	1211.35	1153.68
66	*Biddulphia regia*	968.57	1055.25	969.04
67	*Eucampia zodiacus*	128.66	138.77	131.23
68	*Hemidiscus cuneiformis*	20.67	22.60	13.48
69	*Climacosphenia elongata*	192.03	245.69	232.45
70	*Fragilaria oceanica*	1333.59	1533.34	1401.16
71	*Rhaphoneis amphiceros*	304.37	337.49	322.09
72	*Thalassionema nitzschioides*	33.43	44.84	37.48
73	*Thalassionema* sp.	79.72	136.12	119.55
74	*Thalassiothrix longissima*	146.01	155.80	148.12
75	*Thalassiothrix frauenfeldii*	33.84	41.01	35.50
76	*Thalassiothrix* sp.	55.51	67.06	57.10
77	*Thalassiothrix nitzschioides*	50.55	52.23	49.68
78	*Asterionella japonica*	13.76	21.27	20.44
79	*Asterionellopsis glacialis*	40.56	45.34	43.15
80	*Diatoma vulgare*	93.91	112.66	99.06
81	*Diatoma* sp.	74.77	85.30	74.79
82	*Cocconeis* sp.	119.00	134.40	125.32
83	*Cocconeis scutellum*	137.73	189.10	178.28
84	*Gyrosigma balticum*	187.75	189.70	183.59
85	*Gyrosigma* sp.	340.27	377.75	355.43
86	*Pleurosigma normanii*	95.68	114.92	99.96
87	*Pleurosigma elongatum*	1587.92	1714.37	1621.03
88	*Pleurosigma directum*	1557.62	1776.47	1699.63
89	*Diploneis smithii*	136.13	154.14	142.05
90	*Navicula longa*	308.23	334.48	319.63
91	*Navicula rhombica*	233.25	241.39	232.90

(continued)

Table 5.9 (continued)

S. no.	Species	Pond 1 (0.5 ppm)	Pond 2 (1.5 ppm)	Pond 3 (2.5 ppm)
92	*Navicula pennata*	227.74	240.86	232.90
93	*Navicula* sp.	232.61	249.85	229.27
94	*Navicula cancellata*	448.99	465.09	414.40
95	*Cymbella marina*	37.39	52.61	50.09
96	*Nitzschia sigma*	37.14	43.23	39.72
97	*Nitzschia longissima*	45.49	56.22	50.44
98	*Nitzschia closterium*	59.76	52.20	61.07
99	*Nitzschia striata*	66.15	71.68	67.58
100	*Nitzschia seriata*	55.21	59.76	55.39
101	*Pseudo-nitzschia pungens*	80.32	106.51	92.79
102	*Pseudo-nitzschia australis*	156.86	194.94	177.16
103	*Pinnularia alpina*	52.46	71.54	56.99
104	*Pinnularia* sp.	36.04	37.95	35.48
105	*Pinnularia trevelyan*	125.51	180.65	143.30
106	*Bacillaria paradoxa*	245.87	292.15	236.70
107	*Bacillaria paxillifer*	102.47	138.68	132.49
108	*Triceratium spinosum*	37.01	49.23	42.77
109	*Phaeocystis* sp.	139.13	197.77	184.07
110	*Amphora hyalina*	246.87	289.42	279.22
111	*Amphipleura* sp.	36.55	44.56	36.16
112	*Amphiprora constrica*	114.50	127.88	121.82
113	*Halosphaera viridis*	29.27	37.25	29.05
114	*Halosphaera* sp.	137.76	197.44	183.40
115	*Corethron crinophyllum*	47.57	51.69	49.71
116	*Guinardia flaccida*	59.23	65.30	63.08
117	*Hemiaulus sinensis*	72.39	101.75	77.26
118	*Melosira nummuloides*	156.05	190.86	175.39
119	*Melosira granulata*	469.97	572.42	529.77
120	*Melosira variance*	424.37	468.53	439.24
121	*Melosira* sp.	58.74	72.07	64.70
122	*Bellerochea malleus*	17.18	21.27	18.92
123	*Cerataulina pelagica*	30.15	45.34	36.68
124	*Leptocylindrus danicus*	37.04	37.98	36.50
125	*Leptocylindrus minimus*	96.47	116.18	100.00
126	*Asteromphalus* sp.	44.82	51.56	48.79
127	*Hyalodiscus* sp.	36.99	80.08	43.04
128	*Actinocyclus octanarius*	77.76	101.72	84.66
129	*Actinoptychus* sp.	112.06	143.29	132.07
130	*Eunotia* sp.	51.49	59.04	50.95
131	*Tabellaria* sp.	338.21	469.31	376.09
132	*Grammatophora marina*	286.43	334.44	286.17
133	*Closterium* sp.	71.59	103.52	92.97
134	*Striatella* sp.	188.26	222.58	196.85
135	*Striatella unipunctata*	213.60	287.37	232.00
136	*Diplosalis* sp.	65.74	74.84	64.20
137	*Licmophora ehrenbergii*	47.49	59.09	50.74

Figures in bracket indicate the applied doses of ferrous sulphate in the pond water

Table 5.10 Mean cell carbon (pg) in dinoflagellates in the selected ponds

S. no	Species	Pond 1 (0.5 ppm)	Pond 2 (1.5 ppm)	Pond 3 (2.5 ppm)
1	*Prorocentrum gracile*	191.72	209.48	199.88
2	*Prorocentrum micans*	222.10	285.34	244.85
3	*Prorocentrum concavum*	283.62	319.94	289.68
4	*Ceratium furca*	250.19	330.74	303.16
5	*Ceratium fusus*	232.70	277.78	237.29
6	*Ceratium teres*	208.10	268.86	233.35
7	*Ceratium minutum*	133.36	145.06	140.40
8	*Ceratium tripos (Ceratium tripos)*	364.48	397.36	379.63
9	*Ceratium trichoceros*	123.09	144.62	126.23
10	*Ceratium horridum*	223.01	288.97	254.24
11	*Ceratium inflatum*	215.32	244.51	229.46
12	*Ceratocorys horrida*	115.17	178.01	161.67
13	*Protoperidinium* sp.	284.67	374.73	335.23
14	*Protoperidinium crassipes*	388.04	447.63	427.88
15	*Protoperidinium depressum*	282.79	339.18	296.02
16	*Protoperidinium ovatum*	185.42	226.48	209.52
17	*Protoperidinium pellucidum*	174.83	188.74	181.00
18	*Protoperidinium conicum*	385.78	482.45	444.10
19	*Pyrocystis fusiformis*	215.14	406.88	369.97
20	*Pyrocystis* sp.	93.48	114.95	101.04
21	*Dinophysis caudata*	964.13	981.46	969.91
22	*Dinophysis acuta*	446.50	508.41	482.58
23	*Dinophysis norvegica*	335.92	373.71	361.08
24	*Dinophysis* sp.	272.82	350.38	305.65
25	*Alexandrium* sp.	1062.41	1333.56	1071.41
26	*Preperidinium meunieri*	1567.70	1888.41	1688.55

Figures in bracket indicate the applied doses of ferrous sulphate in the pond water

Annexure

List of Phytoplankton Identified from the Inshore Waters of Bay of Bengal

Researchers involved in sampling and documentation process of phytoplankton during 2006 to 2012 are Dr. Subhra Bikash Bhattacharyya, Dr. Kakoli Banerjee, Dr. Amitava Aich, Dr. Nibedita Mukhopadhyay, Dr. Harekrishna Jana, Dr. Sufia Zaman, Dr. Aftaf Alam, Mr. Subhasmita Sinha, Ms. Kasturi Sengupta, Ms. Mahua Roy-chowdhury, Mr. Kunal Mondal and Mr. Saumya Kanti Ray. This documentation is a spin-off product of the project entitled '**… Study on carbon stock and response of estuarine phytoplankton to iron fertilization…' funded by Department of Science and Technology (DST), Govt. of India.**

List of Diatoms Identified from the Inshore Waters of Bay of Bengal

S. no.	Species	Microscopic view
1	*Coscinodiscus excentricus*	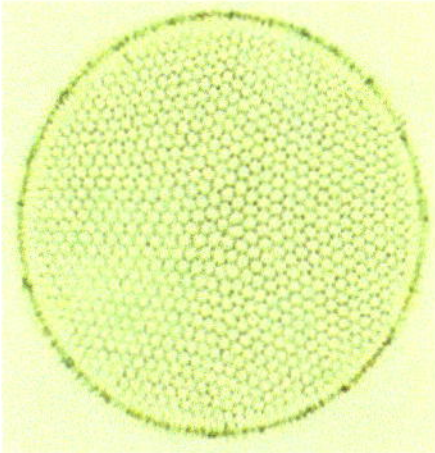
2	*Coscinodiscus jonesianus*	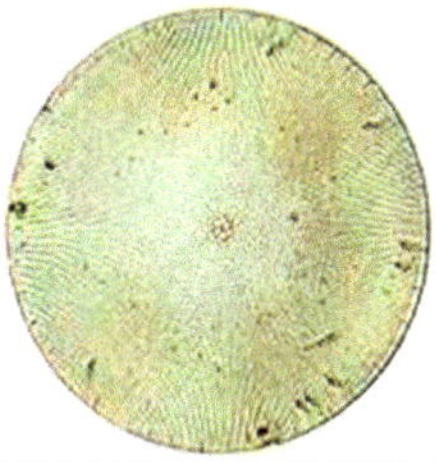
3	*Coscinodiscus lineatus*	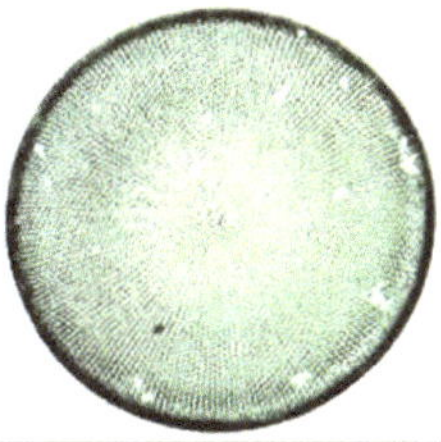
4	*Coscinodiscus radiates*	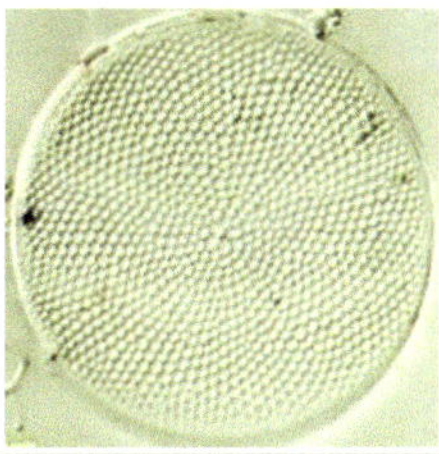
5	*Coscinodiscus gigas*	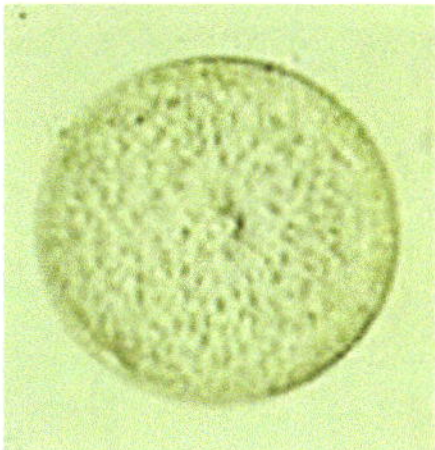

(continued)

(continued)

S. no.	Species	Microscopic view
6	*Coscinodiscus oculus-iridis*	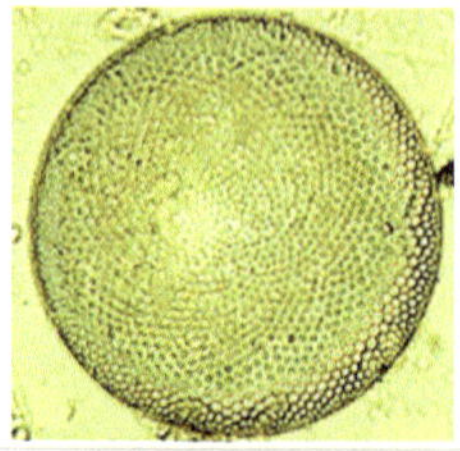
7	*Coscinodiscus concinnus*	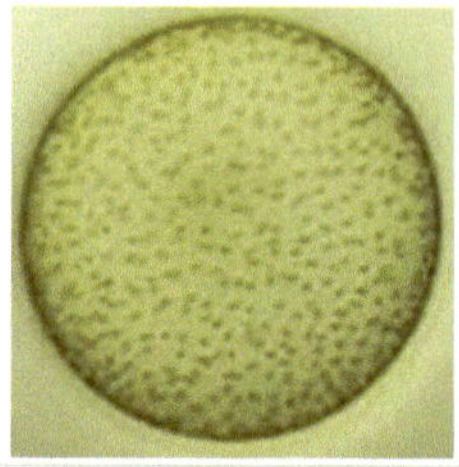
8	*Coscinodiscus perforates*	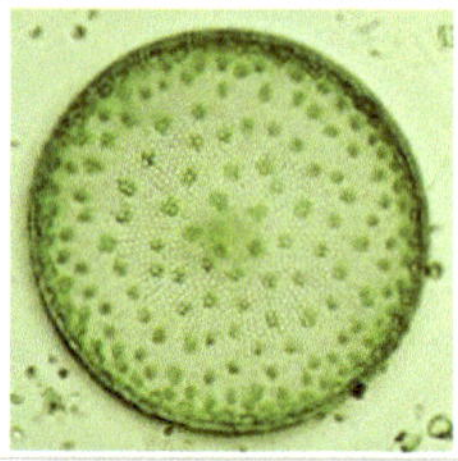
9	*Coscinodiscus asteromphalus*	
10	*Coscinodiscus thorii*	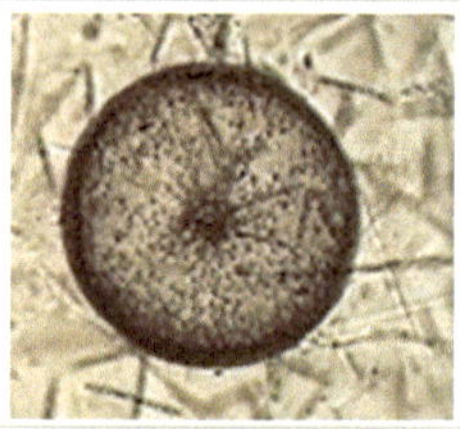
11	*Coscinodiscus granii*	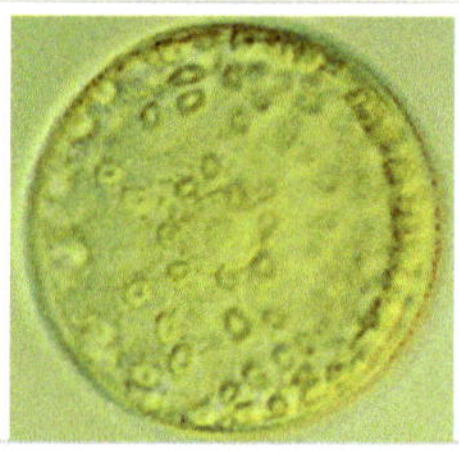

(continued)

(continued)

S. no.	Species	Microscopic view
12	*Cyclotella* sp.	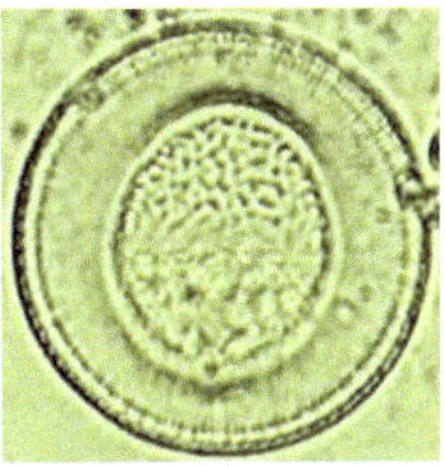
13	*Cyclotella striata*	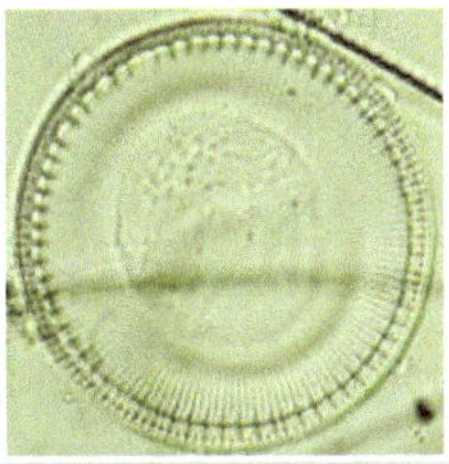
14	*Cyclotella stylorum*	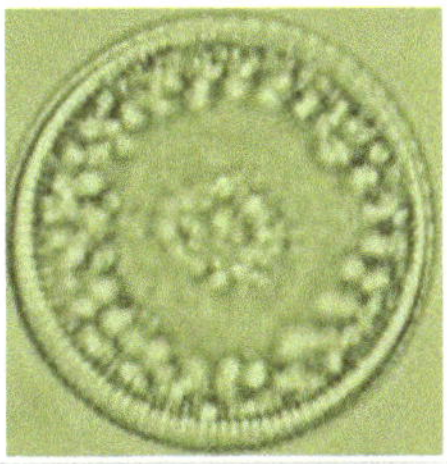
15	*Thalassiosira subtilis*	
16	*Thalassiosira* sp.	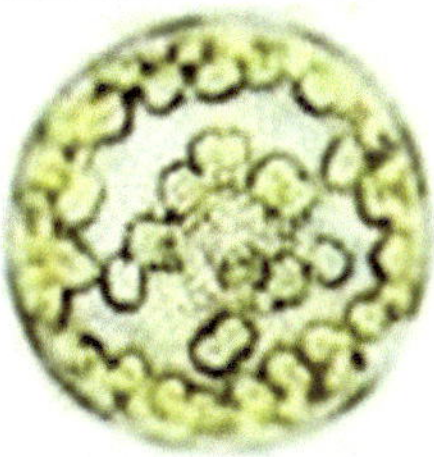
17	*Thalassiosira decipiens*	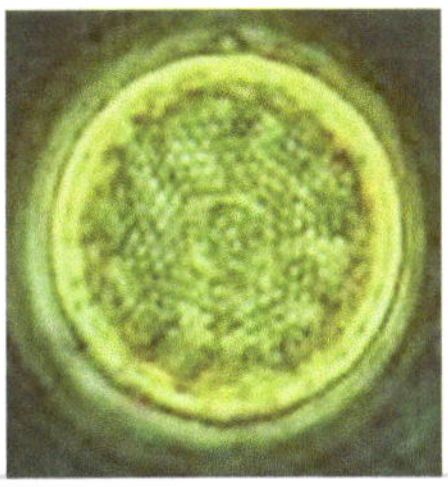

(continued)

(continued)

S. no.	Species	Microscopic view
18	*Thalassiosira punctigera*	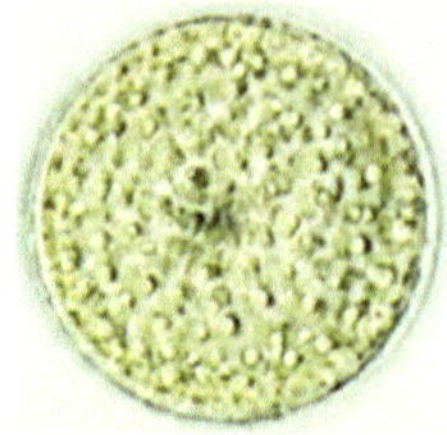
19	*Thalassiosira hyaline*	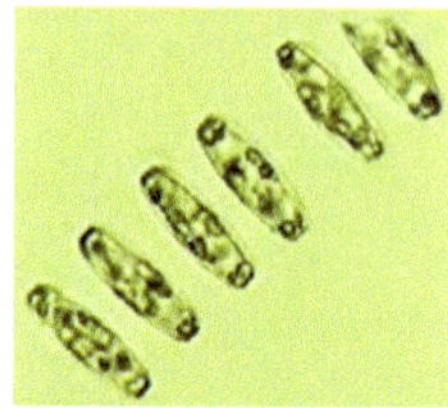
20	*Thalassiosira eccentric*	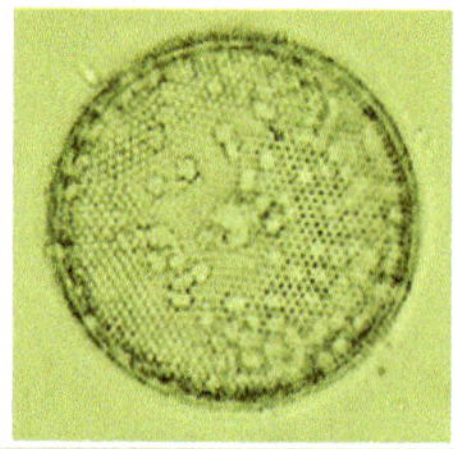
21	*Skeletonema costatum*	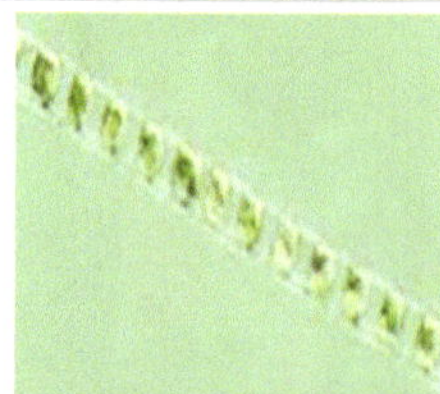
22	*Paralia sulcata*	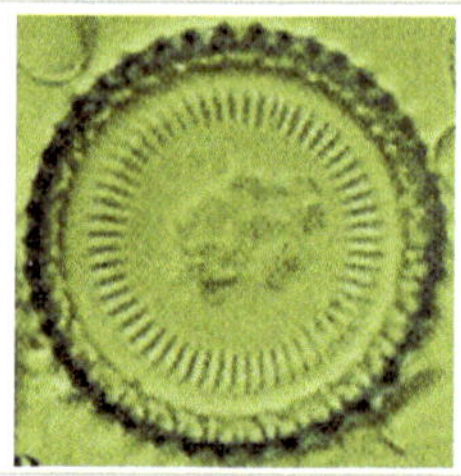
23	*Planktoniella sol*	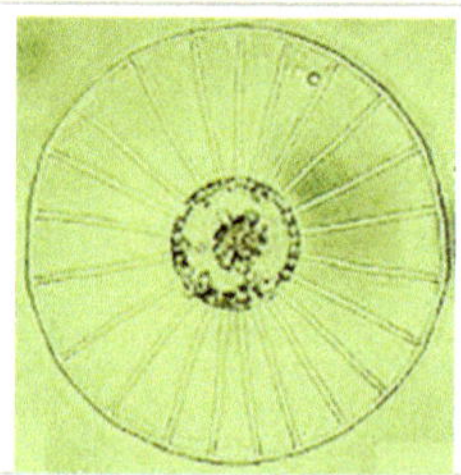

(continued)

(continued)

S. no.	Species	Microscopic view
24	*Planktoniella blanda*	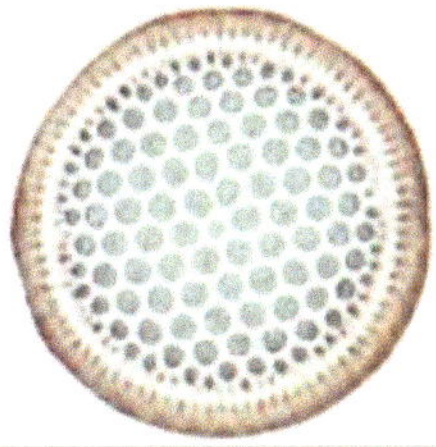
25	*Rhizosolenia setigera*	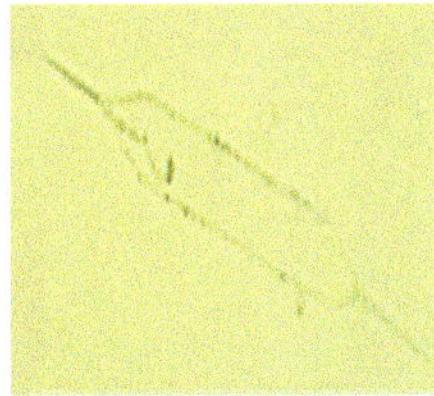
26	*Rhizosolenia alata*	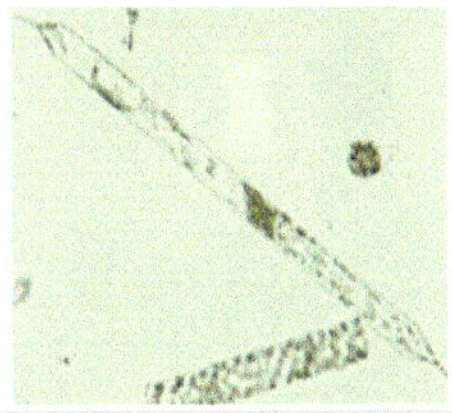
27	*Rhizosolenia hebetata*	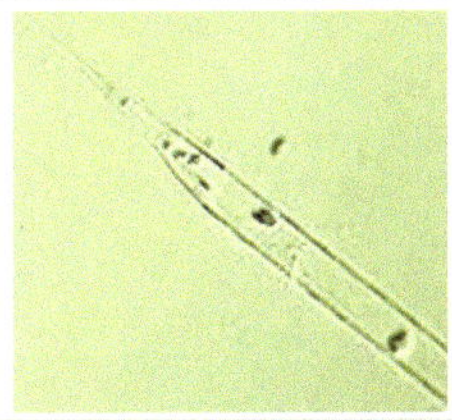
28	*Rhizosolenia styliformis*	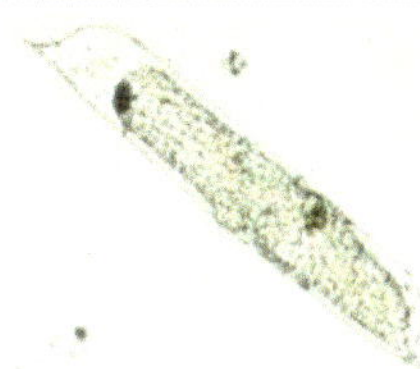
29	*Rhizosolenia robusta*	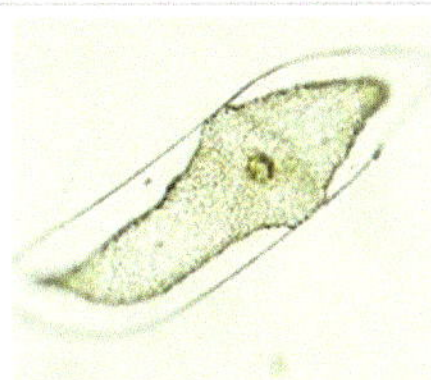
30	*Rhizosolenia stolterfothii*	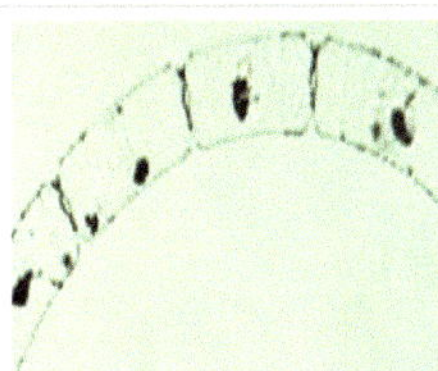

(continued)

(continued)

S. no.	Species	Microscopic view
31	*Rhizosolenia cylindrus*	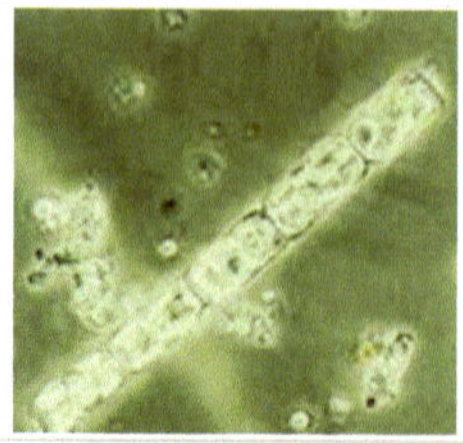
32	*Rhizosolenia shrubsolei*	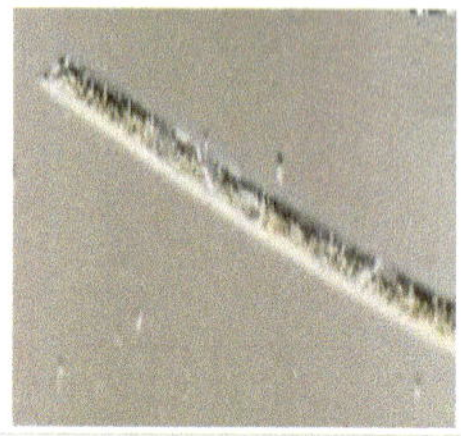
33	*Rhizosolenia imbricata*	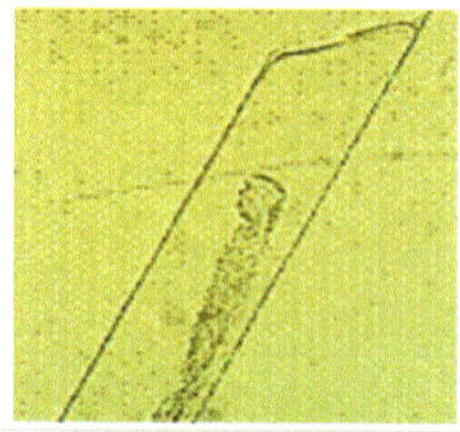
34	*Lauderia annulata*	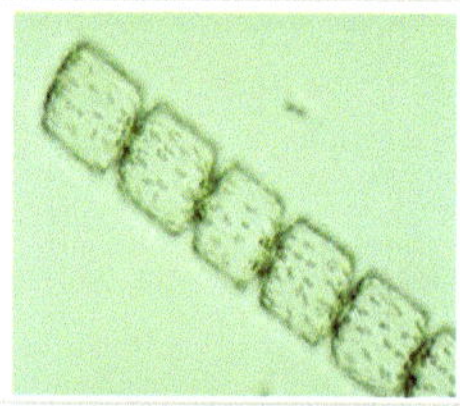
35	*Bacteriastrum* sp.	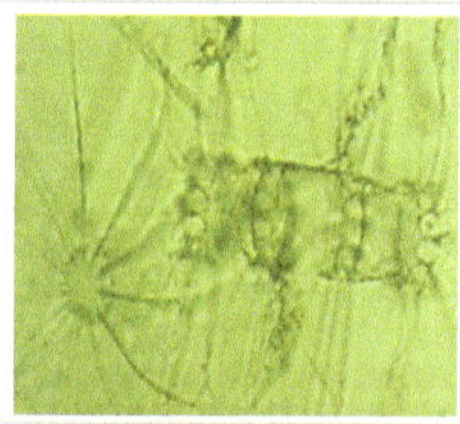
36	*Bacteriastrum delicatulum*	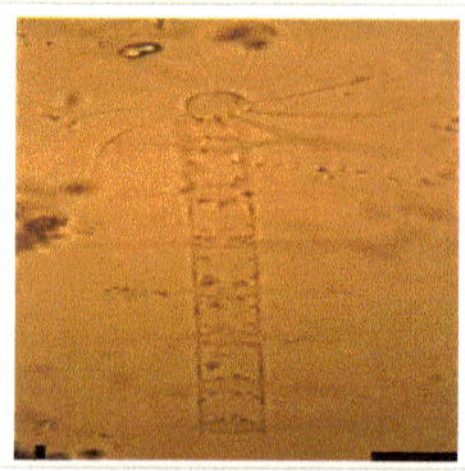
37	*Bacteriastrum varians*	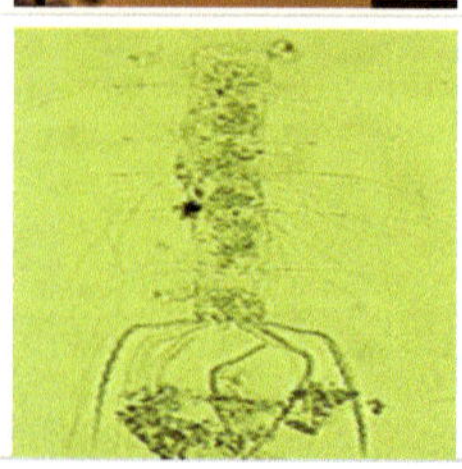

(continued)

(continued)

S. no.	Species	Microscopic view
38	*Bacteriastrum comosum*	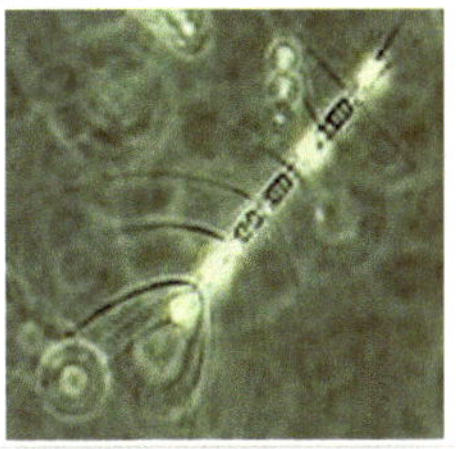
39	*Bacteriastrum hyalinum*	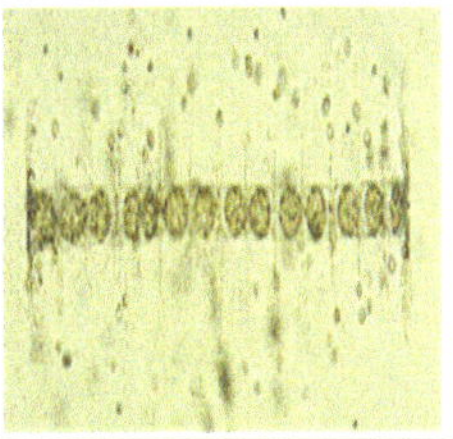
40	*Chaetoceros didymus*	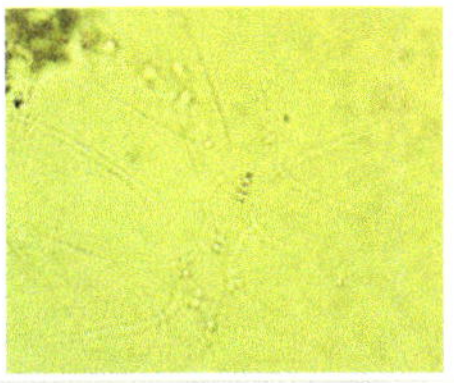
41	*Chaetoceros curvisetus*	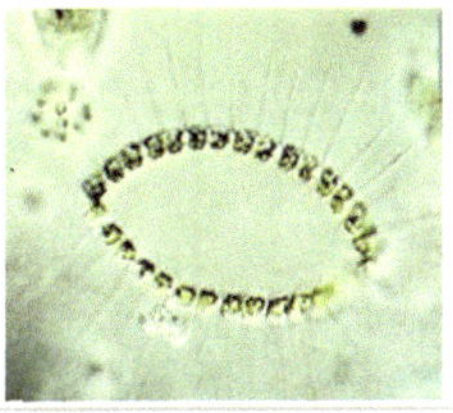
42	*Chaetoceros diversus*	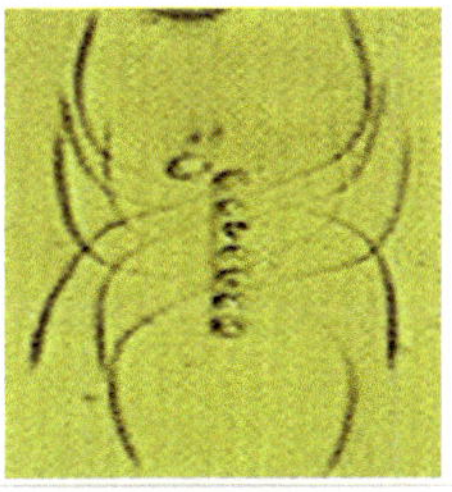
43	*Chaetoceros messanensis*	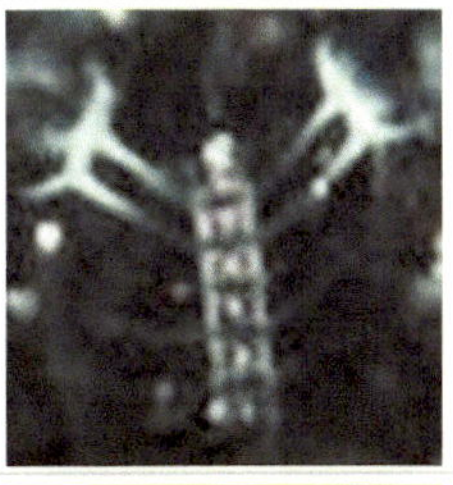
44	*Chaetoceros peruvianus*	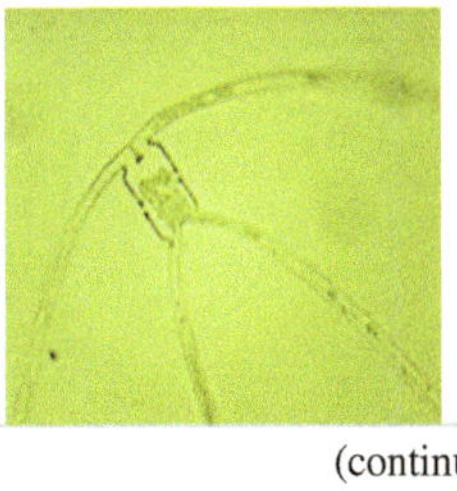

(continued)

(continued)

S. no.	Species	Microscopic view
45	*Chaetoceros eibenii*	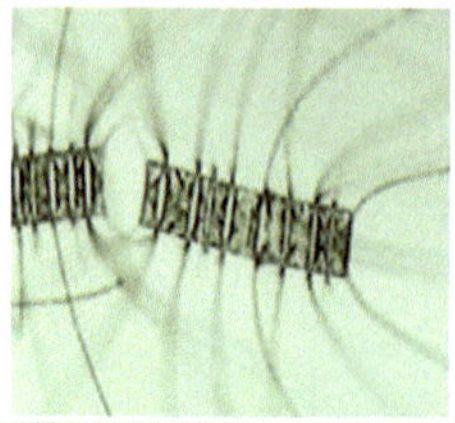
46	*Chaetoceros lorenzianus*	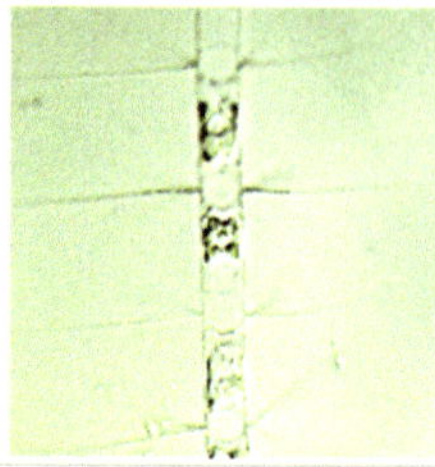
47	*Chaetoceros compressus*	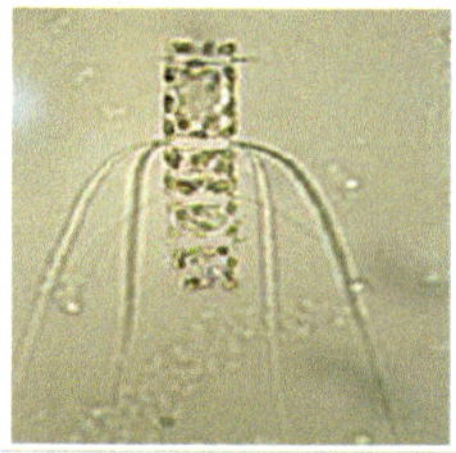
48	*Chaetoceros decipiens*	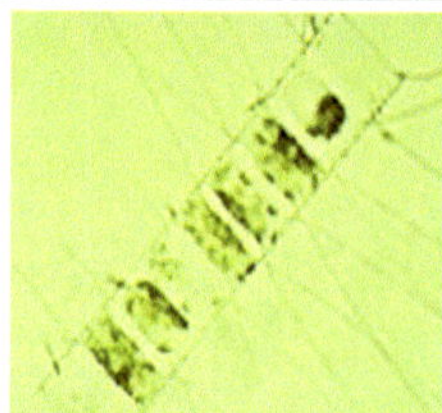
49	*Chaetoceros atlanticus*	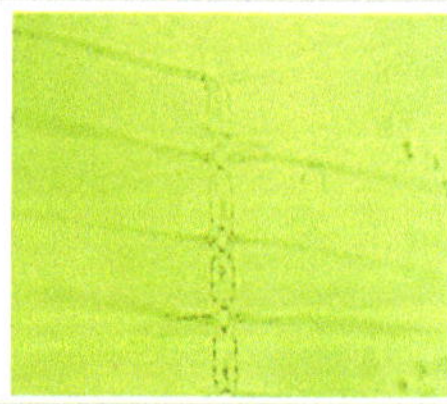
50	*Chaetoceros subtilis*	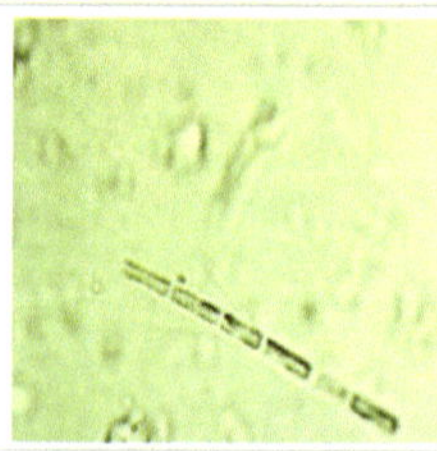
51	*Chaetoceros convolutus*	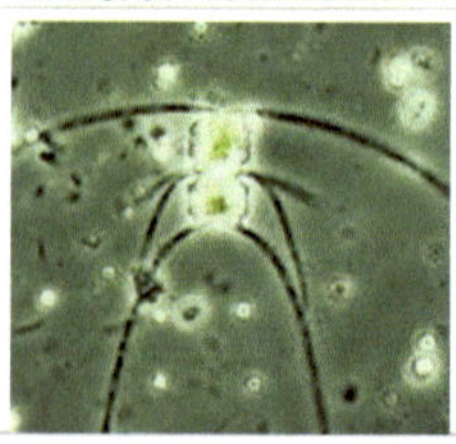

(continued)

(continued)

S. no.	Species	Microscopic view
52	*Chaetoceros holsaticum*	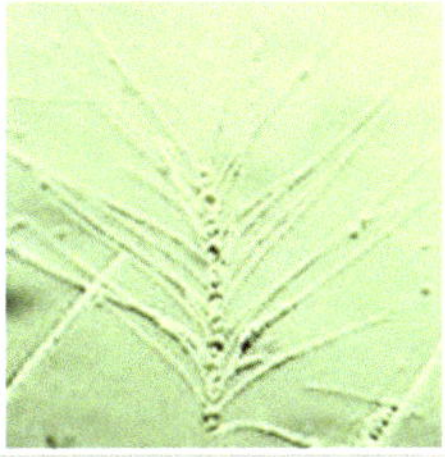
53	*Chaetoceros gracile*	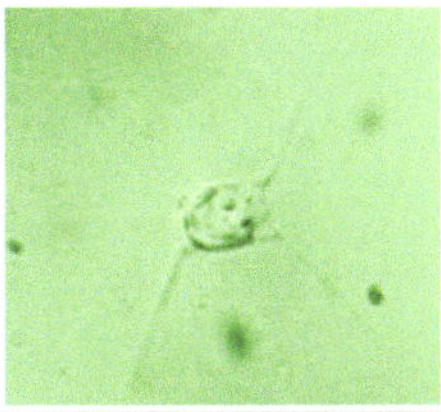
54	*Chaetoceros cinctum*	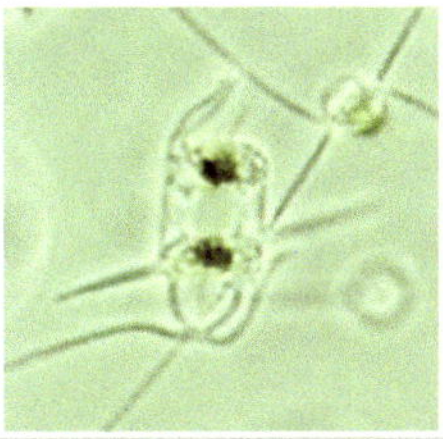
55	*Chaetoceros affinis*	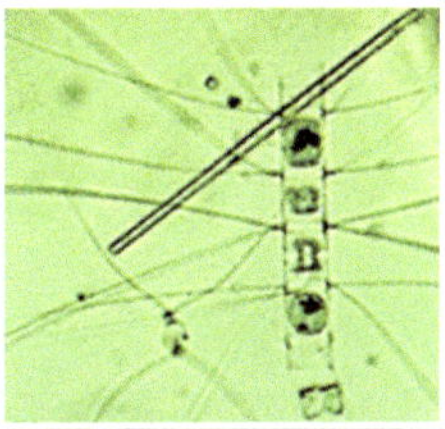
56	*Chaetoceros danicus*	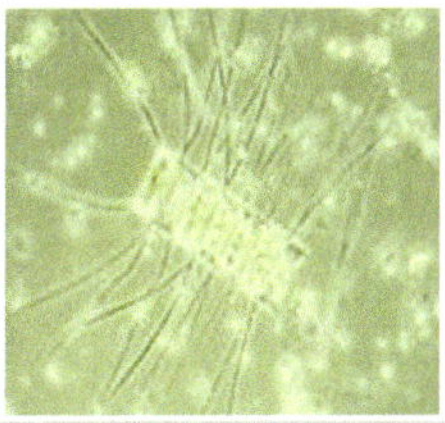
57	*Chaetoceros constrictus*	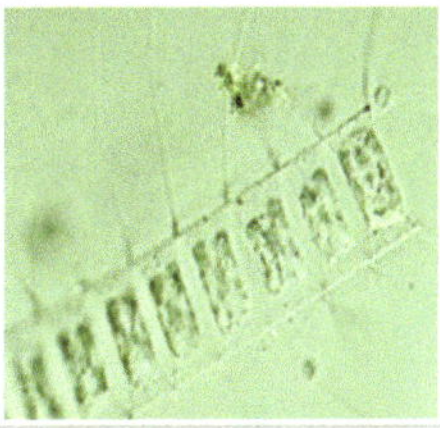
58	*Ditylum sol*	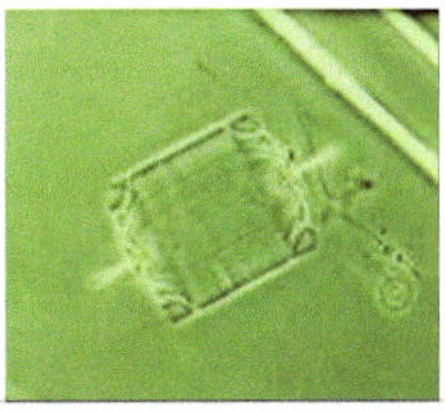

(continued)

(continued)

S. no.	Species	Microscopic view
59	*Ditylum brightwellii*	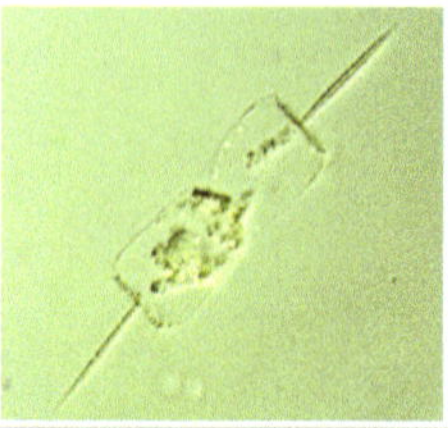
60	*Lithodesmium undulatum*	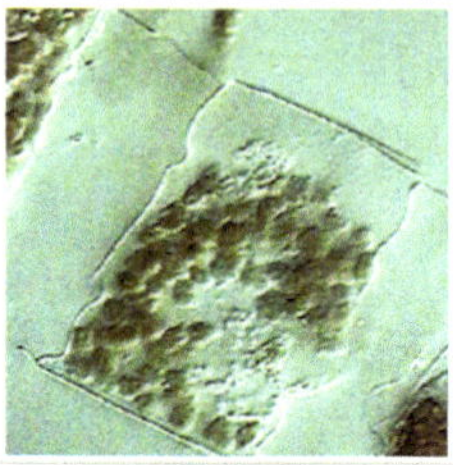
61	*Triceratium favus*	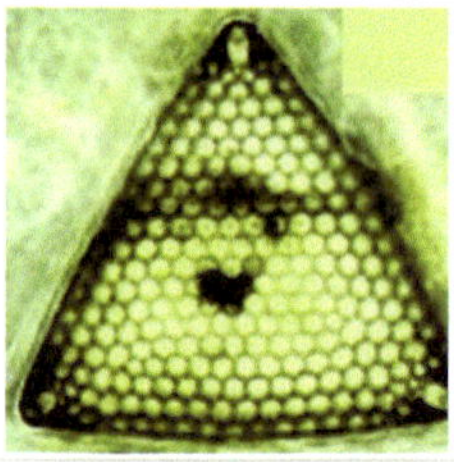
62	*Triceratium reticulatum*	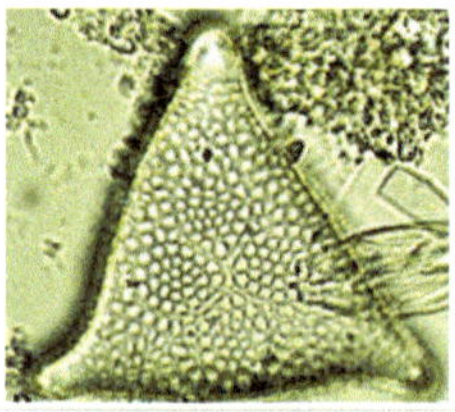
63	*Triceratium* sp.	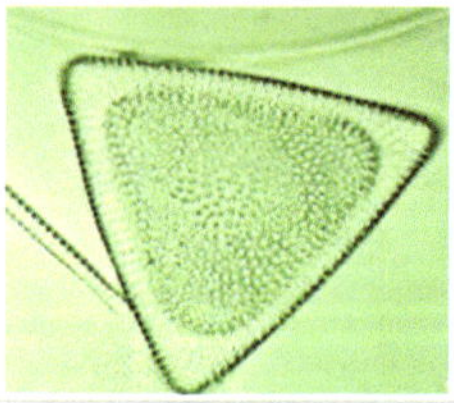
64	*Biddulphia sinensis*	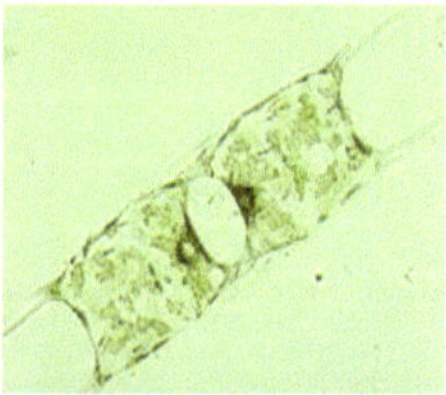
65	*Biddulphia mobiliensis*	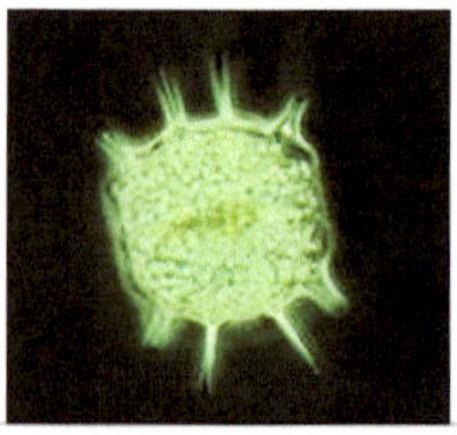

(continued)

(continued)

S. no.	Species	Microscopic view
66	*Biddulphia regia*	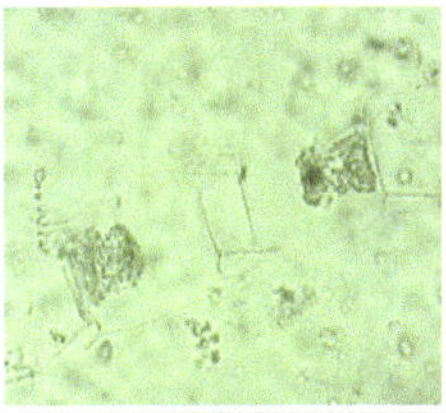
67	*Eucampia zodiacus*	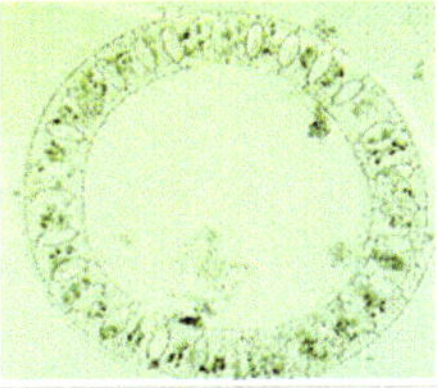
68	*Hemidiscus cuneiformis*	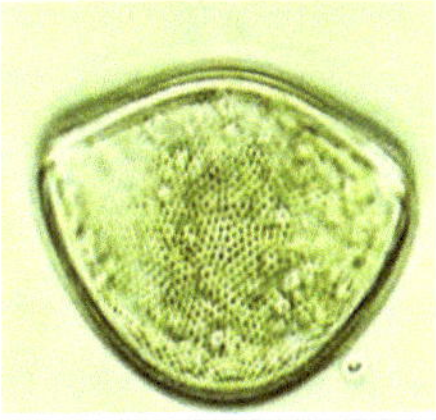
69	*Climacosphenia elongata*	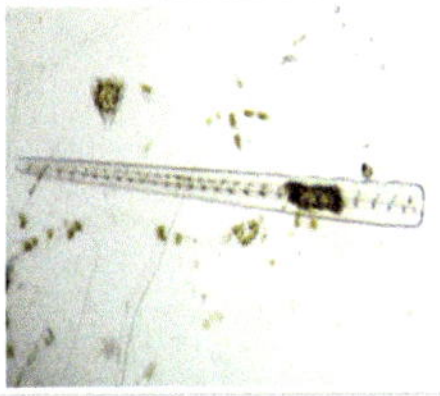
70	*Fragilaria oceanica*	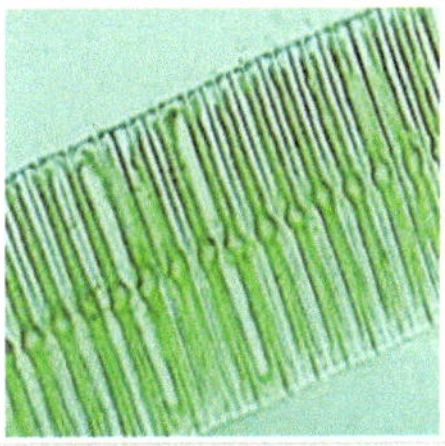
71	*Rhaphoneis amphiceros*	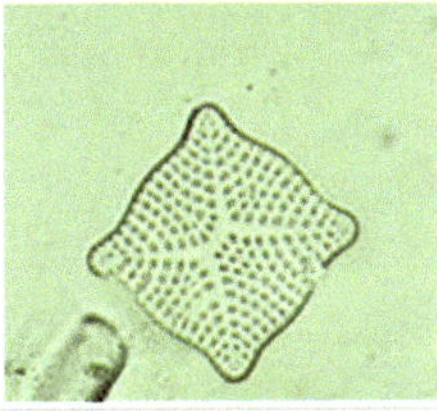
72	*Thalassionema nitzschioides*	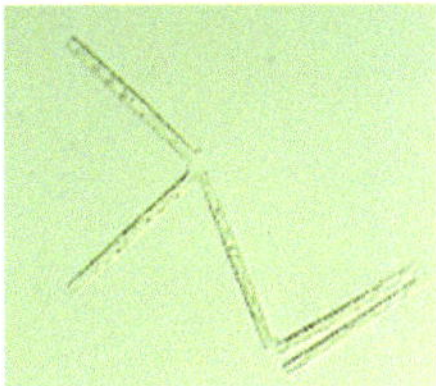

(continued)

(continued)

S. no.	Species	Microscopic view
73	*Thalassionema* sp.	
74	*Thalassiothrix longissima*	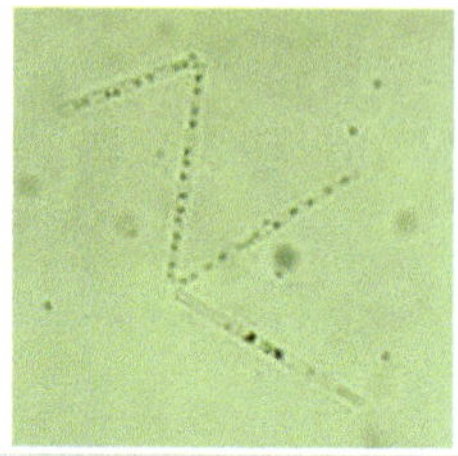
75	*Thalassiothrix frauenfeldii*	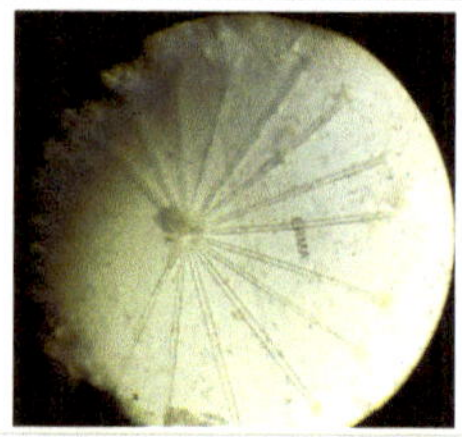
76	*Thalassiothrix* sp.	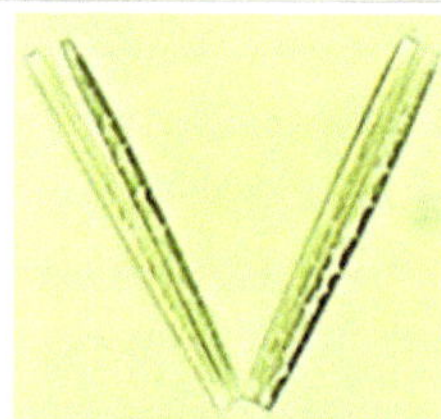
77	*Thalassiothrix nitzschioides*	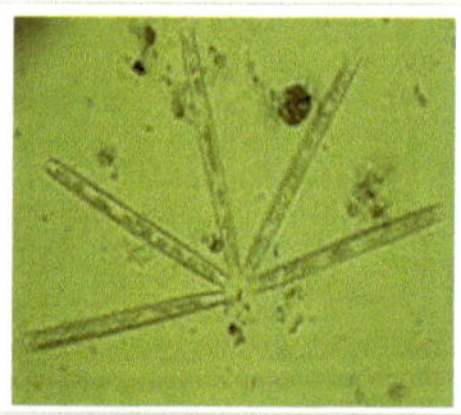
78	*Asterionella japonica*	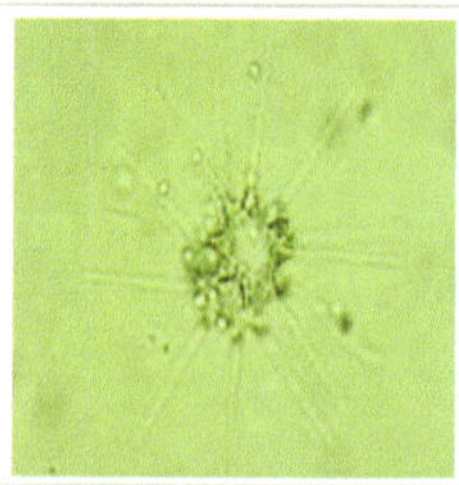
79	*Asterionellopsis glacialis*	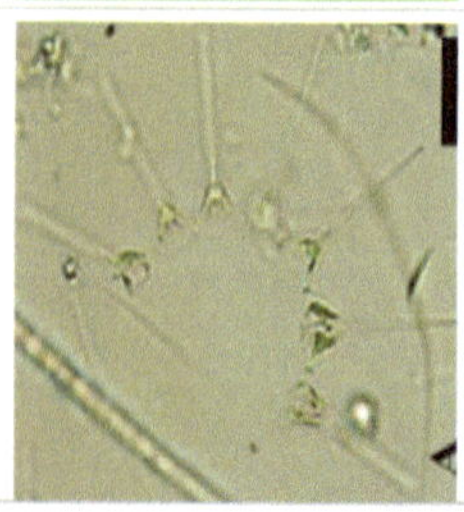

(continued)

(continued)

S. no.	Species	Microscopic view
80	*Diatoma vulgare*	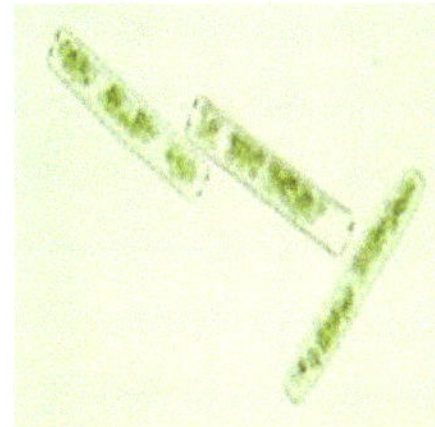
81	*Diatoma* sp.	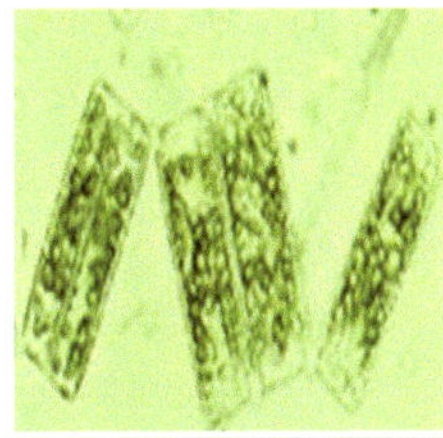
82	*Cocconeis* sp.	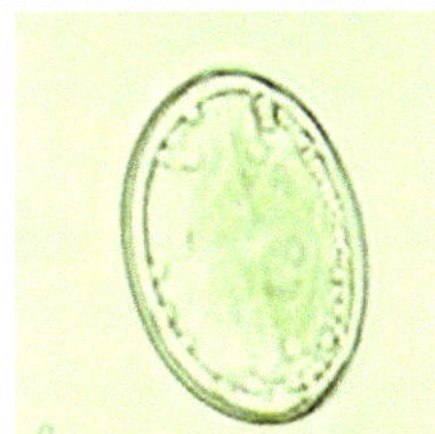
83	*Cocconeis scutellum*	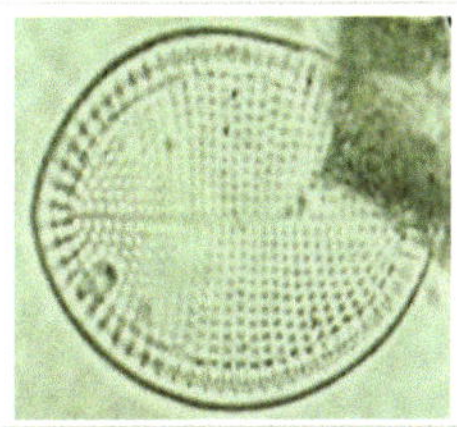
84	*Gyrosigma balticum*	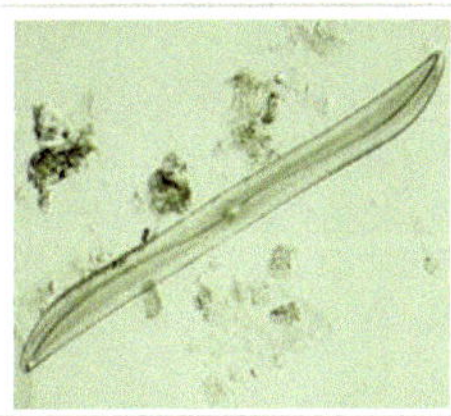
85	*Gyrosigma* sp.	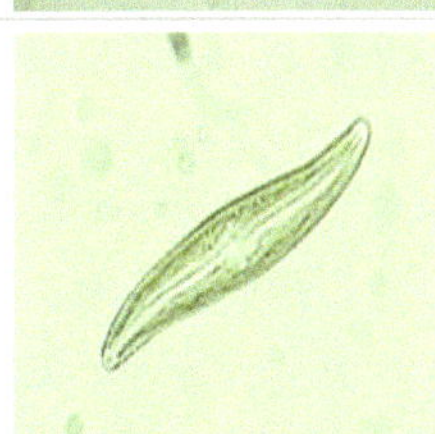
86	*Pleurosigma normanii*	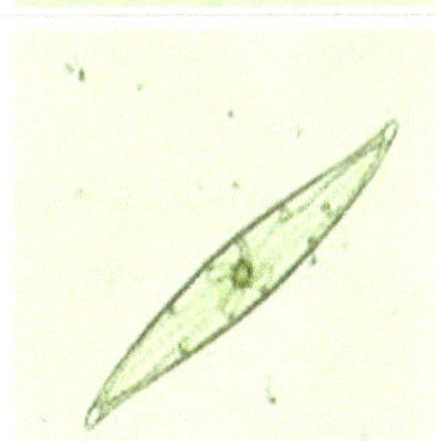

(continued)

(continued)

S. no.	Species	Microscopic view
87	*Pleurosigma elongatum*	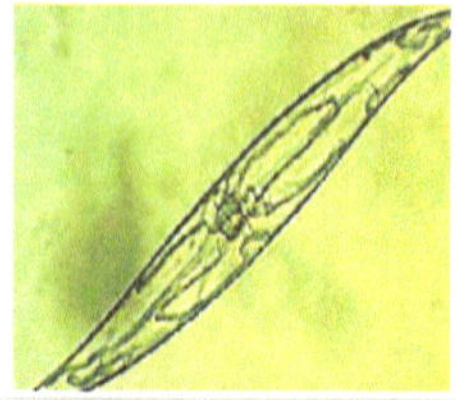
88	*Pleurosigma directum*	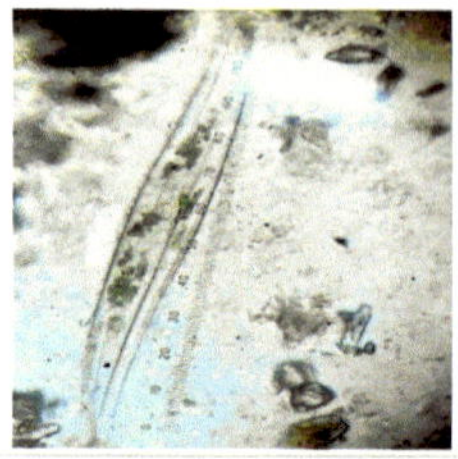
89	*Diploneis smithii*	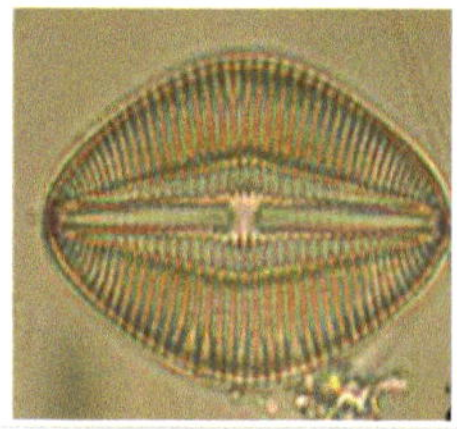
90	*Navicula longa*	
91	*Navicula rhombica*	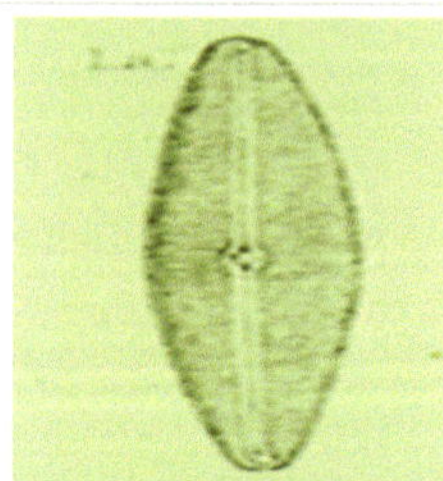
92	*Navicula pennata*	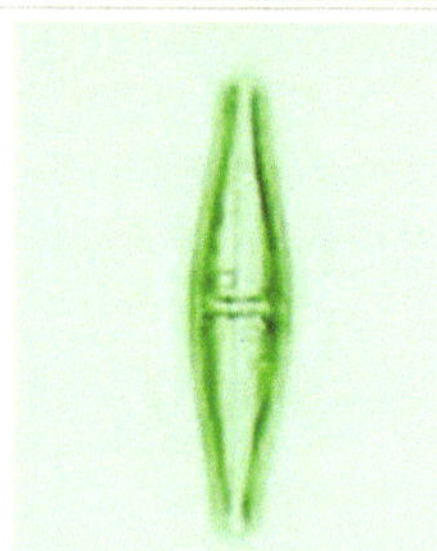

(continued)

(continued)

S. no.	Species	Microscopic view
93	*Navicula* sp.	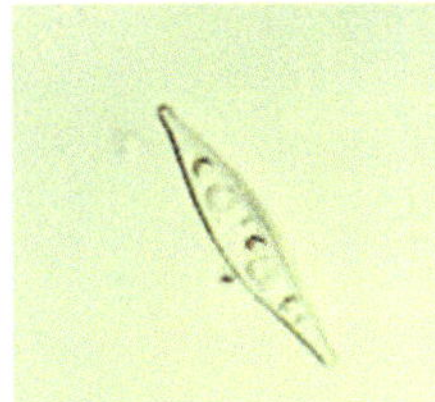
94	*Navicula cancellata*	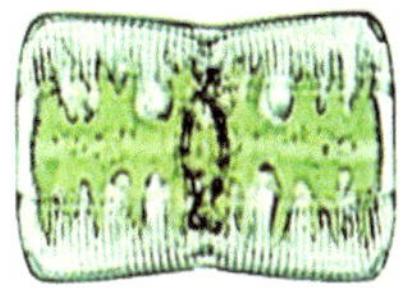
95	*Cymbella marina*	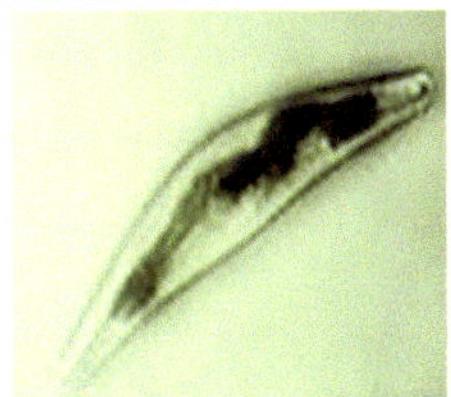
96	*Nitzschia sigma*	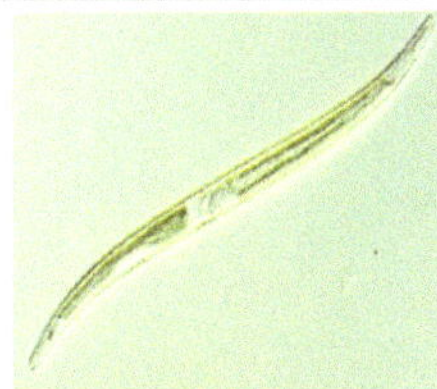
97	*Nitzschia longissima*	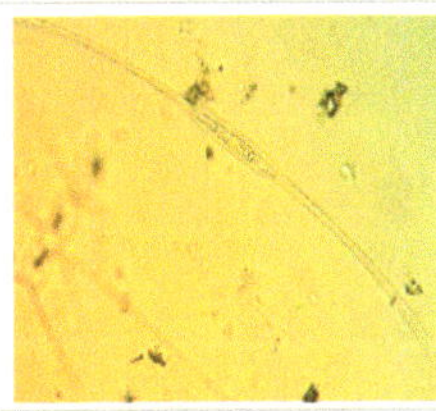
98	*Nitzschia closterium*	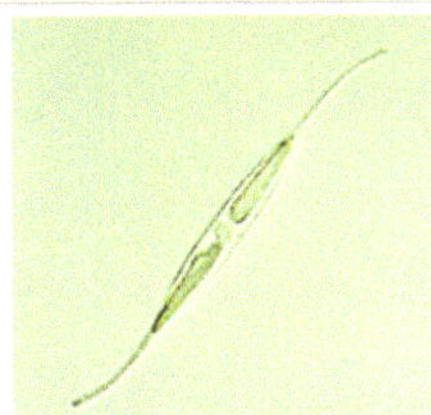
99	*Nitzschia striata*	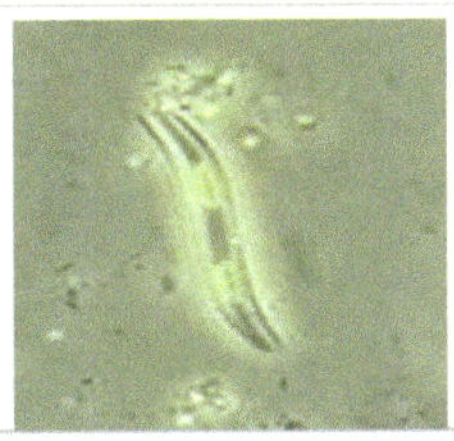

(continued)

(continued)

S. no.	Species	Microscopic view
100	*Nitzschia seriata*	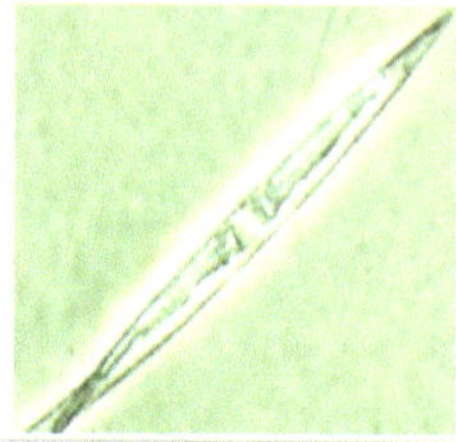
101	*Pseudo-nitzschia pungens*	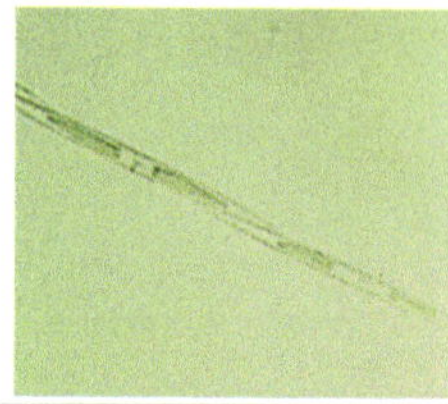
102	*Pseudo-nitzschia australis*	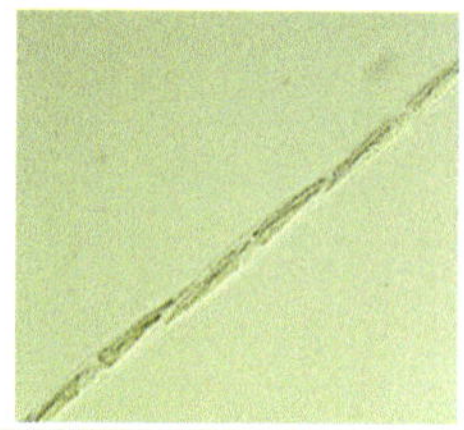
103	*Pinnularia alpina*	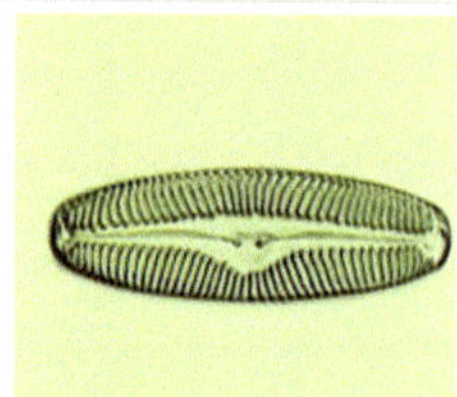
104	*Pinnularia* sp.	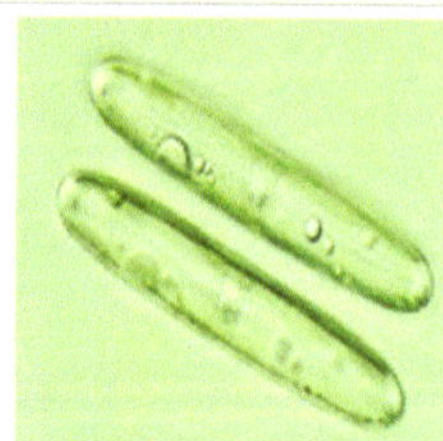
105	*Pinnularia trevelyan*	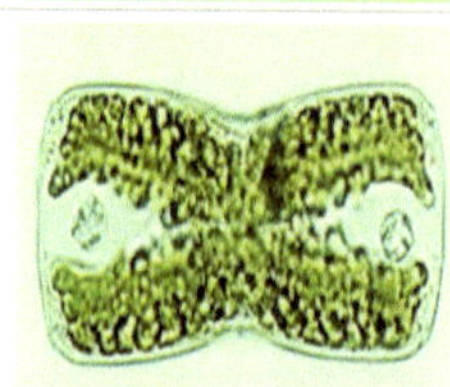
106	*Bacillaria paradoxa*	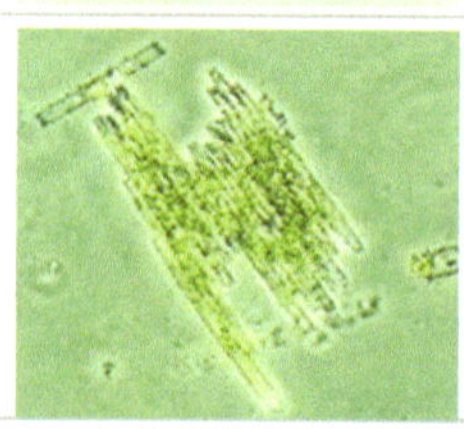

(continued)

(continued)

S. no.	Species	Microscopic view
107	*Bacillaria paxillifer*	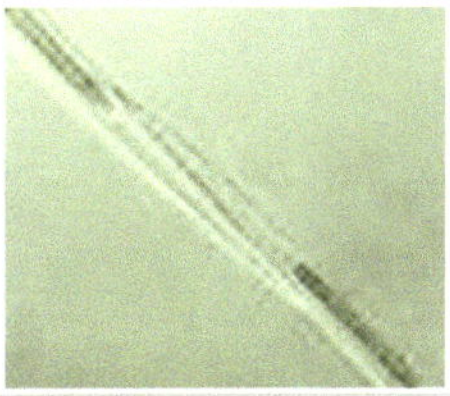
108	*Triceratium spinosum*	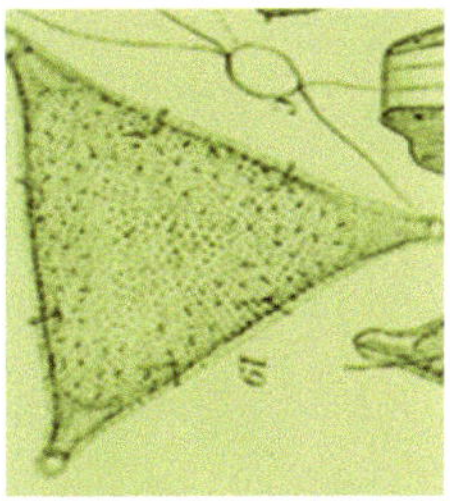
109	*Phaeocystis* sp.	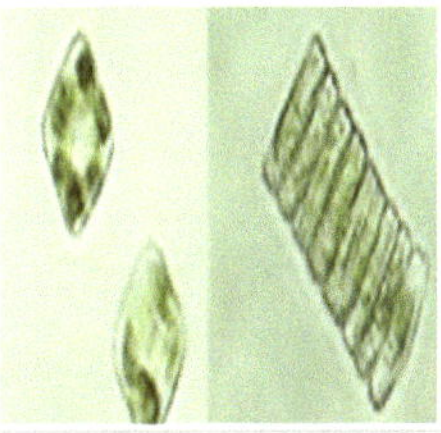
110	*Amphora hyalina*	
111	*Amphipleura* sp.	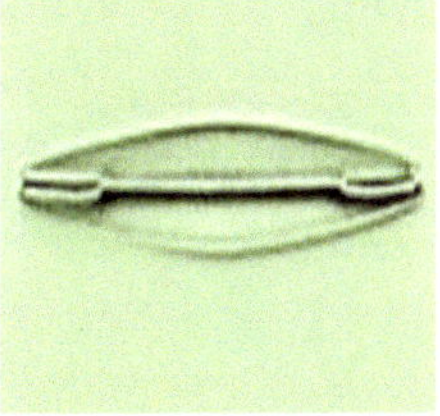
112	*Amphiprora constrica*	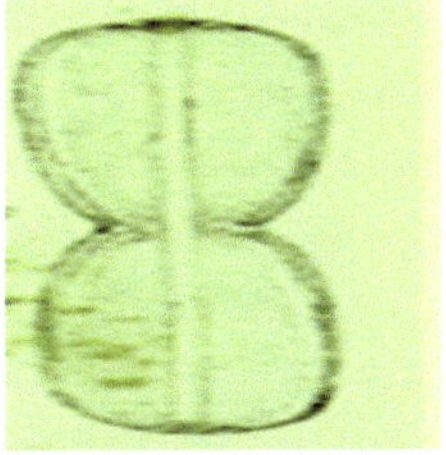
113	*Halosphaera viridis*	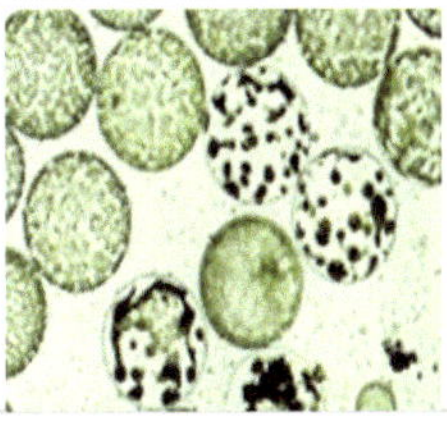

(continued)

(continued)

S. no.	Species	Microscopic view
114	*Halosphaera* sp.	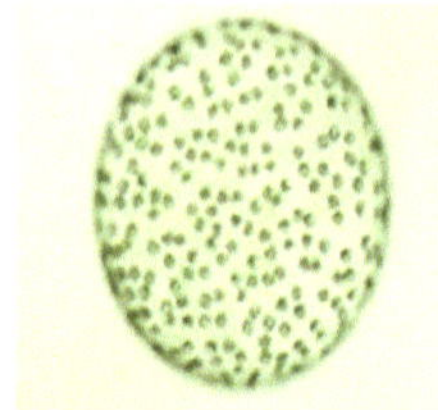
115	*Corethron crinophyllum*	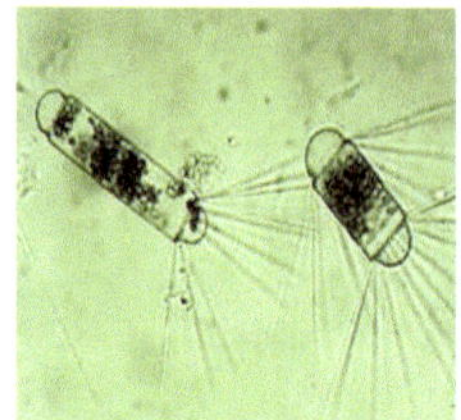
116	*Guinardia flaccida*	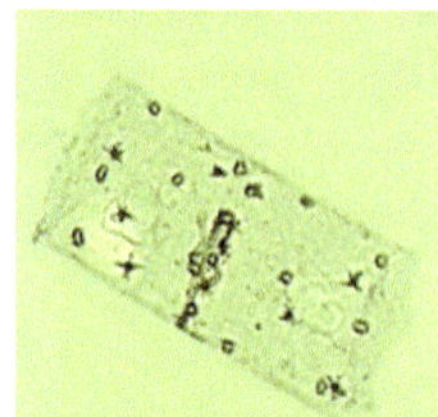
117	*Hemiaulus sinensis*	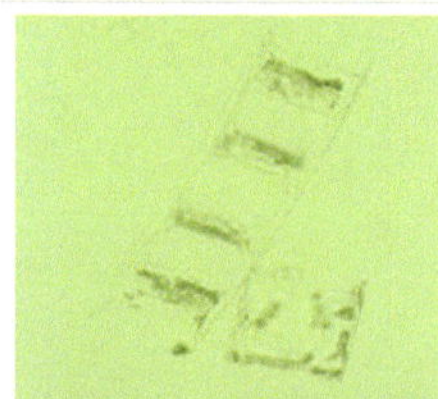
118	*Melosira nummuloides*	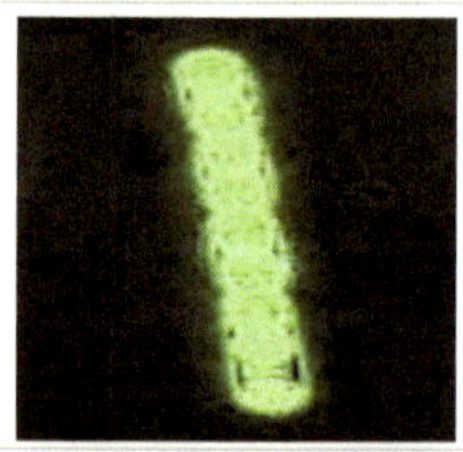
119	*Melosira granulata*	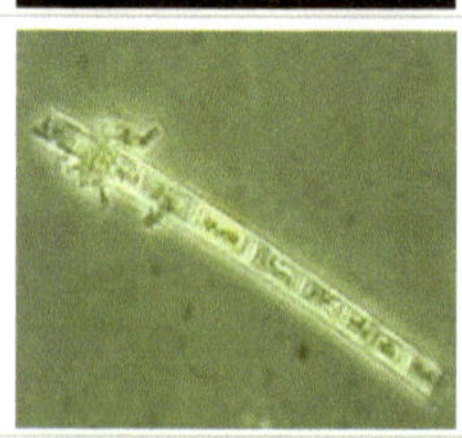
120	*Melosira variance*	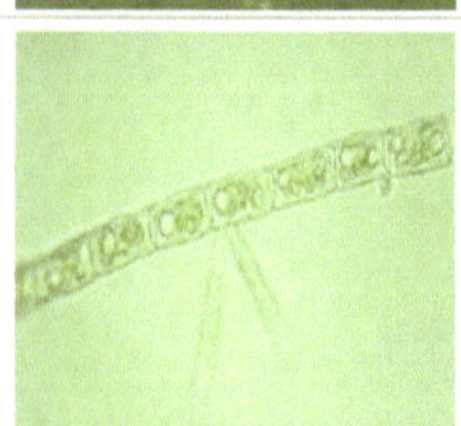

(continued)

(continued)

S. no.	Species	Microscopic view
121	*Melosira* sp.	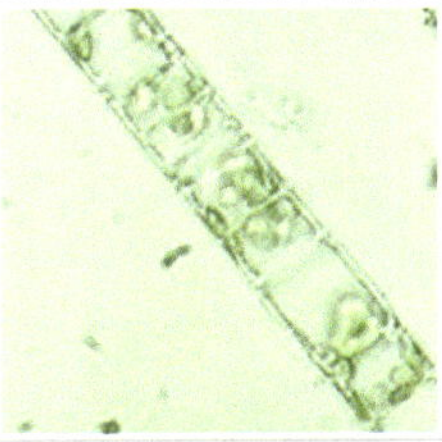
122	*Bellerochea malleus*	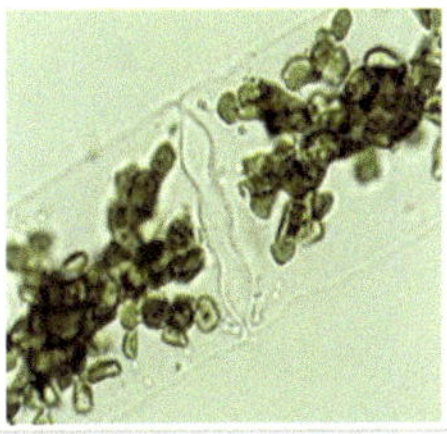
123	*Cerataulina pelagica*	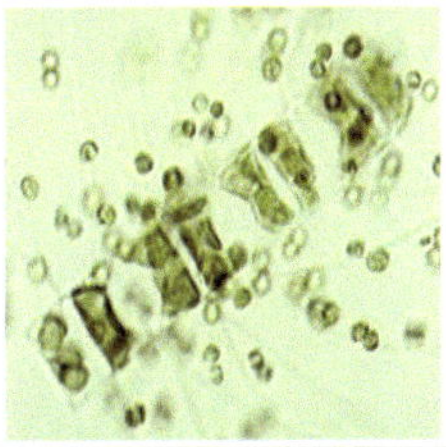
124	*Leptocylindrus danicus*	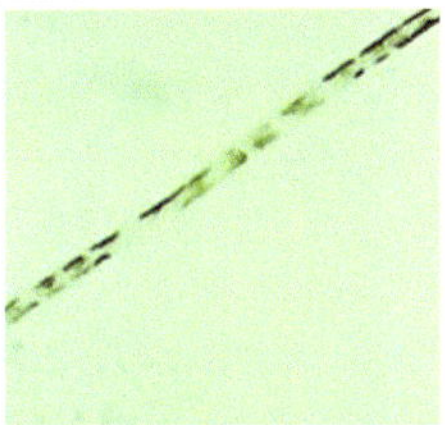
125	*Leptocylindrus minimus*	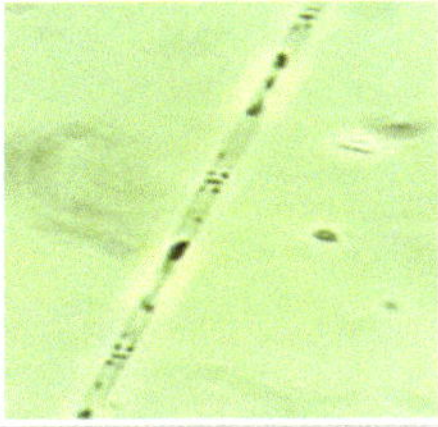
126	*Asteromphalus* sp.	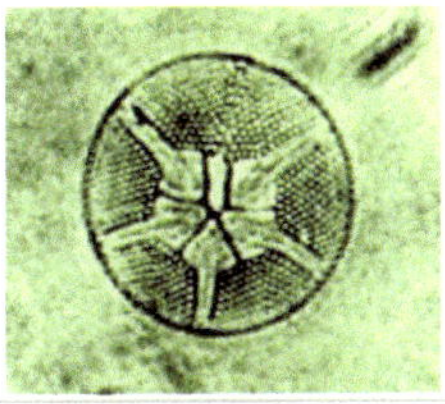
127	*Hyalodiscus* sp.	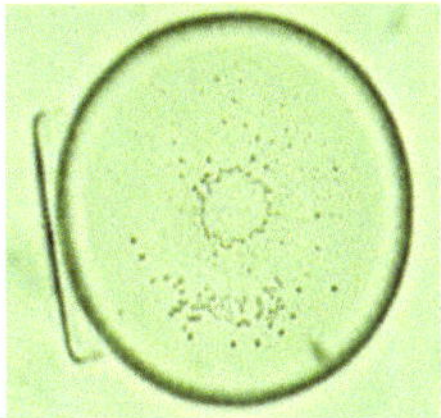

(continued)

(continued)

S. no.	Species	Microscopic view
128	*Actinocyclus octanarius*	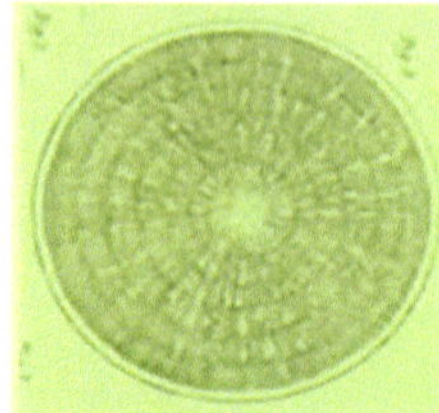
129	*Actinoptychus* sp.	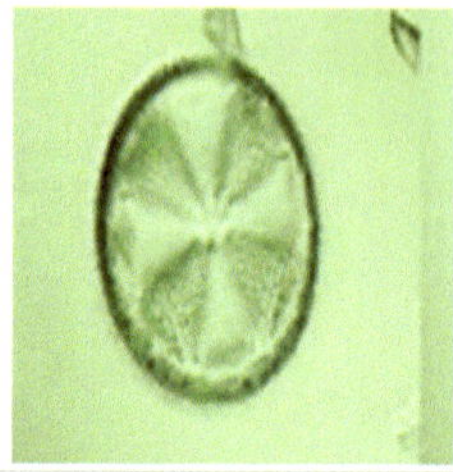
130	*Eunotia* sp.	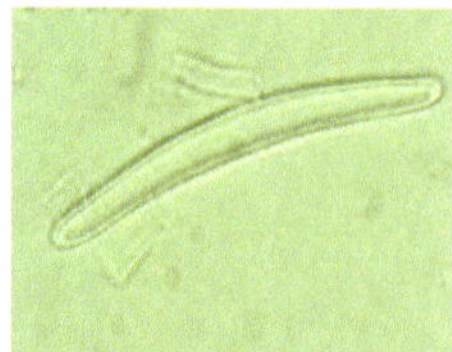
131	*Tabellaria* sp.	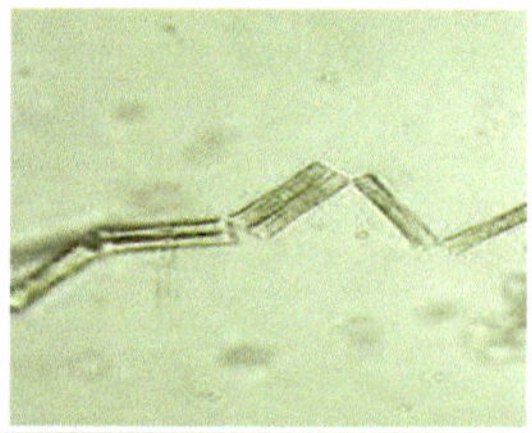
132	*Grammatophora marina*	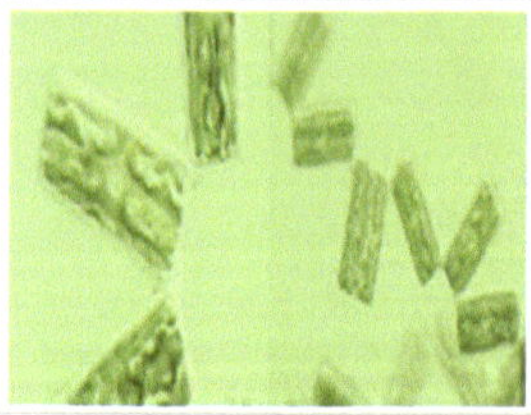
133	*Closterium* sp.	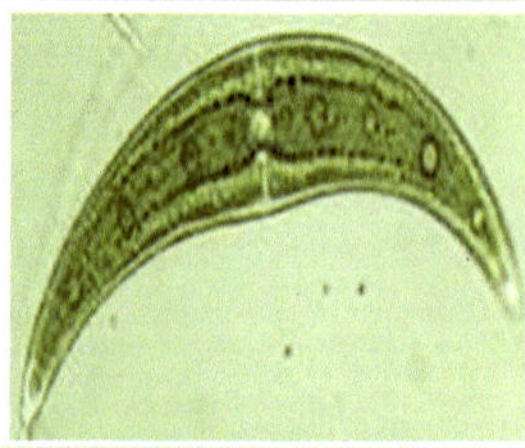
134	*Striatella* sp.	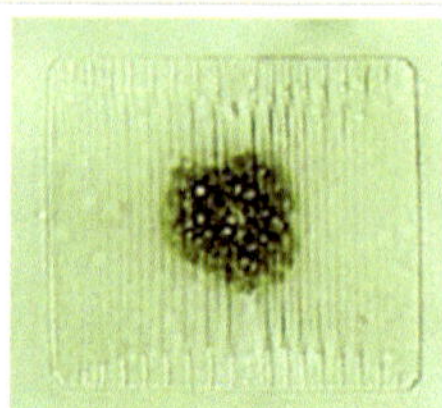

(continued)

(continued)

S. no.	Species	Microscopic view
135	*Striatella unipunctata*	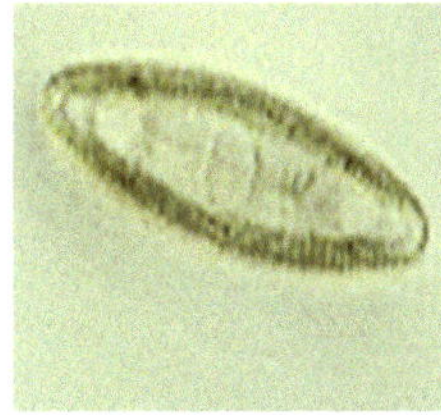
136	*Diplosalis* sp.	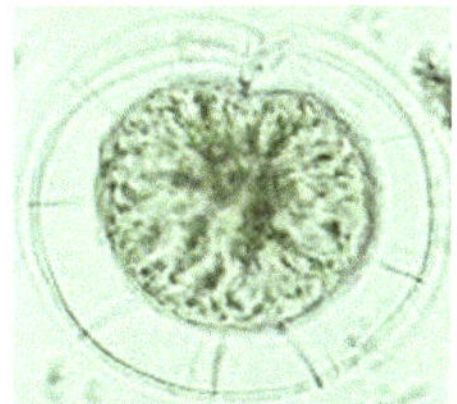
137	*Licmophora ehrenbergii*	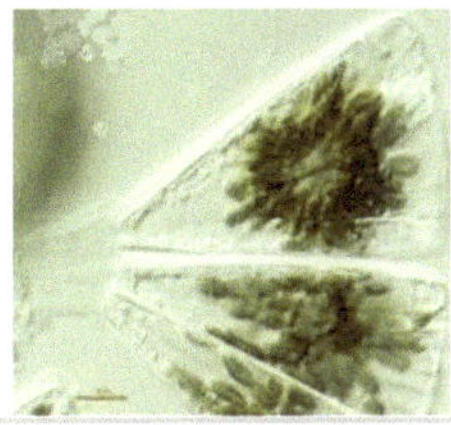

List of Dinoflagellates Identified from the Inshore Waters of Bay of Bengal

S. no.	Species	Microscopic view
1	*Prorocentrum gracile*	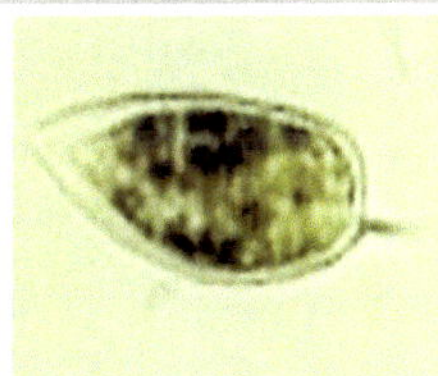
2	*Prorocentrum micans*	

(continued)

(continued)

S. no.	Species	Microscopic view
3	*Prorocentrum concavum*	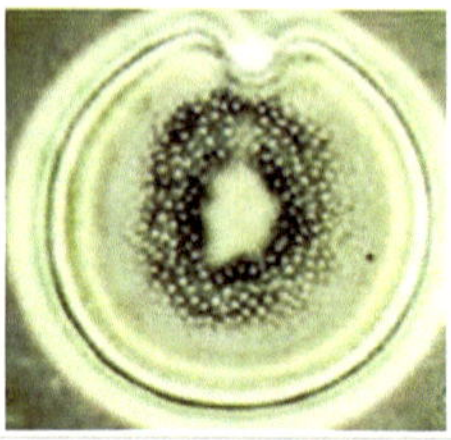
4	*Ceratium furca*	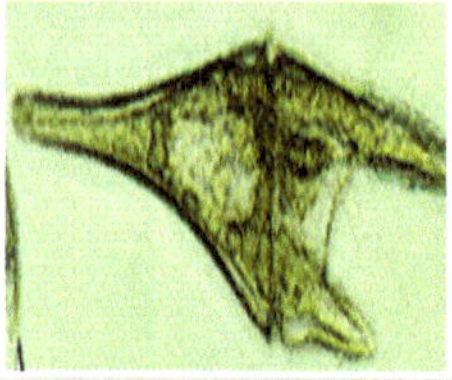
5	*Ceratium fusus*	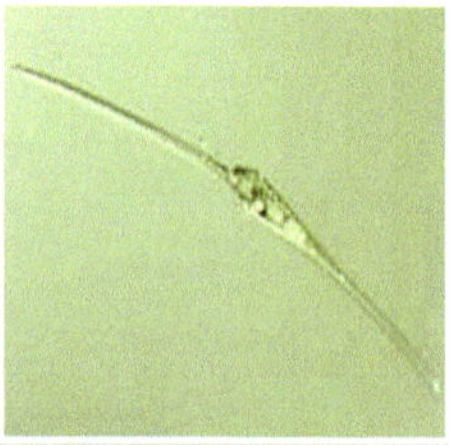
6	*Ceratium teres*	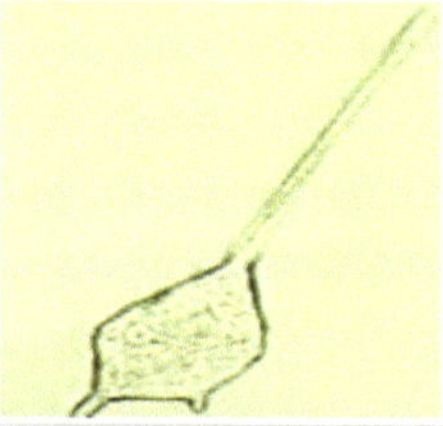
7	*Ceratium minutum*	
8	*Ceratium tripos*	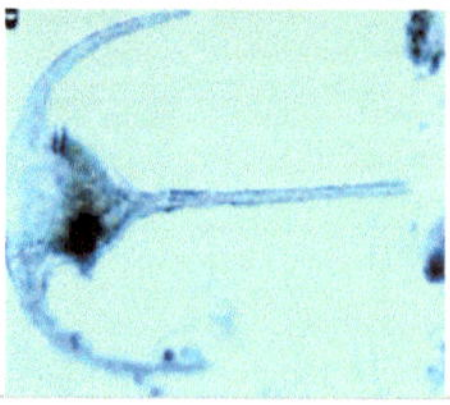

(continued)

(continued)

S. no.	Species	Microscopic view
9	*Ceratium trichoceros*	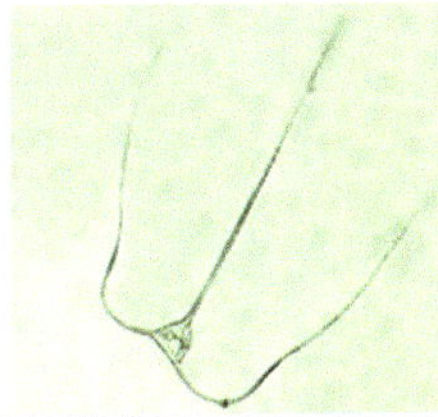
10	*Ceratium horridum*	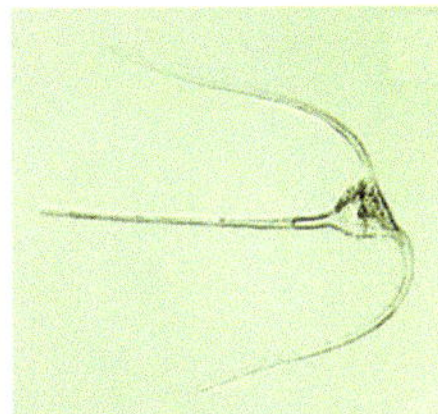
11	*Ceratium inflatum*	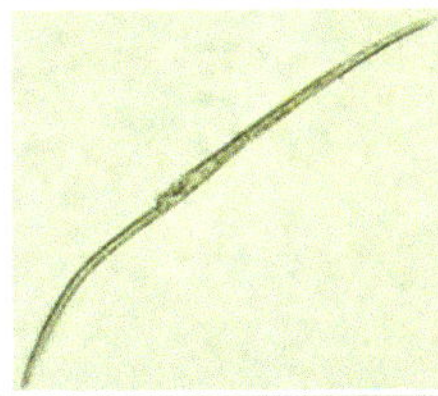
12	*Ceratocorys horrida*	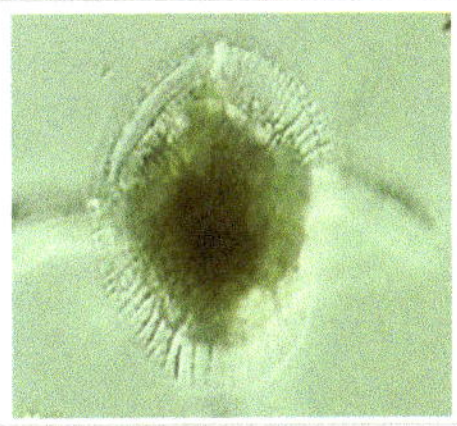
13	*Protoperidinium* sp.	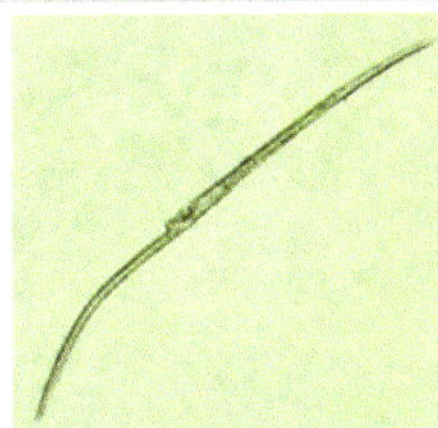
14	*Protoperidinium crassipes*	

(continued)

(continued)

S. no.	Species	Microscopic view
15	*Protoperidinium depressum*	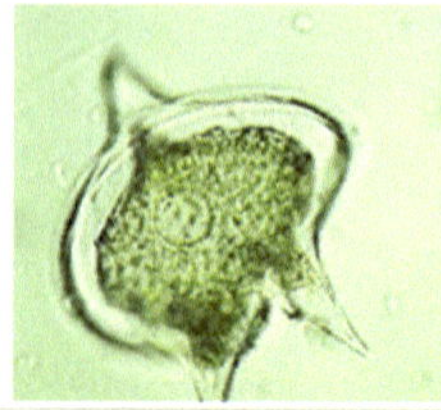
16	*Protoperidinium ovatum*	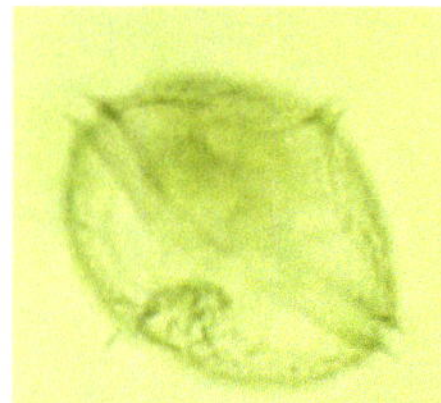
17	*Protoperidinium pellucidum*	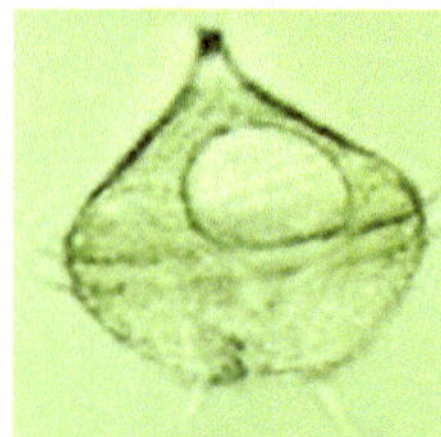
18	*Protoperidinium conicum*	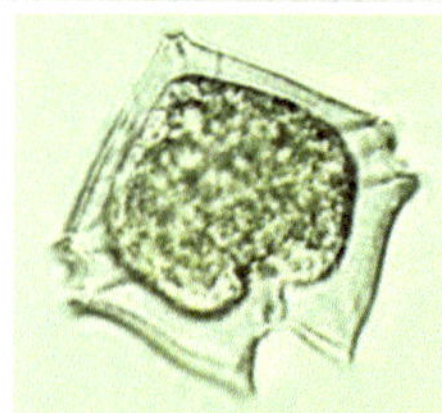
19	*Pyrocystis fusiformis*	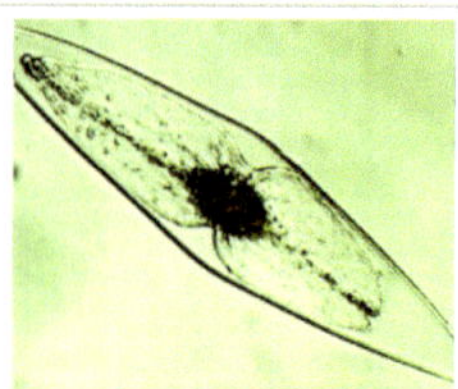
20	*Pyrocystis* sp.	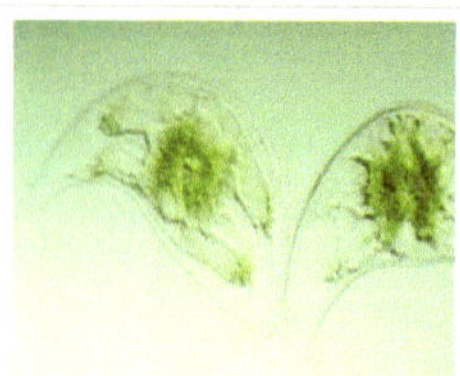

(continued)

(continued)

S. no.	Species	Microscopic view
21	*Dinophysis caudata*	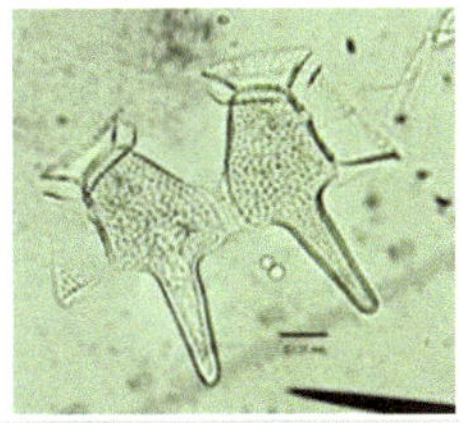
22	*Dinophysis acuta*	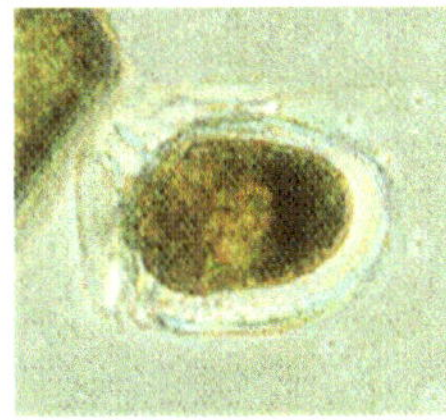
23	*Dinophysis norvegica*	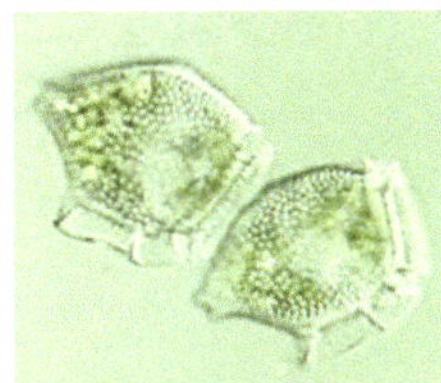
24	*Dinophysis* sp.	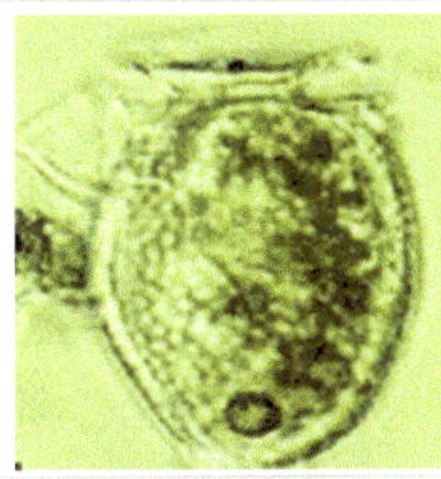
25	*Alexandrium* sp.	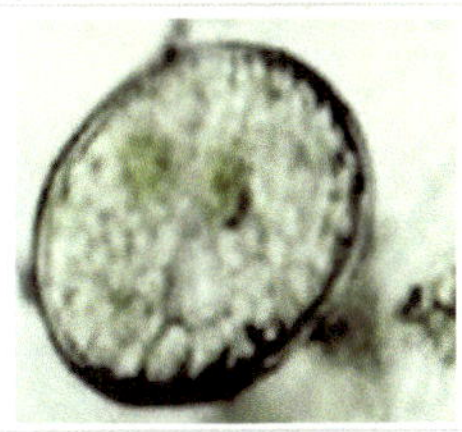
26	*Preperidinium meunieri*	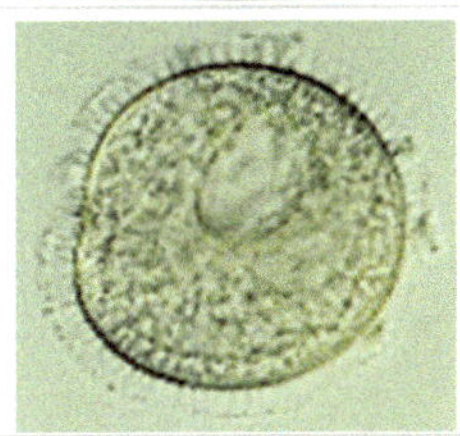

List of Other Algae Identified from the Inshore Waters of Bay of Bengal

S. no.	Species	Microscopic view
1	*Trichodesmium* sp.	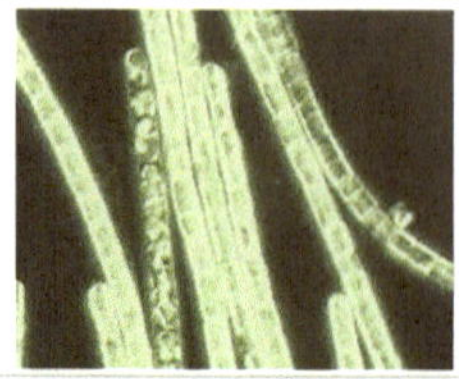
2	*Trichodesmium thiebautii*	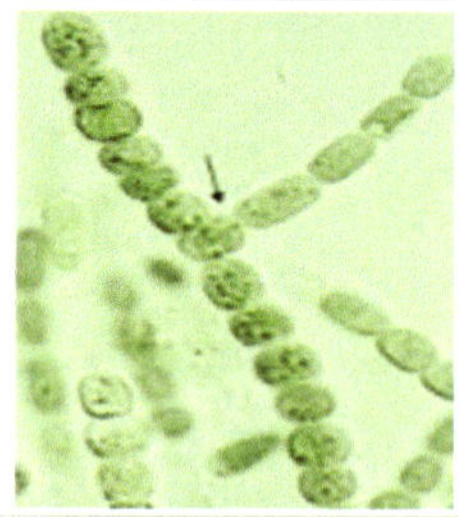
3	*Oscillatoria* sp.	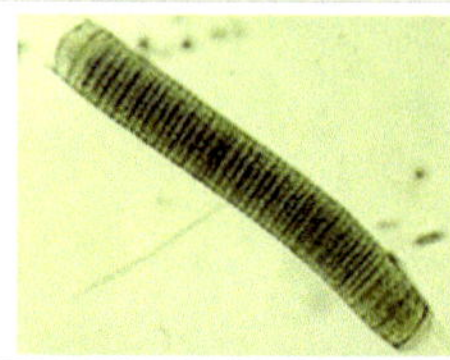
4	*Oscillatoria limosa*	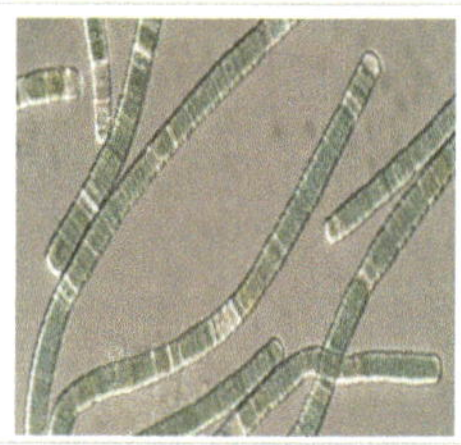
5	*Gloeocapsa* sp.	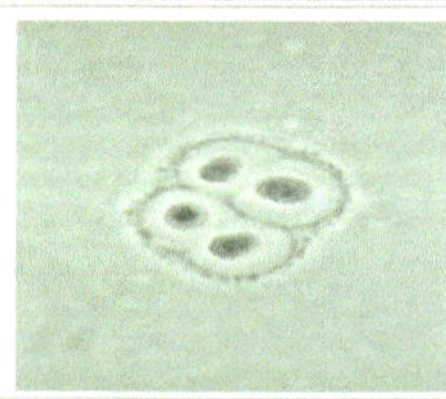
6	*Stigonema* sp.	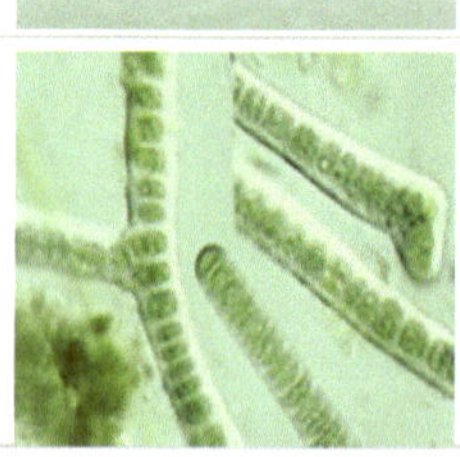

(continued)

(continued)

S. no.	Species	Microscopic view
7	*Cylindrospermopsis raciborskii*	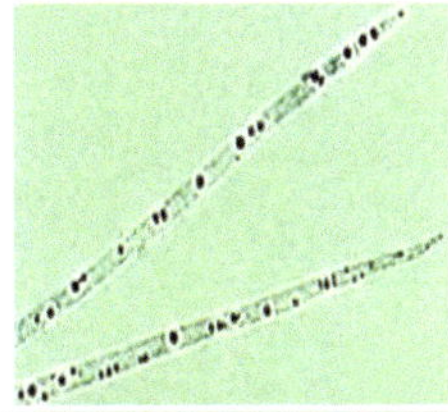
8	*Anabaena* sp.	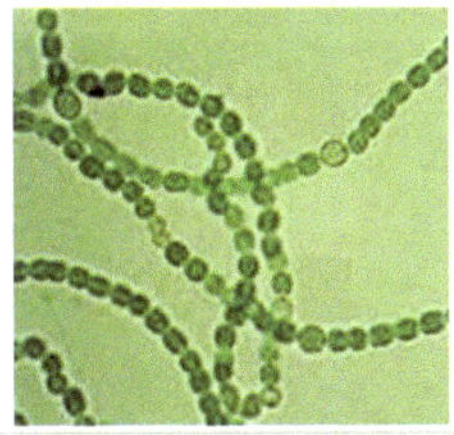
9	*Dictyocha* sp.	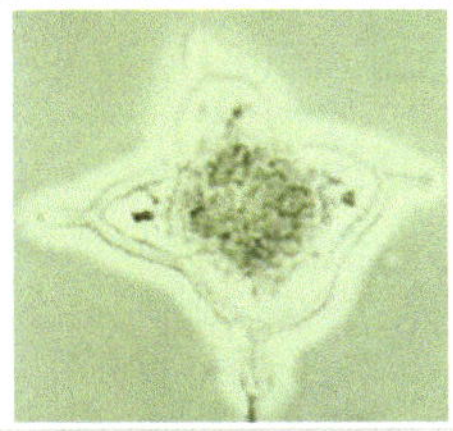
10	*Ditylum sol*	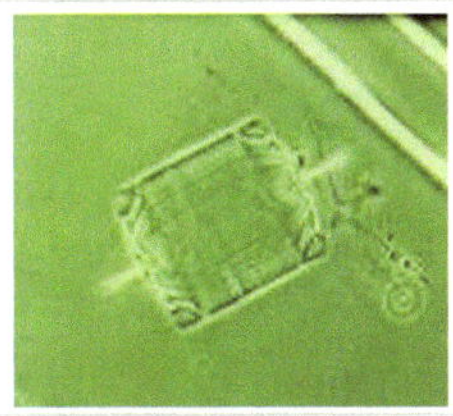
11	*Chlorella salina*	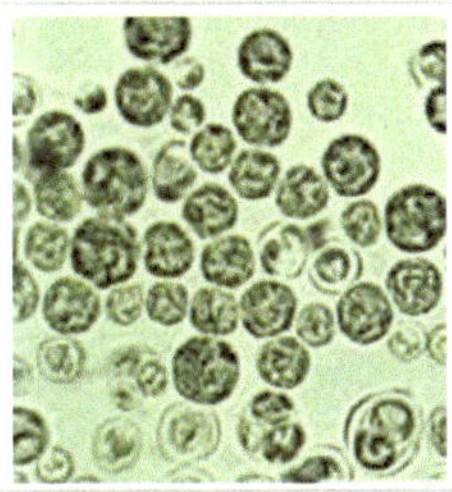
12	*Amphora hyalina*	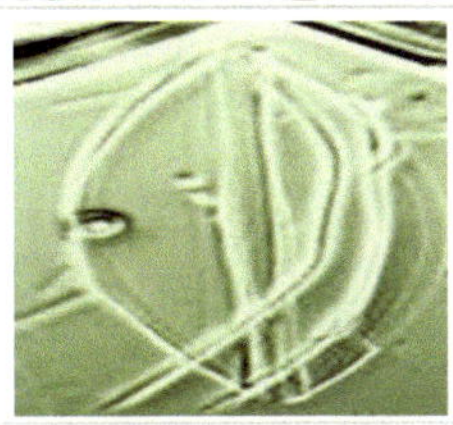
13	*Netrium* sp.	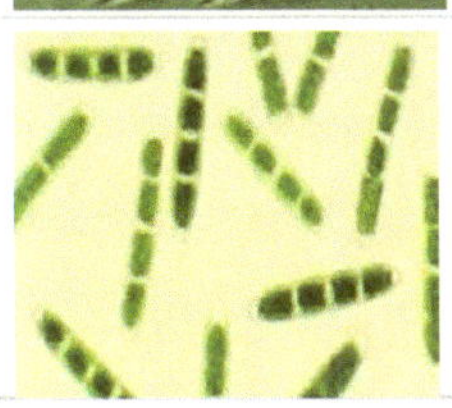

(continued)

(continued)

S. no.	Species	Microscopic view
14	*Draparnaldia* sp.	
15	*Gloeocapsa* sp.	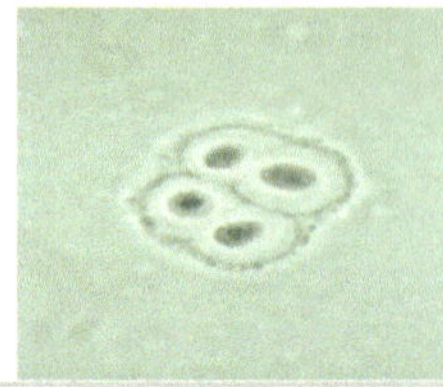

References

Aiken J, Pradhan Y, Barlow R, Lavender S, Poulton A, Holligan P, Hardman-Mountford N (2009) Phytoplankton pigments and functional types in the Atlantic Ocean: a decadal assessment, 1995–2005. Deep Sea Res 56:899–917

Alvain S, Moulin C, Dandonneau Y (2005) Remote sensing of phytoplankton groups in case 1 waters from global SeaWiFS imagery. Deep Sea Res 52:1989–2004

Balch W, Holligan PM, Ackleson SG, Voss KJ (1991) Biological and optical properties of mesoscale coccolithophore blooms. Limnol Oceanogr 36:629–643

Balch W, Kilpatrick KA, Holligan PM, Trees C (1996) The 1991 coccolithophore bloom in the central north Atlantic I—optical properties and factors affecting their distribution. Limnol Oceanogr 41:1669–1683

Bidigare RR, Ondrusek ME, Morrow JH, Kiefer DA (1990) In-vivo absorption properties of algal pigments. Proc Soc Photo-optical Instrum 1302:290–302

Bishop JKB, Wood TJ (2009) Year round observations of carbon biomass and flux variability in the southern ocean. Global Biogeochem Cycles 23. doi:10.1029/2008GB003206

Cairns J Jr, Dixon KL (1971) A simple method for the biological assessment of the effects of waste discharges on aquatic bottom-dwelling organisms. J Watching Pollut Control Fed 43:755–772

Cermeño P, Dutkiewicz S, Harris RP, Follows M, Schofield O, Falkowski PG (2008) The role of nutricline depth in regulating the ocean carbon cycle. Proc Natl Acad Sci USA 105(51):20344–20349

Cullen JT, Lane TW, Morel FMM, Sherrell RM (1999) Modulation of cadmium uptake in photoplankton by seawater CO_2 concentration. Nature 402:165–167

Falkowski PG, Raven JA (1997) Aquatic photosynthesis. Blackwell, Oxford

Garver S, Siegel DA (1997) Inherent optical property inversion of ocean color spectra and its biogeochemical interpretation: 1. Time series from the Sargasso Sea. J Geophys Res 102:18607–18625

González EJ, Matsumura-Tundisi T, Tundisi JG (2008) Size and dry weight of main zooplankton species in Bariri Reservoir (SP, Brazil). Braz J Biol 68(1):69–75

Hein M, Sand-Jensen K (1997) CO_2 increases oceanic primary production. Nature 388:526–527

Hu C, Cannizzaro J, Carder KL, Muller-Karger FE, Hardy R (2010) Remote detection of *Trichodesmium* blooms in optically complex coastal waters: examples with MODIS full-spectral data. Remote Sens Environ 114:2048–2058

Jeffery SW, Humphrey GF (1975) New spectrophotometric equations for determining chlorophylls a, b, c_1, and c_2 in higher plants, algae and natural phytoplankton. Biocheml Physiol Pflanz 167:191–194

Lane TW, Morel FMM (2000a) Regulation of carbonic anhydrase expression by zinc, cobalt, and carbon dioxide in the marine diatom *Thalassiosira weissflogii.* Plant Physiol 123:345–352

Lane TW, Morel FMM (2000b) Regulation of carbonic anhydrase expression by zinc, cobalt, and carbon dioxide in the marine diatom *Thalassiosira weissflogii.* Plant Physiol 123:345–352

Latasa M (2007) Improving estimates of phytoplankton class abundances using CHEMTAX. Mar Ecol Prog Ser 329:13–21

Leonardos N, Richard JG (2005) Elemental and biochemical composition of *Rhinomonas reticulate* (Cryophyta) in relation to light and nitrate-to phosphate supply ratios. J Phycol 41(3):567–576

MacKey MD, MacKey DJ, Higgins HW, Wright SW (1996) CHEMTAX—a program for estimating class abundances from chemical markers: application to

HPLC measurements of phytoplankton. Mar Ecol Prog Ser 144:265–283

McClain C (2009) A decade of satellite ocean color observations. Annu Rev Mar Sci 1:19–42

Mitra A (2000) The northwest coast of the Bay of Bengal and deltaic Sundarbans. In: Sheppard CRC (ed) Seas at the millennium: an environmental evaluation, 145–160. Foundation of Environmental Science. Narendra Publishing House, Delhi

Mitra A (2013) Sensitivity of mangrove ecosystem to changing climate. Springer, Berlin, 323 pp. ISBN: 978-81-322-1508-0

Mitra A, Banerjee K, Gangopadhayay A (2004) Introduction to marine phytoplankton. Daya Publishing House, Delhi, 102 pp

Mitra A, Zaman S, Kanti Ray S, Sinha S, Banerjee K (2012) Inter-relationship between phytoplankton cell volume and aquatic salinity in Indian Sundarbans. Natl Acad Sci Lett. doi:10.1007/s40009-012-0083-1

Moisan JR, Moisan TAH, Linkswiler MA (2011a) An inverse modeling approach to estimating phytoplankton pigment concentrations from phytoplankton absorption spectra. J Geophy Res 116:1–16, C09018. doi:10.1029/2010JC006786

Moisan JR, Moisan TAH, Linkswiler MA (2011b) Estimating phytoplankton pigment concentrations from phytoplankton absorption spectra. J Geophys Res 116:0148–0227

Nair A, Sathyendranath S, Platt T, Morales J, Stuart V, Forget M-HE, Devred E, Bouman H (2008) Remote sensing of phytoplankton functional types. Remote Sens Environ 112:3366–3375

Osborne A, Wanielista P, Yousef A (1976) Benthic fauna species diversity in six central Florida lakes in summer. Hydrobiologia 48:125–129

Raven JA (1991) Implications of inorganic carbon utilization: ecology, evolution, and geochemistry. Can J Bot 69(5):908–924

Raven JA (1997) Inorganic carbon acquisition by marine autotrophs. Adv Bot Res 27:85–209

Riebesell U, Wolf-Gladrow DA, Smetacek V (1993) Carbon dioxide limitation of marine phytoplankton growth rates. Nature 361:249–251

Sabine CL, Richard AF, Nicolas G, Robert MK, Kitack L, John LB, Wanninkhof R, Wong CS, Douglas WR, Wallace BT, Frank JM, Peng T-H, Alexander K, Tsueno O, Aida FR (2004) The oceanic sink for anthropogenic CO_2. Science 305:367–371

Sathyendranath S, Lazzara L, Prieur L (1987) Variations in the spectral values of specific absorption of phytoplankton. Limnol Oceanogr 32(2):403–415

Sathyendranath S, Watts L, Devred E (2004) Discrimination of diatoms from other phytoplankton using ocean-colour data. Mar Ecol Prog Ser 272:59–68

Seiter K, Hensen C, Zabel M (2005) Benthic carbon mineralization on a global scale. Glob Biogeochem Cycles 19. GB1010, doi:10.1029/2004GB002225

Shannon CE, Weiner W (1949) The mathematical theory of communication. University of Illinois Press, Urbanea

Simpson EH (1949) Measurement of diversity. Nature 163:688

Sournia A, Chrdtiennot-Dinetl MJ, Ricard M (1991) Marine phytoplankton: how many species in the world ocean? J Plankton Res 13(5):1093–1099. doi:10.1093/plankt/13.5.1093

Subramanian A, Carpenter EJ (1994) An empirically derived protocol for the detection of blooms of the marine cyanobacterium *Trichodesmium* using CZCS imagery. Int J Remote Sens 158:1559–1569

Subramanian A, Carpenter EJ, Karentz PG, Falkowski PG (1999a) Optical properties of the marine diazotrophic cyanobacteria *Trichodesmium* spp. I. Absorption and spectral photosynthetic characteristics. Limnol Oceanogr 44:608–617

Subramanian A, Carpenter EJ, Falkowski PG (1999b) Optical properties of the marine diazotrophic cyanobacteria *Trichodesmium* spp. II. Reflectance model for remote sensing. Limnol Oceanogr 44:618–627

Tett P, Barton ED (1995) Why are there about 5000 species of phytoplankton in the sea? J Plankton Res 17 (8):1693–1704

Tortell PD, Rau GH, Morel FMM (2000) Inorganic carbon acquisition in coastal Pacific phytoplankton communities. Limnol Oceanogr 45(7):1485–1500

Whittacker RH, Niering WA (1965) Vegetation of the Santa Catalina mountains, Arizona: a gradient analysis of the south slope. Ecology 46:429–452

Wilhm JL, Dorris TC (1968) Biological parameters of water quality. Bioscience 18:477–481

Zou D (2005) Effects of elevated atmospheric CO_2 growth, photosynthesis and nitrogen metabolism in the economic brown seaweed, *Hizikia fusiforme* (Sargassaceae, Phaeophyta). Aquaculture 250:726–735

Internet References

www.helcom.fi/stc/files/Publications/Proceedings/bsep106.pdf

http://www.whoi.edu/conference/ocean/Iron/Fertilization

6 Blue Carbon in Faunal Community

...In the blue soup the bright carbonate shell
Acidification makes them gradually pale...
The Authors

Molluscs constitute of organisms with soft unsegmented body usually covered with calcareous shell. They are the largest marine phylum, comprising of about 23 % of all the recognized marine organisms. About 85,000 species of molluscs have been identified till date. Numerous molluscs also thrive luxuriantly in freshwater and terrestrial habitats. Organisms under phylum mollusca are highly diverse, not only in size and in anatomical structure, but also in behaviour and in habitat. The phylum is typically divided into nine or ten taxonomic classes, of which two are entirely extinct. The two primary molluscan classes are bivalvia and gastropoda (Fig. 6.1). The authors of this book consider molluscan community under the banner of blue carbon primarily because of their habitat in the marshy wetlands, mangroves, seagrass bed, saltmarsh grass bed, coastal zones, estuaries and seas. Carbon is also a major constituent of the calcareous hard shell over the body of molluscan organisms. However, acidification—a strong arm of climate change—has posed a negative impact on this community by way of dissolving and thinning of their calcareous shell.

6.1 Stored Carbon in Bivalves

Filter-feeding oysters, clams or mussels sequester significant amounts of carbon by consuming phytoplankton and absorbing dissolved organic matter. Carbon becomes a primary component of its shell, and it has been observed that for every kilogram of live clams or oysters grown in Florida waters, approximately 114 g of carbon are removed from the water column and benthos. The readers may find interesting to know that how carbon is sequestered by bivalves, as their mind set-up is mostly based on the carbon sequestration by autotrophs. It is true that carbon sequestration is carried out by plants through the process of photosynthesis, in which carbon dioxide is converted into oxygen and plant material (organic compounds) in presence of light energy. The amount of carbon sequestered can vary greatly depending on the species, site and other external factors. The net amount of carbon sequestered by autotrophs is evaluated based on the balance between the changes in aboveground vegetation, below-ground root system, the litter layer and the soil (Hicky 2009; Mitra et al. 2011, 2013; Sengupta et al. 2013). The units used for this type of sequestration are tonnes of carbon per hectare per year. These terrestrial sequestration methods work well, but it is the ocean that stores the vast majority of carbon on the planet as represented in Fig. 6.2. The stored carbon in the marine compartment is used by phytoplankton, coastal vegetations (like mangroves, saltmarshes, seagrasses, etc.) and other shelled organisms.

The shelled organisms preferably the molluscs are important storehouses of carbon. A number of physical, chemical and biological processes carry out carbon sequestration in molluscs, but this section will focus on the biological processes, and more specifically on the processes adopted by a marine/estuarine oyster species. The species *Saccostrea cucullata* (commonly known as rocky oyster) is widely seen occupying hard substrata in the intertidal zone like walls of lighthouse, sluice gates, pillars (Fig. 6.4) of fish

A. Mitra and S. Zaman, *Blue Carbon Reservoir of the Blue Planet*,
DOI 10.1007/978-81-322-2107-4_6, © Springer India 2015

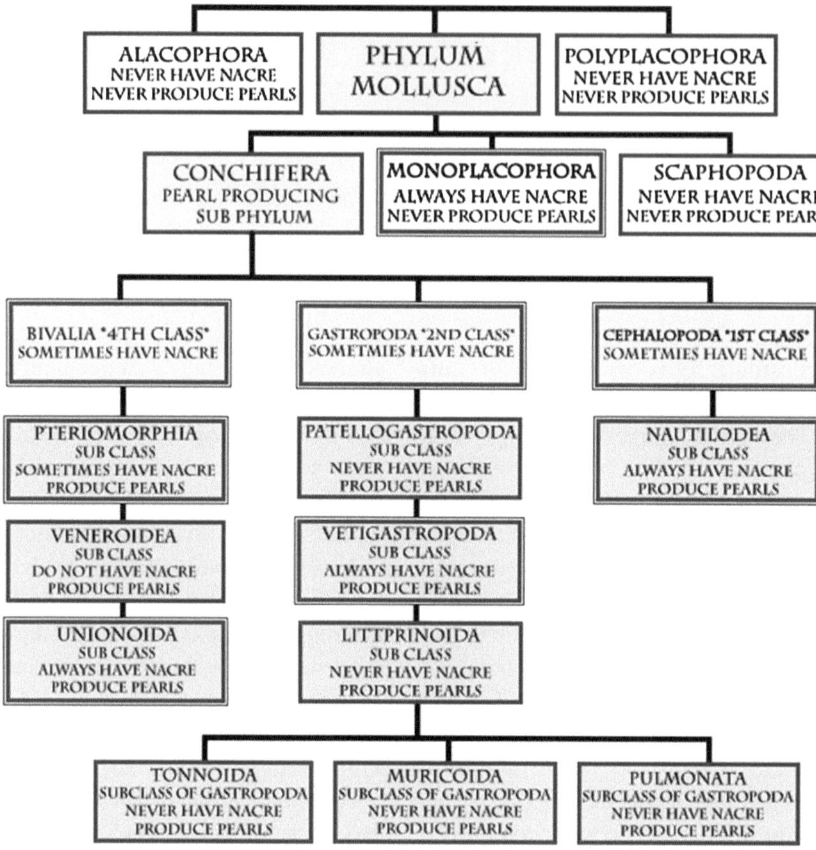

Fig. 6.1 Major divisions and subdivisions of molluscan community

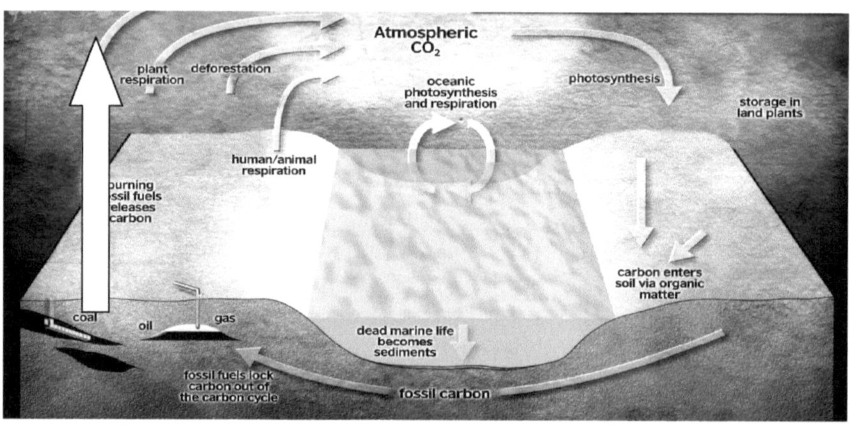

Fig. 6.2 Earth's carbon cycle (*source* NASA Earth Science Enterprise. http://science.nasa.gov/about-us/science-strategy/past-strategy-documents/earth-science-enterprise-plans/earth-science-focus-area-roadmaps-2005/carbon-cycle-ecosystems/)

Fig. 6.3 *Saccostrea cucullata* settling on the mangrove trunk

Fig. 6.4 *Saccostrea cucullata* settling on the supporting pillar of fish landing station

landing stations in the river mouth or estuaries, port/harbour structures and even on the trunk of mangrove trees (Fig. 6.3).

Oysters have the potential of sequestering carbon in their shells, which are made of calcium carbonate. However, through the process of shell generation, the oyster also produces some carbon dioxide.

The potential for oysters to sequester carbon is regulated by the process through which the hard calcareous shell is formed. The chemical process through which this shell is formed is called calcification. The process involves the formation of calcium carbonate as per the equation:

$$Ca^{2+} + 2HCO_3^- \rightarrow CaCO_3 + CO_2 + H_2O$$

The chemical reaction indicates how the oyster converts a calcium ion and two bicarbonate molecules into one calcium carbonate, one carbon dioxide molecule, and water. It may seem that since there is a release of carbon dioxide from this reaction, these bivalves are actually producing carbon dioxide instead of reducing the amount in the water column. However, this is not the complete chemical process. The source of the reactant carbon and fate of the carbon products need to be investigated in order to fully understand the overall effect on the cycle.

The calcium carbonate deposited is accounted for in the shell mass, and the carbon in the shell may be locked up for millennia. Carbon makes up 12 % of calcium carbonate by mass. Oysters offer a medium to capture and store carbon for a much longer time than agricultural products or even forest. Thus, the main question in the process is the source of bicarbonate and the fate of the carbon dioxide. The simple answer to this question is the ambient aquatic medium, but this does not resolve the question of the oyster's role in carbon sequestration. To get a complete answer of the mechanism, one has to critically analyse the carbon cycle within the oceanic compartment.

The main source for oceanic carbon input is from the atmosphere through absorption of carbon dioxide; this process currently absorbs about 2 Gt of carbon per year (Stewart 2005). As atmospheric carbon dioxide production and concentration increase, the amount of carbon dioxide absorbed by the ocean also increases, thus elevating the ocean's carbon dioxide content. However, all the carbon dioxide does not remain in this form in water. As the concentration of carbon dioxide in water increases, a series of reactions occur in order to allow for more carbon dioxide absorption and to balance the different forms of carbon in the water. One of these equilibrium reactions proceeds as follows:

$$CO_2 + CO_3^{2-} + H_2O \rightarrow 2HCO_3^-$$

This shows how carbon dioxide combines with carbonate and water to form two bicarbonates after entering the water. Therefore, increased amounts of carbon dioxide entering the ocean due to increase in atmospheric carbon dioxide would also cause an increase in bicarbonate concentrations (Gattuso 2009). Using these two chemical reactions, it is possible to evaluate the role oyster shell formation that plays in the carbon cycle. Based on these equations, Fig. 6.5 depicts this process. The figure illustrates how the oyster uses calcium (Ca) and carbonate (CO_3) to form its shell and the rest of the reactants and products are recycled. This illustration enables the equation to be simplified and written in Fig. 6.5.

It is now clear that increase in carbon dioxide in the atmosphere leads to an increase in oceanic carbon dioxide absorption. This has profound effects on the seawater carbonate system. It lowers the aquatic pH (ocean acidification), and more importantly, it decreases the availability of carbonate ions. Increases in carbon dioxide push this equation to the right and use up carbonate in the process, which reduces the amount that will be available for calcification. Therefore, the process of calcification actually promotes ocean acidification and reduces the oceans' potential for carbon uptake by locking up carbonate. However, few studies (Gazeau et al. 2007) have shown that oysters are quite resilient in spite of the changes in pH. It is also known that marine animals are, in general, intolerant to changes in

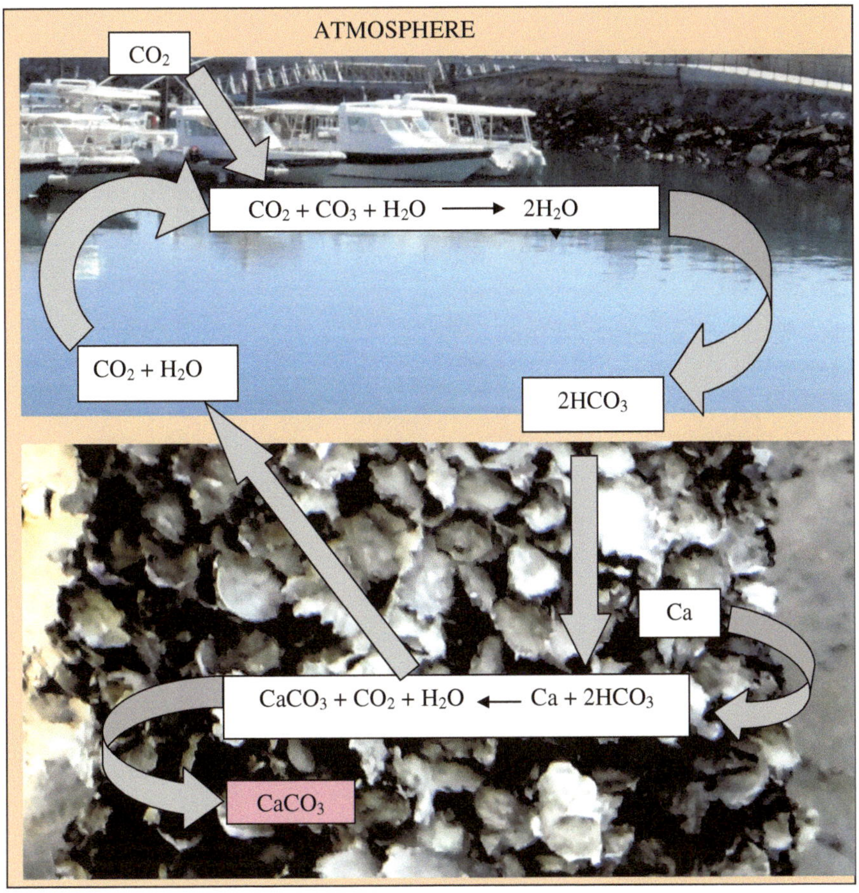

Fig. 6.5 Oyster's role in oceanic carbon cycle

the pH of the water in which they live (Knutzen 1981). But how severe would the pH changes be from carbon dioxide sequestration, and what would be the impact on molluscan community, whose calcareous shell is prone to dissolution at a lower pH? The dissolved inorganic carbon (DIC) concentration in surface ocean waters is already higher than in pre-industrial times as a result of the elevated carbon dioxide concentration in the atmosphere. Calculations show (Haugan and Drange 1996) that there has been a decrease in pH of the ocean surface water of almost 0.1 pH units due to the increase in atmospheric carbon dioxide from pre-industrial times. The 'spring bloom' in the North Atlantic and winter upwelling of deep (CO_2-rich) water in the North Pacific can cause changes in surface water pH of this magnitude, but other ocean areas see much lower seasonal variations. Indeed, below the euphotic zone (the top 100–200 m, where there is enough light to support photosynthesis), the seasonal variation in pH is universally very small. The same calculations show that the current rate of increase in atmospheric carbon dioxide concentration (15 ppm/decade) will cause a decrease in pH of 0.015 units/decade. This, by itself, could have a significant impact on the ecology of the productive surface waters of the oceans in the foreseeable future.

The present authors conducted a study on the aquatic pH value in three sectors of Indian Sundarbans during 1990–2012. The sampling stations are shown in the Fig. 6.6.

It was observed that the aquatic pH decreased by 0.11, 0.10 and 0.06 unit in the western, central and eastern Indian Sundarbans, respectively

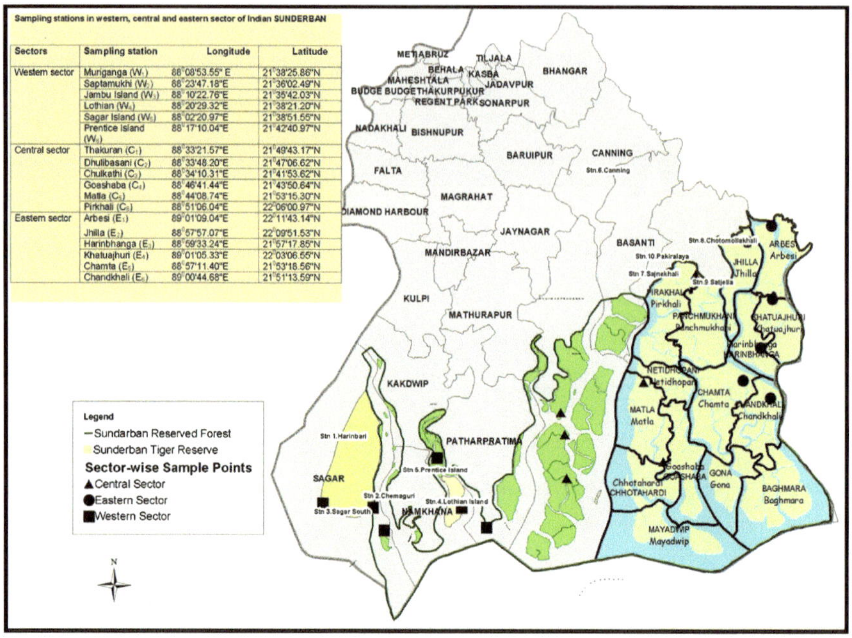

Fig. 6.6 Sampling stations in three sectors of Indian Sundarbans; the stations in the western sector of Indian Sundarbans are marked with *square* marks, and the stations in the central and eastern sectors are represented with *triangular* and *circular* signs, respectively

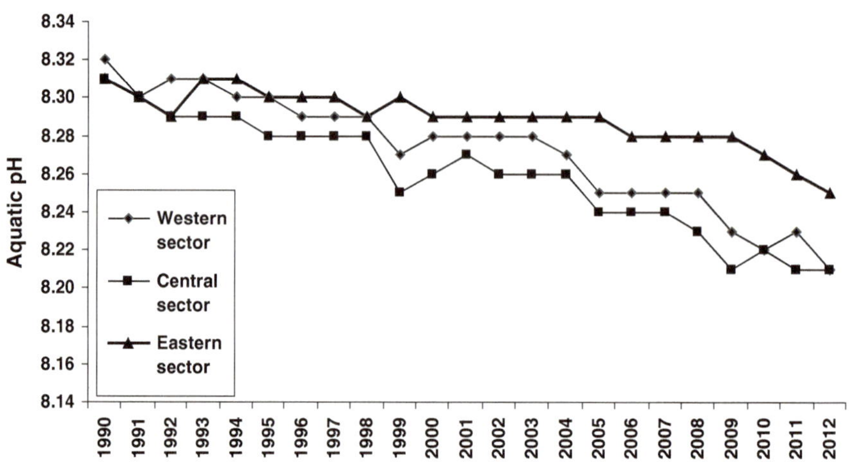

Fig. 6.7 Decreasing trend of aquatic pH in Indian Sundarbans

(Fig. 6.7), which is 0.0047, 0.0043 and 0.0026 unit/year, respectively. In this context, it is extremely interesting to note that decrease in pH (both gross and rate) is lowest in eastern sector compared to other two sectors. This is mainly because the eastern Indian Sundarbans is mostly under Reserve Forest areas and mangroves thrive in this region without any anthropogenic pressure. These halophytes absorb carbon dioxide from the atmosphere and the water column and

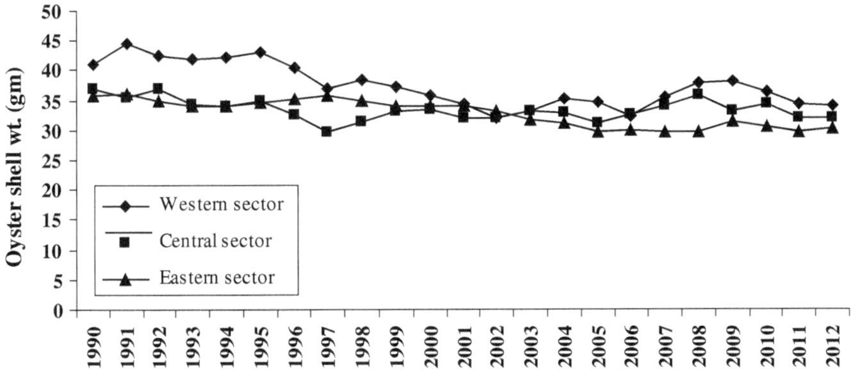

Fig. 6.8 Decreasing trend of oyster shell weight in Indian Sundarbans: is this a signature of marine and estuarine acidification?

store it as carbon in their vegetative and reproductive parts through active participation of the enzyme ribulose-1, 5-biphosphate carboxylase/oxygenase (RuBisco).

The decrease in aquatic pH has been found to exert an adverse impact on shelled organisms. It results in the reduction of the shell weight in oyster species *Saccostrea cucullata* as evidenced by the authors (Fig. 6.8).

The population of oyster is not affected by aquatic pH, but the shell weight is greatly affected by pH. Lowering of pH may cause dissolution of calcareous mass of the shell leading to its thinning, which is reflected through highly significant positive correlation between the oyster shell weight and pH of the ambient water in all the three sectors (except the central sector) of Indian Sundarbans. The degree of positive correlation is masked to a great extent by the results of central Indian Sundarbans (Fig. 6.9 and Table 6.1), where few estuaries like Matla has basically lost its physico-chemical nature in the absence of mixing freshwater from the upstream sources. Instead of considering Matla as an estuary, it may be designated as tide-fed river.

Oyster shell being a storehouse of carbon also exhibits significant positive correlations with aquatic pH in western and eastern sectors of Indian Sundarbans (Table 6.1), but when allometric equation interlinking the two variables is formed, the significant positive interrelationship is diluted as evidenced from the R^2 value in Fig. 6.13. This may again be attributed to the inclusion of central sector in the data set (used for Fig. 6.13), where the correlation between oyster shell carbon and aquatic pH is insignificant (Table 6.1). The central sector of Indian Sundarbans is hypersaline, compared to the western and central Indian Sundarbans (Figs. 6.10, 6.11 and 6.12) (Mitra et al. 2009,

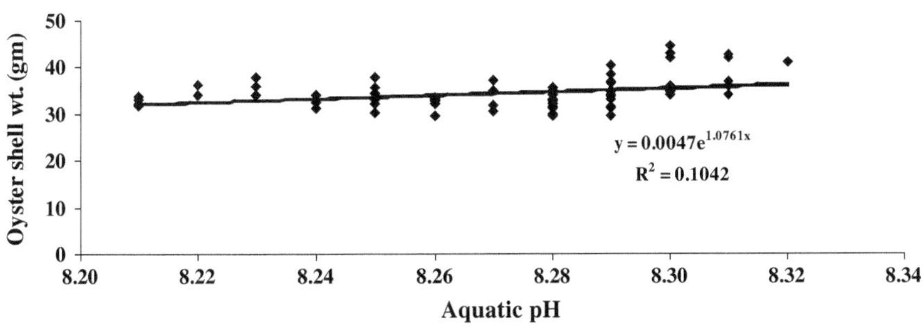

Fig. 6.9 Interrelationship between oyster shell weight and aquatic pH considering all the three sectors of Indian Sundarbans

Table 6.1 Interrelationships between aquatic salinity, pH and oyster variables

Row no.	Combination	Western sector		Central sector		Eastern sector	
		r -Value	*p* -Value	*r* -Value	*p* -Value	*r* -Value	*p* -Value
1	Oyster population × salinity	0.9514	<0.01	0.8430	<0.01	0.8980	<0.01
2	Oyster population × aquatic pH	0.8567	<0.01	−0.8089	<0.01	0.7477	<0.01
3	Oyster shell length × salinity	0.8882	<0.01	−0.2569	IS	0.9561	<0.01
4	Oyster shell length × aquatic pH	0.8567	<0.01	0.2069	IS	0.8792	<0.01
5	Oyster shell weight × salinity	0.7387	<0.01	−0.4558	<0.05	0.8788	<0.01
6	Oyster shell weight × aquatic pH	0.6285	<0.01	0.3339	IS	0.7496	<0.01
7	Oyster flesh weight × salinity	−0.0933	<0.01	0.2495	IS	−0.1104	IS
8	Oyster flesh weight × aquatic pH	−0.1032	IS	−0.1833	IS	−0.1804	IS
9	Oyster shell carbon × salinity	0.7778	<0.01	−0.4657	<0.05	0.8748	<0.01
10	Oyster shell carbon × aquatic pH	0.6886	<0.01	0.3576	IS	0.7541	<0.01
11	Oyster flesh carbon × salinity	−0.0709	IS	0.2762	IS	−0.0793	IS
12	Oyster flesh carbon × aquatic pH	−0.0856	IS	−0.2167	IS	−0.1278	IS
13	Oyster shell weight × oyster shell carbon	0.9578	<0.01	0.9457	<0.01	0.9917	<0.01
14	Oyster flesh weight × oyster flesh carbon	0.9475	<0.01	0.9313	<0.01	0.9506	<0.01

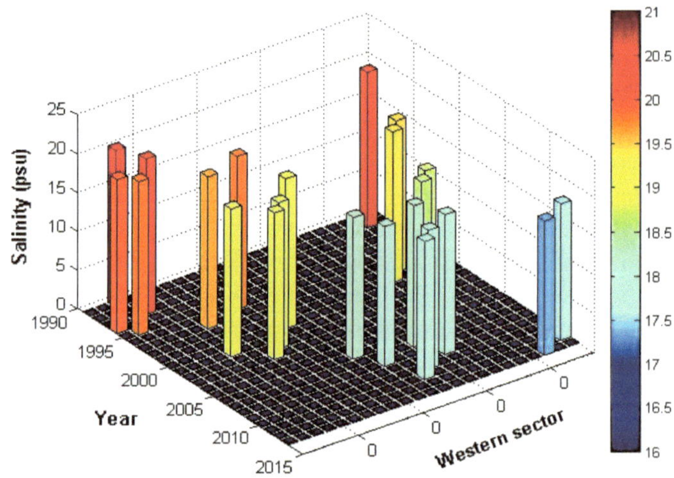

Fig. 6.10 Decreasing trend in aquatic salinity over a period of 23 years (1990–2012) in western Indian Sundarbans (Graph designed in MATLAB by Dr. Nibedita Mukhopadhyay, Environmentalist)

2011, 2012; Sengupta et al. 2013), which may play a regulatory role on the pH and subsequently on the dissolution process of calcareous shell of oyster. Further research needs to be carried out to identify the role of ambient aquatic salinity in the dissolution mechanism of molluscan shell.

The soft mass of the oyster (fleshy part within the shell) is, however, resistant to lowering of pH as evidenced by the insignificant correlation coefficient values between aquatic pH, oyster flesh weight and oyster flesh carbon (Table 6.1).

The authors also conducted a viability study of carbon storage by oyster community and observed that during a span of 23 years (1990–2012), the stored carbon in oyster flesh ranged from 348.85 kg m^{-2} (during 2009 at eastern sector) to 937.90 kg m^{-2} (during 1991 at western sector). This means a sequestration rate of 15.17–40.78 kg m^{-2} year^{-1}. The stored carbon

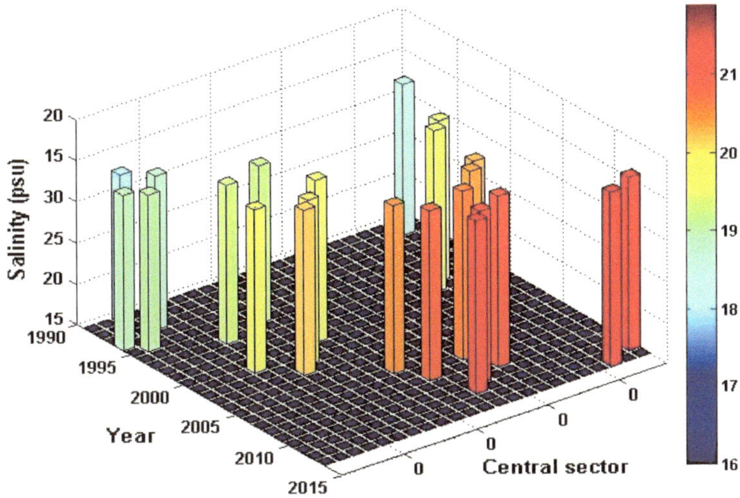

Fig. 6.11 Increasing trend in aquatic salinity over a period of 23 years (1990–2012) in central Indian Sundarbans (Graph designed in MATLAB by Dr. Nibedita Mukhopadhyay, Environmentalist)

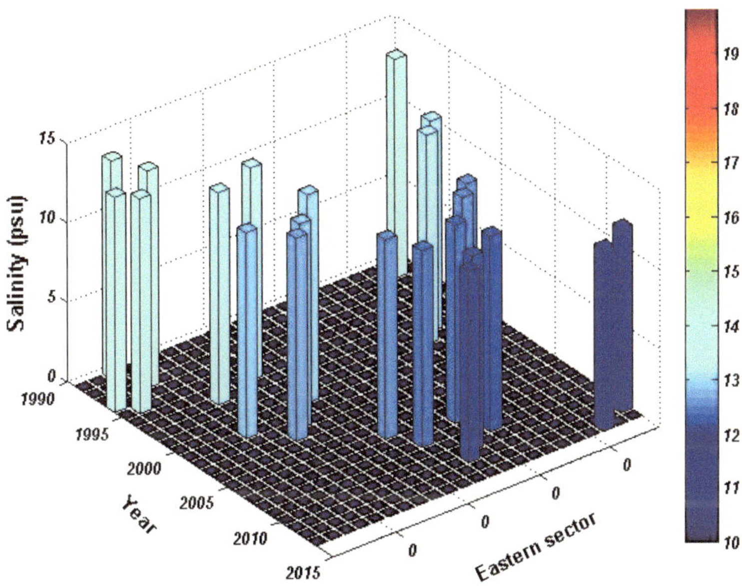

Fig. 6.12 Decreasing trend in aquatic salinity over a period of 23 years (1990–2012) in eastern Indian Sundarbans (Graph designed in MATLAB by Dr. Nibedita Mukhopadhyay, Environmentalist)

in calcareous shell ranged from 1102.62 kg m^{-2} (during 2009 in the eastern sector of Indian Sundarbans) to 3236.90 kg m^{-2} (during 1991 in the western sector), which means a sequestration of 47.94–140.74 kg m^{-2} $year^{-1}$. The values are, however, pH dependent, and therefore, liming the culture area is essential for oyster farming and subsequent carbon sequestration. In the present case study, it was observed that carbon in the soft tissue of oyster is $\sim$40 %, whereas in the calcareous shell, it is $\sim$13 %. Similar findings were also reported by Jurkiewicz-Karnkowska (2005).

Many studies on the role of molluscs in aquatic ecosystems take into account only soft tissues. The organic fraction remains underestimated in molluscan shells which may contain considerable amounts of macroelements.

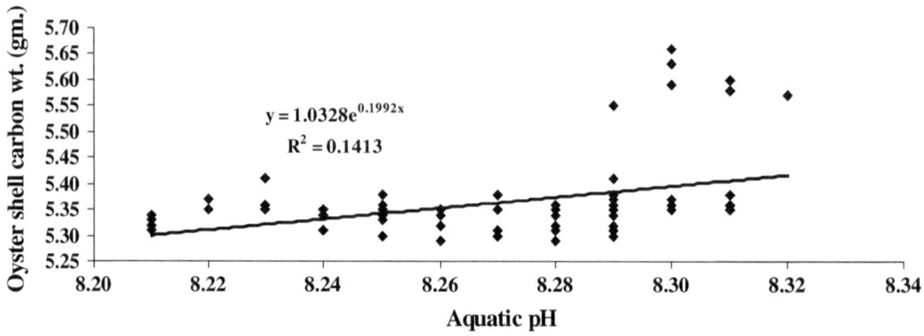

Fig. 6.13 Interrelationship between oyster shell carbon weight and aquatic pH considering all the three sectors of Indian Sundarbans

Cameron et al. (1979) showed a significant contribution of organic shell matter to the total shell weight of unionid bivalves. Their results indicate that measures of standing stock biomass, production or energy flow in bivalve populations can be considerably underestimated unless the organic material contained within the shells is included in the estimates.

Jurkiewicz-Karnkowska (2005) estimated the comparison of carbon concentrations in the soft tissues and shells of molluscan species that is the main component of malacocoenoses of the Zegrzyński Reservoir. An assessment of carbon accumulation in molluscs living in the reservoir and spatial variability of this is also performed. The mean values of carbon content in the soft tissues of the investigated molluscs ranged from about 36 % to about 46 % dry wt., whereas carbon content in shells was similar in all cases on average about 13 % dry wt.

The present chapter depicts few interesting findings on molluscan carbon, which are summarized in points:

1. From the case study conducted by the authors on *Saccostrea cucullata*, a common oyster species widely available in Indian Sundarbans, it is observed that carbon in the oyster flesh is a direct function of flesh weight (*Vide* row 14 in Table 6.1). Similarly, stored carbon in the calcareous shell of the oyster species is directly proportional to shell weight (*Vide* row 13 in Table 6.1). This study is, however, regional and needs further research on global basis. Also site-specific allometric equations need to be developed relating the stored carbon with easily measurable variables like shell weight or flesh weight.
2. In climate science, ocean acidification is a common terminology for a decreasing trend of aquatic pH due to uptake of anthropogenic carbon dioxide (CO_2) from the atmosphere. About 30–40 % of the carbon dioxide released by the human activities into the atmosphere dissolves into the oceans, rivers and lakes. To maintain chemical equilibrium, some of it reacts with the water to form carbonic acid. Some of these extra carbonic acid molecules react with a water molecule to generate a bicarbonate ion and a hydronium ion, thus increasing the ocean's 'acidity' (H^+ ion concentration) as per the expression:

$$CO_{2(aq)} + H_2O \leftrightarrow H_2CO_3 \leftrightarrow HCO_3^- + H^+ \leftrightarrow CO_3^{2-} + 2H^+$$

Between 1751 and 1994, surface ocean water pH is estimated to have decreased from approximately 8.25–8.14, representing an increase of almost 30 % in H^+ ion concentration in the world's oceans. In Indian Sundarbans, at the apex of Bay of Bengal, it is observed that the aquatic pH decreased by 0.11, 0.10 and 0.06 unit in the western, central and eastern sectors, respectively, over a span of 23 years (1990–2012), which is 0.0047, 0.0043 and 0.0026 unit/year, respectively (*Vide* Fig. 6.6).

3. This increasing acidity is believed to have a direct adverse impact on molluscan species. The general norm of lowering shell biomass with the lowering of aquatic pH holds good for western and eastern Indian Sundarbans, where the salinity values range from 18.61 to 20.25 psu (in the western sector) and 12.60 to 13.78 psu (in the eastern sector). However, in the hypersaline central sector, where the salinity value ranges from 18.03 to 20.34 psu, the dissolution of calcareous shell of *Saccostrea cucullata* is not evident (*Vide* row 6 in Table 6.2).
4. The aquatic pH plays no role on stored carbon in oyster flesh in the Indian Sundarbans region (*Vide* row 12 in Table 6.1).
5. Afforestation, reforestation and (avoided) deforestation come under the broad heading of biosequestration. Biosequestration is the general term used to describe activities where plants such as trees are used to 'sequester' or absorb carbon from the atmosphere. Plants achieve this via photosynthesis, which converts carbon dioxide into oxygen and plant material (Drury et al. 2004). The magnitude of sequestered carbon can vary considerably depending on the site, species and other external conditions. The CSIRO states that the 'net amount of atmospheric carbon that is sequestered by a biosequestration project is the balance between changes in above-ground vegetation, roots, the litter layer, and the soil' (www.clw.csiro.au/staff/hairsinep/ffplw19052004.pdf). The amount of carbon sequestered by a stand of vegetation is given by tonnes of carbon per hectare per year (tC ha^{-1} $year^{-1}$). Paul et al. (2003) have calculated a figure of 6–13 tC ha^{-1} $year^{-1}$ for commercial operations involving Tasmanian blue-gum and radiata pine. Hobbs and Bennell (2005) observed that *Eucalyptus socialis* can absorb between 2.5 and 4.2 tC ha^{-1} $year^{-1}$. In the present chapter, the case study on *Saccostrea cucullata* reveals a sequestration rate of 151.70 tC ha^{-1} $year^{-1}$ (during 2009 at eastern sector) to 407.80 tC ha^{-1} $year^{-1}$ (during 1991 at western sector) in the fleshy mass, and in the calcareous shell, the value ranges from 479.40 tC ha^{-1} $year^{-1}$ (during 2009 in the eastern sector of Indian Sundarbans) to 1407.40 tC ha^{-1} $year^{-1}$ (during 1991 in the western sector). The value is much greater than the producer community, but in reality, this does not happen, as the biomass documented in the present case study has not considered the full life cycle of the oyster. If one considers the biomass of oyster spat to adult stage, the sequestration rate will be much lower than the figures stated here. However, the present study suggests that one possible method for achieving the removal of carbon from the ocean and estuaries may be through the practice of shellfish farming.
6. The main advantage of oyster farming is that unlike other ocean sequestration techniques, the oyster shell permanently removes carbon

Table 6.2 Mean surface water salinity in three sectors of Indian Sundarbans; each value is the mean of 6 stations (Fig. 6.10, 6.11 and 6.12) and 3 seasons

Year	Western sector	Central sector	Eastern sector
1990	20.25	18.03	13.78
1991	20.09	18.32	13.76
1992	20.08	18.77	13.65
1993	19.87	18.92	13.51
1994	19.79	19.09	13.50
1995	19.76	19.21	13.38
1996	19.61	19.37	13.33
1997	19.50	19.50	13.25
1998	19.40	19.64	13.16
1999	19.30	19.79	13.08
2000	19.03	19.90	12.95
2001	18.87	20.00	12.87
2002	18.89	20.13	12.77
2003	18.71	20.22	12.65
2004	18.61	20.34	12.60
2005	18.25	20.46	12.55
2006	18.21	20.56	12.48
2007	18.03	20.70	12.40
2008	18.01	20.85	12.31
2009	17.90	20.95	12.21
2010	17.80	21.07	12.08
2011	17.60	21.16	11.58
2012	17.34	21.27	11.44

from the ocean as well as the atmosphere. This aspect will become more important as global warming and ocean carbon capacity affect the amount of carbon absorbed by oceans. A direct comparison between biosequestration activities and shellfish farming may be unrealistic as the former is underpinned by photosynthesis and the latter by respiration. An independent assessment of the shellfish industry is essential to determine the magnitude of carbon sequestration (where all the stages of oyster life cycle need critical analyses). Such assessment may form the basis of the overall costs of generating carbon offsets through oyster farming.

6.2 Stored Carbon in Gastropods

Gastropods play a significant role both in accumulation and circulation of carbon, especially in habitats where they are numerous and their biomass is high, but studies concerning these problems are scarce. In the intertidal mudflats of mangrove-dominated Sundarbans, gastropods like *Telescopium telescopium*, *Nerita articulata*, *Cerithedia cingulata*, *Cymia lacera*, *Littorina scabra scabra* and *Cerithedia obtusa* are abundantly distributed (Fig. 6.14), but no data on their carbon content in shell and fleshy part are available till date.

T. telescopium are often seen grazing on the intertidal mudflats creating unique trail marks (Fig. 6.15). These gastropod species depend on carbon enriched detritus of the intertidal mudflats and form a unique carbon dynamics in the benthic substratum of the coastal zone.

Carbon in gastropod shell carbonate originates from four important sources: atmospheric carbon dioxide, food, water and carbonate rocks. Gastropods incorporate atmospheric carbon dioxide in their shell carbonate via respiration. Respired carbon dioxide is introduced to the bicarbonate pool in the gastropod's haemolymph and passed along to the extrapallial fluid, from which the shell carbonate is ultimately precipitated (Wilbur 1972). Estimates of the contribution of atmospheric carbon dioxide to gastropod shell carbonate vary between negligible (Stott 2002), 16–48 % (Romaniello et al. 2008) and 30–60 % (Goodfriend and Hood 1983).

Carbon from food sources (e.g., living plants, fungi, organic detritus) is incorporated into the extrapallial fluid through two mechanisms: direct digestion and breakdown of urea. Many gastropods (like *Littorina* sp.) feed on mangrove leaves (Fig. 6.16), and even some (like *C. obtusa*) scrap the algae that deposit on the trunk of mangroves (Fig. 6.17).

Fig. 6.14 The dominant gastropod species in mangrove-dominated Indian Sundarbans. **a** *Telescopium telescopium*. **b** *Cerithedia obtusa*. **c** *Cerithedia cingulata*. **d** *Nerita articulata*. **e** *Cymia lacera*. **f** *Littorina* spp

Fig. 6.15 *Telescopium telescopium* abundantly found on the mudflats of Sundarbans

Fig. 6.16 *Littorina* sp. feeding on fresh mangrove leaves

When gastropods consume and digest food, carbon is introduced to the haemolymph and passed along to the extrapallial fluid in the same manner as atmospheric carbon dioxide. There, it mixes with atmospheric carbon before becoming incorporated in the shell carbonate (Wilbur 1972). Carbon derived from urea takes a more indirect pathway. Urea that is not expelled by the gastropod breaks down into carbon dioxide and ammonia via a urease reaction (Stott 2002). The resulting carbon dioxide is then reintroduced directly to the extrapallial fluid and ultimately incorporated into the shell carbonate. Estimates of the amount of carbon derived from plants, either directly or indirectly through urea, vary between 25 and 40 % (Goodfriend and Hood 1983), 36–73 % (Romaniello et al. 2008) and w100 % (Stott 2002).

Terrestrial gastropods ingest water from multiple sources, including dew, soil moisture, standing water and precipitation, all of which contain some amount of dissolved inorganic carbon (DIC). Water is taken up through the foot of the gastropod by contact rehydration (Balakrishnan and Yapp 2004) and introduced to the haemolymph before being passed on to the extrapallial fluid.

The soft fleshy meat and shells of gastropods are storehouse of carbon as evidenced from the concentrations of total carbon (% dry weight) in the soft tissues and shells of gastropods. A comparative study conducted by the present authors

Fig. 6.17 *Cerithedia obtusa* scrapping the algal mat on mangrove trunk

during August 2013 reveals significant spatial differences of stored carbon in gastropod species in Indian Sundarbans, which might be due to carbon derived from algal mat, litter and detritus of mangroves that is more in the eastern sector of Indian Sundarbans compared to the central and western sectors. The stored carbon also exhibited species-wise variation, and the order is *C. obtusa* > *N. articulata* > *T. telescopium* > *C. cingulata* irrespective of locations (Tables 6.3, 6.4 and 6.5). As stated earlier in this section, one of the most important sources of carbon in marine and estuarine gastropods is their diets, which mainly are macroalgae on the mangrove vegetation or detritus on the intertidal mudflats. *C. obtusa* is fully dependent on algal matter for nutrition, and *N. articulata* prefers both algal feed and detritus. *T. telescopium* and *C. cingulata* are normally detritivores (Fig. 6.18).

The spatial order of stored carbon in flesh and shell of the selected gastropod species is eastern sector > central sector > western sector. This spatial difference (as revealed through ANOVA in Tables 6.6 and 6.7) may be due to stored carbon in the food material, which is the main source of carbon in gastropods. The stored carbon in seaweed like *Enteromorpha intestinalis*, *Ulva lactuca* and *Catenella repens* is more in the eastern followed by central and western sectors (Mitra et al. 2014). This trend is also seen in case of organic carbon present in soil and mangrove litter (Mitra 2013).

The carbon content in molluscan species exhibits unique seasonal variation. Compared with the data of Table 6.3, the stored carbon values are relatively lower in the same gastropod species collected from the same sampling

Table 6.3 Concentrations of total carbon (% dry wt.) in the soft tissues and shells of the investigated gastropod species (mean values and standard deviations) collected from eastern sector of Indian Sundarbans

Species	Carbon	
	Tissues	Shells
Telescopium telescopium	37.59 ± 0.60	13.90 ± 0.15
Nerita articulata	37.33 ± 0.58	13.88 ± 0.10
Cerithedia obtusa	39.85 ± 0.89	13.99 ± 0.19
Cerithedia cingulata	37.66 ± 0.50	12.86 ± 0.11

Table 6.4 Concentrations of total carbon (% dry wt.) in the soft tissues and shells of the investigated gastropod species (mean values and standard deviations) in the central Indian Sundarbans

Species	Carbon	
	Tissues	Shells
Telescopium telescopium	36.89 ± 0.70	12.99 ± 0.75
Nerita articulata	37.98 ± 0.95	13.83 ± 0.14
Cerithedia obtusa	39.40 ± 0.55	13.78 ± 0.19
Cerithedia cingulata	35.99 ± 0.77	12.66 ± 0.19

Table 6.5 Concentrations of total carbon (% dry wt.) in the soft tissues and shells of the investigated gastropod species (mean values and standard deviations) in the western Indian Sundarbans

Species	Carbon	
	Tissues	Shells
Telescopium telescopium	34.95 ± 0.78	12.42 ± 0.22
Nerita articulata	35.75 ± 0.93	13.21 ± 0.20
Cerithedia obtusa	37.80 ± 0.90	13.69 ± 0.34
Cerithedia cingulata	34.86 ± 1.02	12.20 ± 0.18

stations during April 2013, the pre-monsoon season in the study area and characterized by high salinity, pH and temperature. A detailed report of the analysis of similar nature was conducted during April 2013.

6.3 Stored Carbon in Coral Reefs

Coral reefs are recognized as one of the major geological features of the earth's surface, although they occupy only about 0.2 % of the world's ocean area (Smith and Jokiel 1978). They support the highest marine biodiversity in the world, containing an estimated 25 % of all marine species. They are often regarded as the oases of the ocean. More than 500 million people worldwide depend on them for food, storm protection, jobs and recreation. Unfortunately, many of the world's coral reefs have been degraded, mainly due to human activities. According to the researchers, about 19 % of the original area of coral reefs have been lost since 1950, 15 % of coral reefs are in a critical state with loss possible

Fig. 6.18 Bed of *Telescopium telescopium* on the mangrove detritus of intertidal mudflats

Table 6.6 ANOVA results showing variations in shell carbon of gastropods between species and sectors

Source of variation	SS	*df*	MS	*F*	*P* -Value	F_{crit}
Between species	2.86	3	0.95	12.56	0.005	4.76
Between sectors	1.21	2	0.61	8.01	0.020	5.14

Table 6.7 ANOVA results showing variations in tissue carbon of gastropods between species and sectors

Source of variation	SS	*df*	MS	*F*	*P* -Value	F_{crit}
Between species	12.99	3	4.33	13.06	0.005	4.76
Between sectors	8.72	2	4.36	13.16	0.006	5.14

within the next 10–20 years and a further 20 % are seriously threatened with loss predicted within 20–40 years.

The Great Barrier Reef, the largest coral reef of the planet Earth, stretches more than 2,000 km from New Guinea southwards along the east coast of Australia. The reefs are constructed and promoted by corals, which are colonial animals and whose individual members are called polyps (Fig. 6.19).

A coral polyp is very similar to a tiny sea anemone possessing stinging tentacles, but unlike the anemone, a coral polyp extracts calcium carbonate from the water and builds within its tissues a skeletal cup of calcareous material. Large numbers of polyps grow together in colonies of delicately branched forms or rounded masses (Figs. 6.19 and 6.20).

Corals are very slowly growing organisms; some species grow less than 1 cm in a year, while others add up to 5 cm each year. The same coral may be found in different shapes and sizes, depending on the depth and wave action of the area. Within the tissues of the coral polyps, single-cell dinoflagellate algae called zooxanthellae are present. Polyps and zooxanthellae have symbiotic relationship in which the coral provides the algal cells with a protected environment, carbon dioxide and nutrients (like nitrate and phosphate) and the algal cells by the process of photosynthesis return oxygen, remove wastes and produce carbon compounds, which help to nourish the coral. Some coral species receive as much as 60 % of their nutrition from the algae. Zooxanthellae also cause the coral to produce more calcium carbonate and increase the growth of their calcareous skeleton.

Fig. 6.19 View of coral polyp

Fig. 6.20 Cluster of coral polyps

The polyps feed actively at night by extending their tentacles for catching the zooplankton, but during the day, their tentacles are contracted, exposing their outer layer of cells containing zooxanthellae to the sunlight. The reef-building corals of tropical regions have specialized requirements like warm, shallow, clean water and a firm substratum to which they can attach. The growth of the corals is restricted to tropical waters between 30°N and 30°S where the temperature usually ranges between 20 and 25 °C. Waters at depths greater than 50–100 m are too cold for considerable secretion of calcium carbonate. Perhaps because of this reason, most of the Caribbean corals are found in the upper 50 m of lighted zone, whereas Indian and Pacific corals are found at depths of 150 m in the more transparent water of these oceans.

Coral reefs are ecologically important ecosystems with rich biodiversity (Table 6.8). They support complex assemblages of many different types of plants and animals having intense competition for space and food. Globally, coral reefs comprise little more than 1 % of total continental shelf area, but they are among the most diverse of the world's ecosystems. It is estimated that about 793 species of scleractinian corals are present worldwide, of which 719 occur in the Indo-West Pacific region. Very few species of any taxa are found in both the Indo-Pacific coral reef area and Caribbean reef area. It has been reported that the coral species richness of Indian Ocean rim countries is very similar to those in the Pacific Ocean with both of these regions having significantly greater coral diversity than the Atlantic. Countries with large coral diversities such as Indonesia (443), Australia (428) and Thailand (238) border both the Indian and Pacific Oceans, thus increasing the apparent diversity of both. The contribution of the Indian Ocean to this diversity is significant with six

Table 6.8 Species richness of coral reef-associated faunas

Taxon	Number of species	References
General Indo-Pacific reef faunas		
Fishes	>2,200	Sale (1980)
Corals	390	Sebens (1994)
Nudibranchs	>500	Gosliner (1993)
Estimates of faunas of single reefs		
Fishes	860	Sale et al. (1994)
Molluscs	150–500	Sorokin (1993)
Crustaceans	100–250	Sorokin (1993)
Polychaetes	100–200	Sorokin (1993)
Echinoderms	50–100	Sorokin (1993)
Sponges	>50	Sorokin (1993)

other countries spread from Seychelles in the west to India in the North to Malaysia in the east, all having over 200 species of coral recorded.

Coral reef is the housing complex of a wide spectrum of fauna. It has been documented that as many as 300 animal species may live together on a single reef. The giant clam (*Tridacna* sp.) measuring up to a metre in length and weighing over 150 kg is an important species of reef ecosystem, but overharvesting and poaching have greatly reduced their numbers. These clams also possess zooxanthellae in large numbers in the colourful tissues that line the edges of the shell. The bubblegum corals are another surprise to the reef community. They are octocorals belonging to the phylum cnidaria, which also includes stony corals and jelly fish. Despite forming enormous tree-like structures, the trunk and branches of bubblegum corals are not corneous or calcareous like most branching octocorals. Instead, the branches are accumulation and near fusion of microscopic sclerites made of calcite.

Crabs, moray eels, colourful reef fish, poisonous stone fish, spiny sea urchins, seahorses, shrimps, lobsters, sponges and various life forms are found in the coral reef ecosystem. Global pattern of coral reef fish diversity has been examined in detail. The biodiversity of coral reef fishes covers a wide array of species. Reef fishes show patterns of distribution and diversity at all scales ranging from global to regional to local. At the largest scale, diversity appears to be principally controlled by the interaction of tectonic and geomorphological events with evolution and dispersion processes. At a regional level, water parameter like turbidity and current strength, perhaps requiring differential adaptation at the larval stage, may control the species distribution pattern. At a local level, diversity is influenced by the habitat characteristics such as depth, heterogeneity and complexity. Because of the vagaries of larval distribution and recruitment, the assemblages to be found at a site are quite variable, although the extent of these variations probably differs among species and areas.

In the Indian subcontinent, coral reefs are concentrated in regions like Gulf of Mannar, Andaman and Nicobar Islands, Lakshadweep Islands, etc. Coral reefs occur in and around 20 Islands between Tuticorin and Pamban in the Gulf of Mannar and Palk Bay.

6.3.1 Types of Coral Reefs

On the basis of the structure and their relationship to the underlying geologic features, coral reefs can be divided into three categories. They are fringing reefs, barrier reefs and atolls (Fig. 6.21).

1. *Fringing reefs*: These are found very close to or near the newly formed volcanic islands as well as surrounding the continental landmasses. They are located directly offshore and project outward to the sea, e.g., the Caribbean reef.
2. *Barrier reefs*: These reefs act as barrier separated by a lagoon from the landmass, with which they are associated. Those surrounding volcanic islands are formed from subsiding islands with fringing reefs. The lagoon forms between the island and the reef. The largest barrier reef is the Great Barrier Reef of Australia, which runs over 2,000 km along the north-eastern coast of Australia to New Guinea. It is so large that it is visible from space shuttle. The second largest barrier reef is located in the Caribbean Sea off the coast of Belize.
3. *Atolls*: These reefs are somewhat circular in shape with a centrally located lagoon. According to Darwin, they are formed in areas where barrier reefs subside below sea level, whereas according to Alexander Agassiz, they are formed on the top of the cones of submerged volcanoes, leaving only the reef around a central lagoon. But in any of the case, the lagoon remains connected to the open sea by breaks in the reef. Over time, the reefs may become eroded or exposed to wind action or waves. These physical processes sometimes give rise to formation of islands surrounding the lagoons, which is perhaps the depositional feature of sand carried by the wind or wave action. More than 300 atolls are present in and around Pacific and Indian Ocean, whereas 10 atolls are present in Atlantic Ocean as the water is too cool or turbid.

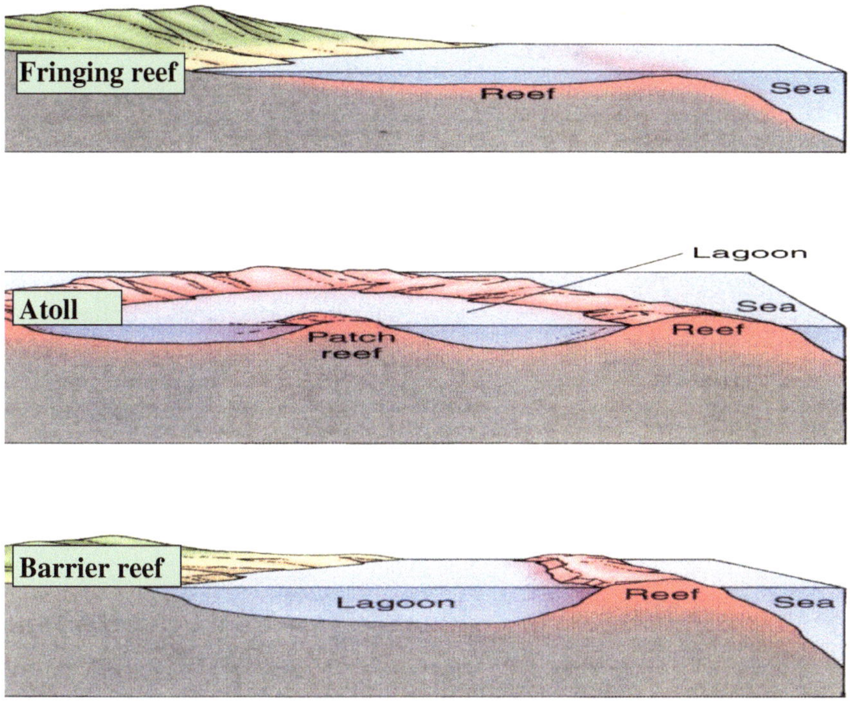

Fig. 6.21 Types of coral reefs

Most atolls have north-east–south-west orientation with an island on the east, a broad well-developed reef on the west and a lagoon in between. Reefs of all atolls are widest on the south-west side.

In addition to these three main reef types, there are several special reef arrangements. These are **table reefs** and **patch reefs**. Table reefs are small reefs found in the open ocean that have no central islands or lagoons. Patch reefs occur numerously in patches located in lagoons associated with atolls and barrier reefs.

There are two important biogeochemical processes in a coral reef system: photosynthesis and calcification:

6.3.2 Photosynthesis

Coral reefs' primary productivity is very high, typically producing 5–10 g of carbon per square metre per day. Two vital requirements for photosynthesis are solar energy (which is a direct function of aquatic transparency) and availability of nutrients. One reason for the unusual clarity of tropical waters is they are deficient in nutrients and drifting plankton. The primary cause behind this is the year-round solar radiation in the tropics, warming the surface layer, making it less dense than subsurface layers. The warmer water is separated from deeper, cooler water by a stable thermocline, where the temperature makes a rapid change. This keeps the warm surface waters floating above the cooler, deeper waters inhibiting the process of exchange between these layers. Organisms that die in aquatic environments generally sink to the bottom, where they decompose, which releases nutrients in the form of nitrogen (N), phosphorus (P) and potassium (K). These nutrients are necessary for plant growth, but in the tropics, they do not directly return to the surface because of thermocline.

Coral reefs often depend on surrounding habitats, such as seagrass meadows and mangrove forests, for nutrients. Seagrass and mangroves supply dead plants and animals which are rich in nitrogen and also serve to feed fish and animals from the reef by supplying wood and vegetation. Reefs, in turn, protect mangroves and seagrass from waves and produce sediment in which the mangroves and seagrass can root. The input of nutrients through continental run-off influences the productivity of coral reef ecosystem. Suzuki et al. (2001) estimated the riverine C:N:P molar ratio in the GBR region as 1,120:30:1, based on their carbon budget calculation and annual riverine nitrogen and phosphorous input estimated by Furnas et al. (1995). These ratios are much higher than the Redfield ratio for plankton (106:16:1; Redfield et al. 1963), and the ratio of benthic plants in coral reefs (550:30:1; Atkinson and Smith 1983). Therefore, even if all discharged nutrients are utilized by biological processes of both planktonic organisms and benthic communities, surplus carbon must remain in the reef water as oxidized *DIC*. Although nutrient inputs may stimulate organic carbon production in a reef area, net oxidation of organic matter rather than net organic carbon fixation would be expected in the GBR lagoon. A similar influence was also recognized in the Shiraho fringing reef of the Ryukyu Islands (Kawahata et al. 2000). Land-derived freshwaters, including river water and groundwater, make a relatively large contribution to the reef's circulation system. These terrestrial waters exhibit extremely high pCO_2, up to 6,400 μatm, reflecting enrichments in *AT* and *DIC* due to dissolution of carbonate rocks and decomposition of organic matter in the soil of subtropical island. These terrestrial influences may enhance carbon dioxide degassing from the coastal zone.

The variations of seawater carbon dioxide system and organic and inorganic carbon production of coral reefs were investigated by Suzuki and Kawahata (2004) with respect to topographic types and oceanographic settings. Because of dominant carbonate production in coral reef ecosystems, most coral reefs are likely to act as a net or at least a potential carbon dioxide source to the atmosphere. The comparison of the seawater carbon dioxide system parameters (pH, total alkalinity, dissolved inorganic carbon and partial pressure of carbon dioxide; pCO_2) between a reef lagoon and the surrounding ocean allowed the researchers to evaluate the system-level performance of the carbon cycle in the particular reef system. Surface pCO_2 in the lagoons of some atolls and barrier reefs in the western Pacific were consistently higher than those of their offshore waters. The alkalinity decrease in the lagoon water was attributed to calcification of reef organisms. Reef topography, especially residence time of lagoon water, affects the carbon budget of coral reefs to some extent. The offshore–lagoon differences in pCO_2 from several reefs showed a tendency to increase with the longer residence time of reef water. Another important factor controlling carbon turnover in coral reefs is proximity to land: terrestrial carbon and nutrient inputs were clearly observed in the northern Great Barrier Reef lagoon as well as a fringing reef of the Ryukyus.

6.3.3 Calcification

Corals secrete their skeleton through the calicoblastic layer, which is the outer layer of skin or ectoderm located at the underside of all polyps. This layer contains specialized cells which continuously secrete calcium (Ca^{2+}) and bicarbonate ions (HCO_3^-) to the external environment. Eventually, this leads to the deposition of a matrix of calcium carbonate ($CaCO_3$ or aragonite). Pumping these ions against a gradient is quite energy demanding, and this energy is provided by the coral's symbiotic zooxanthellae. Many species are able to grow over 1 cm (0.4 in.) a month this way. According to estimations, tropical stony corals are able to deposit about 10 kg of aragonite per m^2 per year. The entire process is directly driven by solar radiation through photosynthesis by the algae. Because of this, the process is commonly referred to as 'light-enhanced calcification'. From several experiments, it has indeed become clear that this statement is accurate. Corals deposit significantly more calcium carbonate during the day, and for the stony coral *Stylophora pistillata,* this rate is about four times in the daytime compared to the night-time.

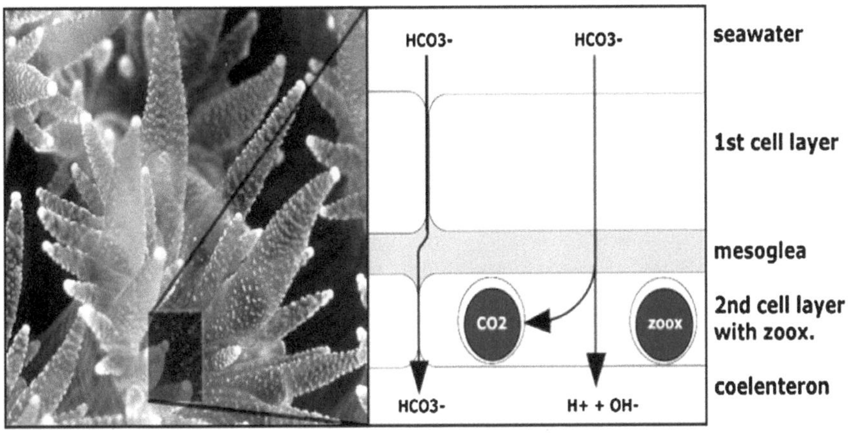

Fig. 6.22 The uptake of bicarbonate ions from seawater by the ectoderm of a coral polyp. The ions diffuse through both cellular layers and the mesoglea, after which they enter the coelenteron. A part of the ions splits into CO_2 and hydroxide ions (OH^-). A major portion of the CO_2 molecules is taken up by the zooxanthellae (ZOOX). The hydroxide ions help stabilize the pH of the coelenteron by reacting with protons (H^+ ions). *Source* Furla et al. (2000)

Building a coral skeleton is a complex process and can be divided into two major steps:

Step 1. Uptake of calcium and bicarbonate ions

The ion uptake occurs from the water column through the oral pore and through the ectoderm. Next, the ions enter the gastrovascular cavity or coelenteron. This process is diffusion dependent and therefore passive, where no energy is required (Fig. 6.22).

A part of the ions splits into carbon dioxide and hydroxide ions (OH^-). A major portion of the carbon dioxide molecules is taken up by the zooxanthellae. The hydroxide ions help stabilize the pH of the coelenteron by reacting with protons (H^+-ions) into water (H_2O).

Step 2. Growth of skeletal mass

The next step is to transport the bicarbonate ions to the so-called calicoblastic fluid; this is the stagnant layer of water located directly beneath a coral polyp, where aragonite deposition takes place. This process is energy demanding, which requires an energy carrier. This is provided by very commonly used molecule by all life on our planet, and it is called adenosine tri phosphate (ATP). ATP itself is produced in the power reactors of a living cell, the mitochondria. The calicoblastic cells in the outer skin layer are highly enriched with these cell organelles (Fig. 6.22), and work hard on a daily basis to allow corals to build their skeleton. ATP is produced by oxidizing carbohydrates and fatty acids, and the yielded energy is used to transport mainly calcium ions over the calicoblastic layer (Fig. 6.23). The needed carbohydrates are produced by the zooxanthellae and provide up to 95 % of the energy budget. The transport of bicarbonate ions takes place by the exchange of negatively charged molecules from the external environment (represented as A^-). This principle is called antiport. The pumping of calcium ions over the cellular membrane is also carried out by an antiport system; the only difference is that the calcium ions as well as the protons (H^+) have to be transported across a gradient. ATP, in the end, is the energy carrier for this process allowing the calcium/H^+-pump to perform its task.

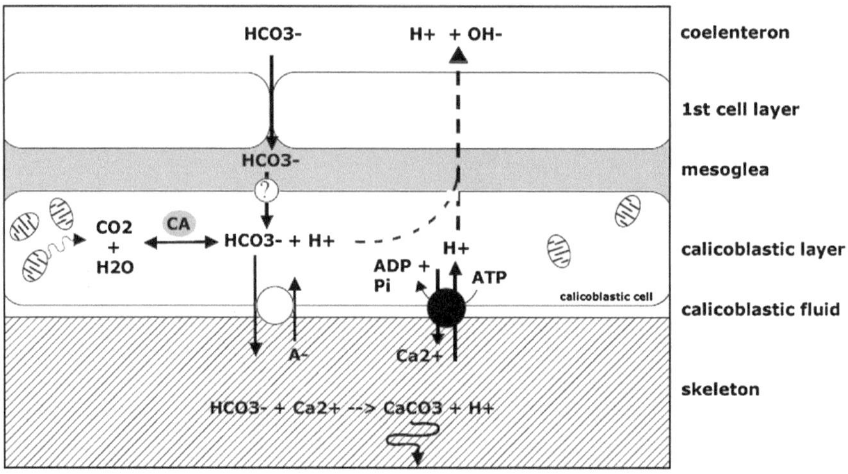

Fig. 6.23 Deposition of calcium carbonate by the outer skin layer or ectoderm at the basal plate of a coral polyp

The bicarbonate ions again diffuse through the mesoglea, although this process is not yet completely understood. The next step is, however, not passive, but active; bicarbonate ions are pumped into the calicoblastic fluid by an antiporter system which uses up negatively charged ions (A^-). Calcium ions (Ca^{2+}) are also translocated to the calcifying layer, and H^+-ions are pumped into the calicoblastic cells at the same time. As this process works against a chemical gradient, it uses up energy, which is provided by the hydrolysis of ATP to ADP and inorganic orthophosphate (Pi). Note that most bicarbonates originate from the coral cell's own metabolism; the mitochondria exhale carbon dioxide, after which the enzyme carbonic anhydrase (CA) catalyses the reaction to generate bicarbonate ions (HCO_3^-). Up to 75 % of the available bicarbonate originates from the coral itself, and not from the water column. Eventually, the bicarbonate and calcium together precipitates as calcium carbonate ($CaCO_3$). The released protons (H^+) are constantly pumped back into the coral cells to ensure a continuous high pH value in the calicoblastic layer. This remains about 9.3 during the day and drops to about 8 at night. This means that stony corals grow mostly during the day.

Eventually, the bicarbonate and calcium together precipitate as calcium carbonate ($CaCO_3$). The released protons (H^+) are constantly pumped back into the coral cells to ensure a continuous high pH value in the calicoblastic layer. This has to do with a very important chemical equilibrium in seawater:

$$CO_2 + H_2O \leftrightarrows H_2CO_3 \leftrightarrows \overleftarrow{HCO_3^- + H^+ \leftrightarrows CO_3^{2-}} + 2H^+$$

As pH levels drop, more carbonate ions are converted into bicarbonate ions. This provides more room for new carbonate ions, such as those from the coral skeleton. For this reason, corals maintain high calicoblastic fluid pH levels to prevent newly produced skeleton from redissolving. The amount of free carbonate ions is called the aragonite saturation state. During the day, calicoblastic fluid pH levels lie around 9.3 and around 8 at night. At these pH levels, calcium carbonate cannot dissolve properly and precipitates as aragonite. Without this high pH level, coral skeleton would dissolve quickly. As pH levels drop, more carbonate ions are converted into bicarbonate ions. This provides more room for new carbonate ions to dissolve, such as those from the coral skeleton. For this reason, corals maintain high calicoblastic fluid pH levels to prevent newly produced skeleton from redissolving.

6.3.4 Coral Reefs: A Source or Sink of Carbon?

Researchers have indicated the dominance of inorganic carbon in the reef carbon sequestration domain. This is consistent with the observation that the calcium carbonate percentage in reef sediments is close to 100 % (so inorganic carbon percentage is close to 12 % by mass), while the organic carbon percentage is typically ~0.5 %. These proportions imply that inorganic carbon accounts for about 95 % of carbon burial in reef sediments. Thus, if the reef area is considered to be 0.6×10^{12} m^2 and the calcium carbonate burial to be ~1,200 g m^{-2} $year^{-1}$, then the contemporary accumulation on of calcium carbonate in coral reefs is ~700 Tg $year^{-1}$. Based on mass contribution of carbon to the molecular weight of calcium carbonate (12/100), inorganic carbon burial is about 80 Tg $year^{-1}$. Recalling that calcium carbonate precipitation causes gas evasion and applying the '0.6 rule' for the ratio of precipitated inorganic carbon to evaded carbon dioxide implies that coral reefs are a source for approximately 50 Tg C $year^{-1}$. This is close to the value of 43 Tg $year^{-1}$ as observed in the model estimate derived by Kleypas (1997).

References

Atkinson MJ, Smith SV (1983) C:N:P ratios of benthic marine plants. Limnol Oceanogr 28:568–574

Balakrishnan M, Yapp CJ (2004) Flux balance models for the oxygen and carbon isotope compositions of land snail shells. Geochim Cosmochim Acta 68:2007–2024

Cameron CJ, Cameron IF, Peterson CG (1979) Contribution of organic shell matter to biomass estimates of unionid bivalves. Can J Zool 57:1666

Drury CF, Yang XM, Reynolds WD, Tan CS (2004) Influence of crop rotation and aggregate size on carbon dioxide production and denitrification. Soil Tillage Res 79:87–100

Furla P, Galgani I, Durand I, Allemand D (2000) Sources and mechanisms of inorganic carbon transport for coral calcification and photosynthesis. J Exp Biol 3445–3457(203)

Furnas MJ, Mitchell AW, Skuza M (1995) Nitrogen and phosphorous budgets for the central Great Barrier Reef Shelf. Great Barrier Reef Marine Park Authority research report no. 36, 194 pp., Great Barrier Reef Marine Park Authority, Townsville

Gattuso JP (2009) Ocean acidification. Congressional Research Service

Gazeau F, Quiblier C, Jansen JM, Gattuso JP, Middelburg JJ, Heip CHR (2007) Impact of elevated carbon dioxide on shellfish calcification. Geophys Res Lett 34

Goodfriend GA, Hood DG (1983) Carbon isotope analysis of land snail shells: implications for carbon sources and radiocarbon dating. Radiocarbon 25:810–830

Gosliner TM (1993) Biodiversity of tropical opisthobranch gastropod faunas. Proc 7th Int Coral Reef Symp 2:702–709

Haugan PM, Drange H (1996) Effects of carbon dioxide on the ocean environment. Energy Convers Manage 37:1019–1022

Hicky JP (2009) Carbon sequestration potential of shellfish. School of Natural and Built Environs, University of South Australia

Hobbs TJ, Bennell M (2005) Plant biometrics and biomass productivity in the River Murray dry land corridor. A report for the SA Centre for Natural Resource Management. Flora search series. SA water, land and biodiversity conservation, Adelaide

Jurkiewicz-Karnakowska E (2005) Pol J Environ Stud 14 (2):173–177

Kawahata H, Yukino I, Suzuki A (2000) Terrestrial influence on the Shiraho fringing reef, Ishigaki Island, Japan: high carbon input relative to phosphate. Coral Reefs 19:172–178

Kleypas JA (1997) Modeled estimates of global reef habitat and carbonate production since the last glacial maximum. Paleoceanography 12:533–545

Knutzen J (1981) Effects of decreased pH on marine organisms. Mar Pollut Bull 12:25–29

Mitra A (2013) Sensitivity of mangrove ecosystem to changing climate. Springer, Berlin. ISBN:978-81-322-1508-0 (Print) 978-81-322-1509-7 (Online)

Mitra A, Banerjee K, Sengupta K, Gangopadhyay A (2009) Pulse of climate change in Indian Sundarbans: a myth or reality? Nat Acad Sci Lett 32:1–2

Mitra A, Sengupta K, Banerjee K (2011) Standing biomass and carbon storage of above ground structures in dominant mangrove trees in the Sundarbans. Forest Ecol Manage 261:1325–1335

Mitra A, Banerjee K, Zaman S, Roy Chowdhury M, Sengupta K, Sett S (2012) Matrix effects of natural and anthropogenic factors on the soil organic carbon status in the Indian Sundarbans. J Coast Environ 3 (1):41–54

Mitra A, Raha AK, Ghosh R, Banerjee K (2013) A baseline study of temporal and spatial variations of physico-chemical variables in Gangetic delta. In: Rao DV, Kosygin L, Dash S (eds) Estuaries of India. Nature Books, India, pp 315–328

Mitra A, Zaman S, Pramanick P, Bhattacharyya SB, Raha AK (2014) Stored Carbon in Dominant Seaweeds of Indian Sundarbans Peternika. J Trop Agri Sci 37 (2):263–274

Paul EA, Morris SJ, Six J, Paustian K, Gregorich EG (2003) Interpretation of soil carbon and nitrogen dynamics in agricultural and afforested soils. Soil Sci Soc Am J 67:1620–1628

Redfield AC, Ketchum BH, Richards FA (1963) The influence of organisms on the composition of sea-water. In: Hill MN (ed) The sea, vol 2. Wiley, New York, pp 26–77

Romaniello L, Quarta G, Mastronuzzi G, D'Elia M, Calcagnile L (2008) 14C age anomalies in modern land snails shell carbonate from Southern Italy. Quat Geochronol 3:68–75

Sale PF (1980) The ecology of fishes on coral reefs. Oceanogr Mar Biol Annu Rev 18:367–421

Sale PF, Guy JA, Steel WJ (1994) Ecological structure of assemblages of coral reef fishes on isolated patch reefs. Oecologia 98:83–99

Sebens KP (1994) Biodiversity of coral reefs: what are we losing and why? Am Zool 34:115–133

Sengupta K, Roy Chowdhury M, Bhattacharya SB, Raha A, Zaman S, Mitra A (2013) Spatial variation of stored carbon in *Avicennia alba* of Indian Sundarbans. Discov Nat 3(8):19–24

Smith SV, Jokiel PL (1978) Water composition and biogeochemical gradients in the Canton Atoll lagoon. Atoll Res Bull 221:15–53

Sorokin YI (1993) Coral reef ecology. Springer, Berlin, p 465

Stewart RR (2005) Oceanography in the 21st Century. Department of Oceanography, Texas A&M University.unfccc.int. 2009. Kyoto Protocol

Stott LD (2002) The influence of diet on the ^{d13}C of shell carbon in the pulmonate snail *Helix aspersa*. Earth Planet Sci Lett 195:249–259

Suzuki A, Kawahata H (2004) Reef water CO_2 system and carbon production of coral reefs: topographic control of system-level performance. In: Shiyomi M et al. (eds) Global environmental change in the ocean and on land. pp 229–248. Copyright by TERRAPUB

Suzuki A, Kawahata H, Ayukai T, Goto K (2001) The oceanic CO_2 system and carbon budget in the Great Barrier Reef, Australia. Geophys Res Lett 28: 1243–1246

Wilbur KM (1972) Shell formation in molluscs. In: Florkin M, Scheer BT (eds) Chemical zoology, mollusca, vol 7. Academic Press, New York, pp 103–145

Internet References

www.clw.csiro.au/staff/hairsinep/ffplw19052004.Pdf

http://science.nasa.gov/about-us/science-strategy/past-strategydocuments/earth-science-enterprise-plans/earth-science-focus-area-roadmaps-2005/carbon-cycle-ecosystems/)

7 Threats to Blue Carbon Reservoir

...The blue carbon is under threat
Let us seek the conservation gate...
The Authors

The reservoir of blue carbon is perhaps the most valuable of all natural assets as it is not only the source of several bioactive molecules to run the wheel of human civilization smoothly, but also is the source of future fuel and energy. Several ecosystem services of blue carbon are still under the carpet. How many of us know that silver nanoparticles can even be synthesized from the extracts of seaweeds such as *Sargassum plagiophyllum*, *Ulva reticulata* and *Enteromorpha compressa*?

The blue carbon community such as mangroves acts as the primary line of defence against natural calamities such as tidal surges, cyclone and wave actions. Some species of salt marsh grass (such as *Porteresia coarctata*) stabilize a newly formed island in the estuaries or river mouth. Threats to such a super-priced natural vault and its subsequent destruction may pose a negative effect on mankind, if not proper care and measures are taken in right time with high precision.

It is difficult to segregate the natural- and human-induced threats on coastal ecosystems as both of them go hand in hand. Often the threats are synergistic in nature. The example of deterioration of mangrove vegetation is very relevant in this context. The cumulative effects of natural and anthropogenic pressures make mangrove wetlands one of the most threatened natural communities worldwide. Roughly, 50 % of the global area has been lost since 1900 and 35 % of the global area has been lost in the past two decades, primarily due to human activities such as conversion for aquaculture (IUCN 1989; Secretariat 2001; Valiela et al. 2001). Between 56 and 75 % of different Asian mangroves have been lost during the twentieth century. This is primarily due to overuse and conversion for aquaculture (Primavera 1997; Smith et al. 2001). There are roughly 17 million ha of mangroves worldwide (Valiela et al. 2001; FAO 2003). The global average annual rate of mangrove loss is about 2.1 %, exceeding the rate of loss of tropical rainforests (0.8 %) (Valiela et al. 2001; Wells 2006). In some regions, natural factors such as massive siltation and rise of salinity (resulting from tectonic shift) have also reduced the growth of mangroves as seen in central sector of Indian Sundarbans (Mitra et al. 2011; Sengupta et al. 2013; Banerjee et al. 2013). The mangrove trees in the central Indian Sundarbans have become considerably stunted owing to hypersalinity.

The blue carbon reservoir also experiences human-induced threats such as cutting of coastal vegetation for fuel and fodder, absence of natural regeneration, poor growth and survival due to decreased freshwater flow (which may be due to construction of dam in the upstream zone) and clearance of large tracts of mangrove forests for development activities. It has already been stated that the domain of threats on blue carbon community is full of 'noise' and it is extremely difficult to isolate the natural causes from the artificial causes from this noisy matrix. However, for the convenience of the readers, we have attempted to segregate the natural threats from the anthropogenic influences.

A. Mitra and S. Zaman, *Blue Carbon Reservoir of the Blue Planet*,
DOI 10.1007/978-81-322-2107-4_7, © Springer India 2015

7.1 Natural Threats

The natural factors that affect the process of erosion and accretion of habitats that sustain coastal vegetation as a whole may be broadly divided into five heads.

A. Changes in the Earth's orbit
B. Alteration of atmospheric carbon dioxide
C. Volcanic activities
D. Variations in solar output
E. Plate tectonics
F. Natural disasters and extreme weather events.

A. Changes in the Earth's Orbit

How it is possible that variation in the Earth's orbit regulates the distribution and growth of coastal vegetation? This is a general question that will peep out from the mind of readers. The answer is simple. The variation in the Earth's orbit, which affects the intensity of solar energy reaching the surface of the Earth, regulates the vegetation pattern. It has been documented that regions with considerable solar energy coincide with higher floral diversity.

There are several factors that regulate the alteration of the Earth's orbit. The first cyclical variation, known as eccentricity, controls the shape of the Earth's orbit around the Sun. The orbit gradually changes from being elliptical to being nearly circular and then back to elliptical in a period of about 100,000 years. The greater the eccentricity of the orbit (i.e. the more elliptical it is), the greater is the variation in solar energy received at the top of the atmosphere between the Earth's closest (perihelion) and farthest (aphelion) approach to the Sun. Currently, the Earth is experiencing a period of low eccentricity. The difference in the Earth's distance from the Sun between perihelion and aphelion (which is only about 3 %) is responsible for approximately a 7 % variation in the amount of solar energy received at the top of the atmosphere. When the difference in this distance is at its maximum (9 %), the difference in solar energy received is about 20 %.

The second cyclical variation results from the fact that, as the Earth rotates on its polar axis, it wobbles like a spinning top changing the orbital timing of the equinoxes and solstices (Fig. 7.1). This effect is known as the precession of the equinox. The precession of the equinox has a cycle of approximately 26,000 years. According to illustration (A), the Earth is closer to the Sun in January (perihelion) and farther away in July (aphelion) at the present time. Because of precession, the reverse will be true in 13,000 years and the Earth will then be closer to the Sun in July. This means, of course, that if everything else remains constant, 13,000 years from now seasonal variations in the Northern Hemisphere should be greater than at present (colder winters and warmer summers) because of the closer proximity of the Earth to the Sun.

The third cyclical variation is related to the changes in the tilt (obliquity) of the Earth's axis of rotation over a 41,000 years period. During the 41,000 year cycle, the tilt can deviate from approximately 22.5–24.5°. At the present time, the tilt of the Earth's axis is 23.5°. When the tilt is small, there is less climatic variation between the summer and winter seasons in the middle and high latitudes. Winters tend to be milder and summers cooler. Warmer winters allow for more snow to fall in the high-latitude regions. When the atmosphere is warmer, it has a greater ability to hold water vapour and therefore more snow is produced at areas of frontal or orographic uplift. Cooler summers cause snow and ice to accumulate on the Earth's surface because less of this frozen water is melted. Thus, the net effect of a smaller tilt would be more extensive formation of glaciers in the polar latitudes.

Periods of a larger tilt result in greater seasonal climatic variation in the middle and high latitudes. At these times, winters tend to be colder and summers are warmer. Colder winters produce more snow because of lower atmospheric temperatures. As a result, more snow and ice accumulates on the ground surface. Moreover, the warmer summers produced by the larger tilt provide additional energy to melt and evaporate the snow that fall and accumulate during the winter months. In conclusion, glaciers in the Polar Regions should be generally receding, with other contributing factors constant, during this part of the obliquity cycle.

Fig. 7.1 Modification of the timing of aphelion and perihelion over time

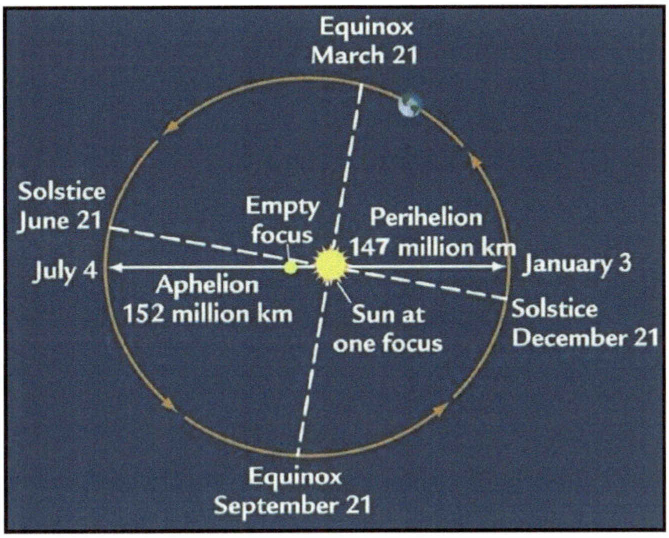

The alteration of temperature and incoming solar radiation has profound influence on the distribution pattern and species diversity of coastal vegetation. Increased surface temperature is expected to affect mangroves (Field 1995; Ellison 2000) by:

1. changing species composition;
2. changing phenological patterns (e.g. timing of flowering and fruiting);
3. increasing mangrove productivity where temperature does not exceed an upper threshold; and
4. expanding mangrove ranges to higher latitudes where range is limited by temperature, but is not limited by other factors, including a supply of propagules and suitable physiographic conditions.

Mangroves reach a latitudinal limit at the 16 °C isotherm for air temperature of the coldest month, and the margins of incidence of ground frost, where water temperatures do not exceed 24 °C (Ellison 2000). The optimum mangrove leaf temperature for photosynthesis is believed to be between 28 and 32 °C, while photosynthesis ceases when leaf temperatures reach 38–40 °C (Clough et al. 1982; Andrews et al. 1984). The frequency, duration and intensity of extreme cold events have been hypothesized to explain the current latitudinal limits of mangrove distribution (Woodroffe and Grindrod 1991; Snedaker 1995). However, the incidence of extreme cold events is not likely to be a factor limiting mangrove expansion to higher latitudes in response to increased surface temperature. The Intergovernmental Panel on Climate Change projects reduced extreme cold events (Solomon et al. 2007), in correlation with projected changes in average surface temperatures. For instance, Vavrus et al. (2006) predicted a 50–100 % decline in the frequency of extreme cold air events in Northern Hemisphere winter in most areas, while Meehl et al. (2004) projected decreases in frost days in the extratropics, where the pattern of decreases will be determined by changes in atmospheric circulation.

The effect of temperature fluctuation due to variation in the Earth's orbit is explained here with mangroves, but for all the species under the domain of blue carbon the impact is considerable. The reproductive biology of seagrass species has attracted naturalists for about two centuries. Flowering of seagrasses is primarily controlled by temperature and often occurs simultaneously across large spatial scales. European seagrass species flower in late spring, and some of them (*Zostera* spp.) throughout the summer as well, when irradiance improves and water temperature increases, except for the Mediterranean species *Posidonia oceanica*, which flowers in the fall (October).

The diversity and abundance of phytoplankton that serve as the storehouse carbon in the aquatic system is also regulated by solar radiation. In estuarine system and deltaic region, where the suspended solid load is usually high, the phytoplankton community exhibits a poor standing stock. This is primarily because the transparency of the water and the intensity of solar radiation are inhibited by suspended solids of the aquatic system.

B. Alteration of Atmospheric Carbon Dioxide

Studies on long-term climate change have revealed a connection between the concentrations of carbon dioxide in the atmosphere and mean global temperature. Carbon dioxide is one of the most important gases responsible for the greenhouse effect. Certain atmospheric gases, such as carbon dioxide, water vapour and methane, are able to alter the energy balance of the Earth by their property to absorb long wave radiation emitted from the Earth's surface. The net result of this process is the rise of Earth's temperature. Without the greenhouse effect, the average global temperature of the Earth would be around −18 °C rather than the present 15 °C.

Research works on climate change have confirmed the findings that the temperature variations are closely correlated with the concentration of carbon dioxide in the atmosphere and variations in solar radiation received by the planet as controlled by the Milankovitch cycles. Measurements indicated that atmospheric carbon dioxide levels were about 30 % lower during colder glacial periods. It was also theorized that the oceans were a major storehouse of carbon dioxide and that they controlled the movement of this gas to and from the atmosphere depending on the temperature. Carbon dioxide is released from the oceans when global temperatures become warmer and diffuses into the ocean when temperatures are cooler. Initial alterations in global temperature were triggered by changes in received solar radiation by the Earth through the Milankovitch cycles. The increase in carbon dioxide then amplified the global warming by enhancing the greenhouse effect. The direct relationship between carbon dioxide concentration and temperature is shown in Fig. 7.2.

The effect of enhanced carbon dioxide on blue carbon community is poorly understood and sometimes contradictory and there is a paucity of research in this area. A direct effect of elevated atmospheric carbon dioxide levels may be increased productivity of some mangrove species (Field 1995; Ball et al. 1997; Komiyama et al. 2007). Mangrove metabolic responses to increased atmospheric carbon dioxide levels are likely to be increased growth rates (Farnsworth et al. 1996) and more efficient regulation of water loss (UNEP 1994). For some mangrove species, the response to elevated carbon dioxide may be sufficient to induce substantial change of vegetation along natural salinity and aridity gradients. Ball et al. (1997) showed that doubled carbon dioxide had little effect on mangrove growth rates in hypersaline areas, and this may combine with reduced rainfall to create some stress. The greatest effect may be under low salinity conditions. Elevated carbon dioxide conditions may enhance the growth of mangroves when carbon gain is limited by evaporative demand at the leaves, but not when it is limited by salinity at the roots. There is no evidence that elevated carbon dioxide will increase the range of salinities in which mangrove species can grow.

In red seaweed, *Gracilaria lemaneiformis,* the growth rate is usually higher when exposed to carbon dioxide-enriched air compared to non-enriched air under intermediate irradiance.

As authors we feel to extend the domain of blue carbon across the boundary of coastal vegetation to shelled organisms thriving in the oceans, estuaries and bays as they are important drivers in regulating the carbon dioxide level both in the oceanic and atmospheric compartments (*Vide* Chap. 6 for detailed mechanism and role of molluscs in storing/sequestering carbon). The increase of atmospheric carbon dioxide and its subsequent dissolution in ocean and estuarine waters is, however, not favourable for molluscan communities as ocean acidification may dissolve their shell. The calcification rates of the edible mussel (*Mytilus edulis*) and Pacific oyster (*Crassostrea gigas*) decline linearly with increasing carbon dioxide levels (Gazeau et al. 2007). Squids are especially sensitive to ocean

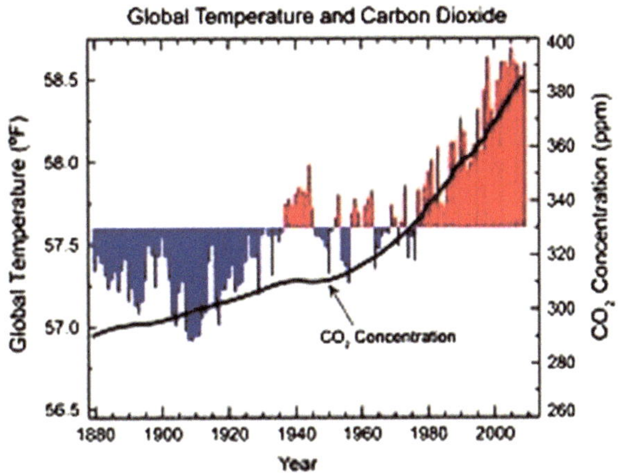

Fig. 7.2 Global average temperature and carbon dioxide concentrations

acidification because it directly impacts their blood oxygen transport and respiration (Portner et al. 2005). Sea urchins raised in lower-pH waters show evidence for inhibited growth due to their inability to maintain internal acid-base balance (Kurihara and Shirayama 2004). Scientists have also seen a reduced ability of marine algae and animals to produce protective carbonate shells (Gattuso et al. 1998; Langdon et al. 2000; Riebesell et al. 2000). These organisms are important food sources for many marine species. One such example is a pteropod, a free-swimming molluscan species that is eaten by organisms ranging in size from tiny zooplankton to whales. In particular, pteropods are an important source for North Pacific juvenile salmon. Mackerel, Pollock, herring and cod are also known to feed on pteropods. Many species of other marine calcifiers, such as coccolithophores (microscopic algae), foraminifera (microscopic protozoans), coralline algae (benthic algae), echinoderms (sea urchins and starfish) and molluscs (snails, clams and squid) also exhibit a general decline in their ability to produce shells with decreasing pH (Kleypas et al. 2006). A very recent study conducted by the present authors exhibits significant inverse relationship between aquatic pH and shell weight of *Saccostrea cucullata* in Indian Sundarbans estuarine water. The detailed study along with the results and interpretations are discussed in Chap. 6. Thus, in conclusion, it can be advocated that increase of atmospheric carbon dioxide may one-way cause a positive impact on coastal vegetation, but on the other hand may pose an adverse impact on shell forming organisms by triggering the process of ocean acidification.

C. Volcanic Activities

Volcanic eruptions are major natural disturbances with varied and complex consequences. Apart from generation of ash particles, volcanic activities also alter the surrounding water chemistry to a great extent. This can cause considerable impact on coastal vegetation and more preferably on phytoplankton. The volcanic activity of Barren Island is found to have considerable impact in chemical wet depositions occurred due to rainwater at Port Blair. It also appears to be a causative factor for higher sulphate and nitrate and the lower-pH values of the ambient aquatic phase. After the first eruption noticed in April 1991 began with hot gases and strong ash emissions with activity continuing through October, another eruption was noticed in December 1994 and January–March 1995, where thick clouds of pale brownish gas, dark ash particles and white steam were observed. Very high values of sulphate ion concentrations in rainwater were observed from January to April in 1995 when predominant winds are northeasterly. The volcanic emission of smoke was also reported during January 2000. The pH values showed a decreasing sulphate and nitrate concentrations exhibited an increasing tendency from the year 1995 onwards. As the volcanic eruption at Barren Island during 1991 almost

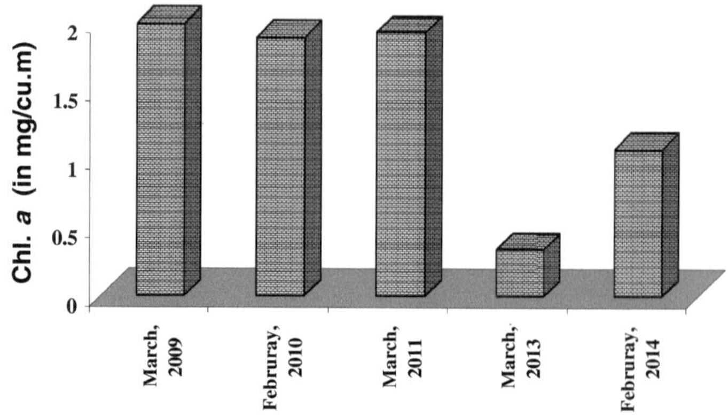

Fig. 7.3 A sudden drop in chlorophyll *a* (expressed in m gm^{-3}) level during March 2013—one month after the volcanic episode

coincided with major volcanic eruption of Mt. Pinatubo in June 1991 in Philippines and Mt. Pinatubo's eruption was of very large magnitude, it is very difficult to differentiate individual contributions during 1991 episode, and hence, it may be considered as a combined effect of both the eruptions. However, such eruptions cause considerable decrease of phytoplankton standing stock. The chlorophyll concentration also reduces drastically. However, very few studies reflect the cause–effect relationship between volcanic eruptions and coastal vegetation.

The present authors observed a significant decrease in chlorophyll *a* (Fig. 7.3) and phytoplankton count (Fig. 7.4) in the aquatic phase around Barren Island after the volcanic episode of 16 February 2013.

Volcanically active islands abound in the tropical Pacific and harbour complex coral communities. Researchers have observed severe adverse effect of deep volcanic ash deposits that often devastate coral communities through burial and smothering. However, little is known about the effect of moderate amounts of small particulate ash deposits on reef communities. Volcanic ash contains a diversity of chemical compounds that can induce nutrient enrichments triggering changes in benthic composition. Two independently collected data sets on the marine benthos of the pristine and remote reefs around Pagan Island, Northern Mariana Islands, reveal a sudden critical transition to cyanobacteria-dominated communities in 2009–2010, which coincides with a period of continuous volcanic ash eruptions.

Pagan is the largest island of the northern Mariana Arc (48 km^2) and the fifth largest of all Mariana Islands. The largest eruption of Pagan in recent history occurred in 1981 and instigated the evacuation of its fifty-some permanent residents. Since then, few temporary residents have lived on the island with lately just one or two official representatives of the mayor of the Northern Mariana Islands. The strong volcanic activity in 1981 was followed by intermittent light

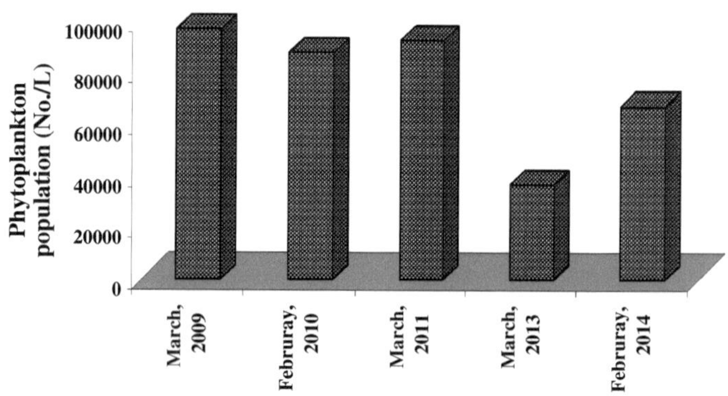

Fig. 7.4 A sudden drop in phytoplankton population (expressed in No./L) during March 2013—one month after the volcanic episode

Fig. 7.5 Emission of gas and ash at Pagan Island

ejection of primarily phreatic ash until 1996. From 1996 onwards no emissions were documented until early December 2006 when ash and gas plumes were reported by local residents and were observed in satellite imagery. This stint of eruptive activity ended later that month, and the volcano remained quiet until mid-April 2009. Although ground-based geophysical instruments are lacking on Pagan, USGS records show that volcanic activity increased in 2009 and frequent observations of gas and ash emissions have been documented up to December 2010 (Fig. 7.5).

Pagan's volcanic rocks mainly consist of silicon dioxide (SiO_2, 50.5 %), aluminium oxide (Al_2O_3, 17.4 %), and iron oxides (Fe_2O_3 and FeO, 11.2 %). The latter compounds are the most likely candidates to stimulate cyanobacterial growth as iron requirements of cyanobacteria are greater than other algal groups because of their high PS I to PS II ratio and the presence of nitrogenase in diazotrophs. The iron limitation of cyanobacteria in oligotrophic coral habitats has been attributed to their evolutionary origin in an anoxic ocean with high bioavailability of iron. When volcanic ash dissolves in sea water, bioactive trace metals such as iron become immediately available to primary producers and experimental studies have demonstrated that this increase in iron concentration can initiate cyanobacterial blooms. Although a great deal of progress has been made in recent years, there are still significant gaps in our understanding which hamper our ability to predict the outcomes of volcanic activities on blue carbon domain. Some of the most important areas in which we lack a general or even basic understanding include (i) the importance of rates, timing, magnitude and duration of environmental change associated with volcanic eruption, (ii) non-additive effects of multiple stressors, (iii) population-level implications of variable environmental impacts among life-history stages, (iv) the scope for population- or species-level adaptation to environmental change caused by volcanic activities and (v) ecological responses at the level of communities and ecosystems, including tipping points and sudden phase shifts. With regard to uncertainties in the nature of environmental forcing, we require additional eco-physiological and eco-mechanical studies—especially ones that move beyond single-factor ANOVA designs—and further development in the emerging field of ecological genomics to identify biological responses to key environmental drivers or combinations of drivers that are associated with volcanic activities.

D. Variations in Solar Output

Several factors and parameters are responsible for variations in solar output that cause temperature difference in the planet Earth. The presence/absence and number of sunspots is one of the important reasons that regulate the magnitude of solar radiation.

Sunspots have magnetic fields of strength up to 3,000 times as great as the average magnetic field of either the Sun or the Earth. Astronomers believe the cause of sunspots is attributed to this fact. According to a standard explanation, the strong magnetic fields of the Sun have the shape of tubes just below the solar surface at the beginning of the sunspot cycle. These tubes lie perpendicular to the Sun's equator. The Sun rotates faster at its equator than at its poles, and so the tubes are stretched out in the east–west direction. Kinks then develop in the magnetic tubes and push through the solar surface. A pair of sunspots appears wherever a kink penetrates, because the kink both leaves and re-enters the surface. The number and size of sunspots show cyclical patterns, reaching a maximum about every 11, 90 and 180 years. The decrease in solar energy observed in the early 1980s corresponds to a period of maximum sunspot activity based on the 11-year cycle. In addition, measurements made with a solar telescope from 1976 to 1980 showed that during this period, as the number and size of sunspots increased, the Sun's surface cooled by about 6 °C. Apparently, the sunspots prevented some of the Sun's energy from leaving its surface. However, these findings tend to contradict observations made on longer timescales.

During periods of maximum sunspot activity, the Sun's magnetic field is strong. When sunspot activity is low, the Sun's magnetic field weakens. The magnetic field of the Sun also reverses every 22 years, during a sunspot minimum. Some researchers believe that the periodic droughts on the Great Plains of the USA are in someway correlated with this 22-year cycle.

Temperature, a direct function of solar output, determines the performance of coastal vegetation, and indeed all organisms, at the fundamental levels of enzymatic processes and metabolic function. Seaweeds have evolved biochemical and physiological adaptations, including variation in the identity and concentration of proteins and the properties of cell membranes, that enable them to optimize their performance with respect to the temperatures they encounter (Eggert 2012). Although seaweeds are generally well adapted to their thermal environment, they nevertheless experience temperatures in nature—particularly during periods of environmental change—that are sufficiently high or low to result in disruptive stress in the form of cellular and subcellular damage (reviewed in Davison and Pearson 1996). Such damage and any reallocation of resources for protection and repair can slow growth, delay development and lead to mortality (Davison and Pearson 1996). In response, seaweeds can produce heat shock proteins that repair or remove damaged proteins (e.g. Vayda and Yuan 1994; Lewis et al. 2001). However, protein thermal physiology is not well understood in macroalgae and the up regulation of heat shock protein production is only one of many transcriptional changes that occur in seaweeds during periods of thermal stress (Collen et al. 2007; Kim et al. 2011). Relevant genomic, transcriptomic and proteomic studies are only just beginning to scratch the surface and most links from gene expression to organismal performance are far from well established. As a result of non-stressful conditions at intermediate temperatures and stress at the extremes, the relationship between temperature and most subcellular, tissue-level, or whole-organism processes is described by a hump-shaped thermal performance curve. From colder to warmer, these curves generally rise exponentially as rates of biochemical reactions increase, peak at some optimum temperature, and then fall rapidly as the biological components of the system become less efficient or damaged (Kordas et al. 2011). When properly parameterized across the full-temperature tolerance range of a species, thermal performance curves have the potential to predict the physiological effects of any given warming or cooling scenario (barring any further acclimatization, adaptation or context-dependent surprises; see below). The effect of a small increase in thallus temperature will be beneficial when the initial temperature is cooler than optimal and detrimental when it is warmer than optimal, and the precise change in performance can be predicted from the starting and ending temperature values along the curve. Unfortunately, the shapes of thermal performance curves and the positions of their optima are poorly described in most seaweed. Although many physiological and ecological studies have linked seaweed performance to temperature, a substantial fraction of these studies do not investigate enough temperatures across a wide enough range to characterize the underlying, nonlinear relationship between the two. Furthermore, various physiological parameters within an organism differ in the shape and optimum temperature of their thermal performance curves, which limits our ability to use an easily measured parameter (e.g. photosynthesis) as a proxy for parameters that may be more ecologically relevant (e.g. growth and reproduction). Indeed, growth rates do tend to peak at lower temperatures than photosynthetic rates, presumably because metabolic rates increase faster than photosynthetic rates at higher temperatures. Much remains to be learned regarding the thermal dependence of the key physiological processes that control growth, reproduction and survival across the full range of temperatures experienced by an individual in its lifetime.

A study conducted by Indian researchers exhibits that the biochemical composition of a red seaweed species in lower Gangetic delta region is greatly influenced by temperature, particularly the carbohydrate content exhibited significant positive correlation with ambient aquatic temperature (Table 7.1). This table indicates that salinity is a vital parameter regulating the biochemical composition of coastal vegetation. High salinity is a major threat to the protein level of the seaweed. The carbohydrate and astaxanthin levels are, however, positively correlated with salinity proving that under salinity stress the seaweeds survive by increasing the astaxanthin content of the thallus. This is a major biochemical adaptation in context to salinity stress that is quite common in the coastal regions.

E. Plate Tectonics

The movement of the Earth's crust is the basis of the theory of plate tectonics. In this theory, the lithosphere is viewed as a series of rigid plates that are separated by the earthquake belts of the world, that is, the trenches, ridges and faults. There are seven major lithospheric plates: the Pacific, Eurasian, African, Australian, North American, South American and Antarctic plates. Each plate is composed of continental and/or oceanic crust.

At the mid-oceanic ridges, where plate boundaries move apart as new lithosphere is formed, divergent plate boundaries occur. Convergent plate boundaries occur at trenches, where plates move towards each other and old lithosphere is destroyed. The plates move past each other at regions known as faults, which represent breaks in the Earth's crust where one plate can move past the other.

A transform fault is a special kind of fault that is found in sections of the mid-ocean ridge. Each side of a transform fault is formed by a different plate, and these plates are moving away from each other in opposite directions. The fault zone produced by this movement is quite active and is the site of frequent earthquakes. The motion of the plates along these faults produces a nearly continuous line of cliffs with sharp vertical drops, known as escarpments. In these regions, there are sudden changes in the ocean depth. Regions where the lithosphere splits, separates and moves apart as new crust is formed are called rift zones. The mid-ocean ridge and rise systems represent the major rift zones at this time. It is generally thought that rifting occurs when rising magma causes enough tension to stretch the overlying crust, creating a sunken rift zone. As this process continues, the rift spreads and the fault deepens and cracks, allowing the magma to seep through and eventually form a ridge. When this happens to the lithosphere under the continents, a sunken rift zone can occur. As the process continues, the fault gets thinner and deeper and can eventually fill with sea water. The Red Sea is an example of this type of formation.

As a plate moves away from the rift zone, it cools and thickens. At the rift, thinning of the crustal plate and increased flow of magma into

Table 7.1 Correlation coefficients between environmental variables and biochemical composition of *Catenella repens* (a red seaweed) at three collection sites from Indian Sundarbans

	Biomass	Protein	Carbohydrate	Lipid	Astaxanthin
Site name: GOSABA					
Salinity	0.147	−0.950*	0.490**	−0.775*	0.944*
pH	0.728*	−0.754*	0.161	−0.393	0.777*
Dissolved oxygen	−0.881*	0.794*	0.058	0.540**	−0.860*
Temperature	−0.465**	−0.464**	0.798*	−0.498**	0.428
Extinction coefficient	−0.603*	0.940*	−0.340	0.760*	−0.940*
Nitrate	−0.484**	0.908*	−0.437	0.796*	−0.891*
Phosphate	−0.695*	0.933*	−0.270	0.740*	−0.947*
Silicate	0.760*	0.915*	−0.217	0.656*	−0.938*
Chl *a*	0.696*	−0.930*	0.257	−0.683*	0.961*
Chl *b*	0.698*	−0.934*	0.283	−0.721*	0.950*
Chl *c*	0.518**	−0.882*	0.432	−0.822*	0.836*
Biomass		−0.504**	−0.394	−0.309	0.557*
Protein			−0.440	0.805*	−0.883*
Carbohydrate				−0.484**	0.262
Lipid					−0.735*
Site name: CHOTOMOLLAKHALI					
Salinity	0.226	−0.860*	0.597*	−0.777*	0.770*
pH	0.739*	−0.596*	0.149	−0.603*	0.719*
Dissolved oxygen	−0.966*	0.319	0.392	0.445**	−0.566*
Temperature	−0.309	−0.497**	0.830*	−0.343	0.297
Extinction coefficient	−0.586*	0.889*	−0.347	0.886*	−0.938*
Nitrate	−0.549**	0.885*	−0.371	0.871*	−0.918*
Phosphate	−0.519**	0.900*	−0.366	0.892*	−0.912*
Silicate	−0.871*	0.736*	0.230	0.843*	−0.876*
Chl *a*	0.214	−0.867*	0.594*	−0.779*	0.763*
Chl *b*	0.700*	−0.924*	0.207	−0.924*	0.960*
Chl *c*	0.883*	−0.892*	0.044	−0.892*	0.971*
Biomass		−0.631*	−0.374	−0.631*	0.803*
Protein			−0.243	0.887*	−0.741*
Carbohydrate				0.022	0.098
Lipid					−0.897*
Site name: BALI					
Salinity	0.654*	−0.920*	0.861*	−0.812*	0.884*
pH	0.188	−0.680*	0.418	−0.913*	0.602*
Dissolved oxygen	−0.102	0.073	−0.122	0.022	−0.147
Temperature	−0.362	−0.230	0.657*	−0.587*	0.450**
Extinction coefficient	−0.449	0.851*	−0.933*	0.877*	−0.901*
Nitrate	−0.744*	0.944*	−0.778*	0.781*	−0.867*
Phosphate	−0.476**	0.787*	−0.937*	0.831*	−0.925*
Silicate	−0.523**	0.865*	−0.653*	0.746*	−0.720*
Chl *a*	0.665*	−0.928*	0.785*	−0.793*	0.836*
Chl *b*	0.664*	−0.888*	0.822*	−0.750*	0.851*

(continued)

Table 7.1 (continued)

	Biomass	Protein	Carbohydrate	Lipid	Astaxanthin
Chl *c*	0.465**	−0.766*	0.850*	−0.855*	0.909*
Biomass		−0.670*	0.417	−0.250	0.580*
Protein			−0.686*	0.811*	−0.746*
Carbohydrate				−0.681*	0.891*
Lipid					−0.743*

$*p < 0.01$; $**p < 0.05$

the rift causes the land masses to separate. A low-lying region of oceanic basalt is formed as well.

Coastal drainages and strandlines are sensitive to land-level changes by tectonic movements and deformations. In the case of slow deformation as during the inter-seismic intervals, the coastlines and the drainage systems adjust slowly to the changes. The adjustments, however, may be abrupt in response to the fast, almost instantaneous (in geological perspective) coseismic land-level changes associated with morphogenic earthquakes (i.e. the earthquakes that produce recognizable surface deformation). Coseismic subsidence induces landward shift of the coastline resulting in drowning of the coasts and consequent distress, whereas uplift induces seaward migration of the coastline, exposing stretches of the seafloor. Moreover, the land-level changes lead to a change in the configuration of the intertidal zones (tidal swamps/mudflats/wetlands/estuaries), which may also shrink (Fig. 7.6) or expand due to ground uplift and subsidence, respectively.

The case study of North Andaman is very relevant in this context where the dimension of blue carbon reservoir has undergone a change due to tectonic movement. The western coastline of North Andaman is indented by several tidal marshes/swamps, bays, estuaries and mudflats. Vast stretches of intertidal wetlands covered by luxuriant mangrove forests (i.e. mangrove swamps) and traversed by a network of tidal creeks occur along the western coast.

During the event of Tsunami on 26 December 2004, the estimated 0.3–1.5 m uplift and its easterly tilt in the North Andaman Island

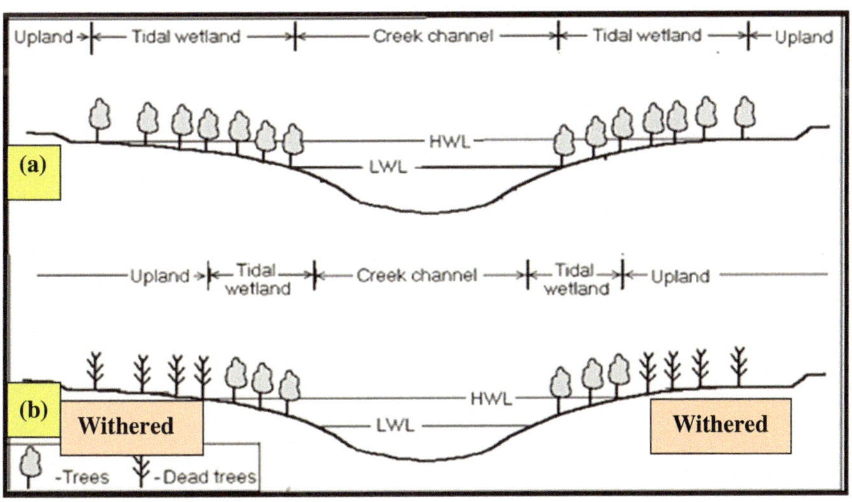

Fig. 7.6 Rise and fall of relative sea level due to ground uplift; **a** ground uplift and consequent fall of sea level **b** distribution after the earthquake and seismic uplift. The plants in the intertidal zone that have been shifted above the high-tide level slowly wither away because of unavailability of sea water

have caused many changes in the coastal geomorphic pattern. The strandline has moved seaward exposing vast stretches of seabed and the fringing coral reefs. Upper reaches of the tidal creeks and streams that flow from the hilly interior of the island to the sea have dried up, leaving dry streambeds with stagnant pools of water. The boundary of the tidal zones has migrated towards the central drainage channels, converting the upper intertidal swamps to subaerial coastal uplands (Fig. 7.7). It was documented that after 2 years, the vegetation of the supra-tidal zone died and there was mass mortality in blue carbon reservoir due to non-availability of tidal water.

There are also instances of complete blockage of freshwater supply due to plate tilting that not only caused massive alteration of coastal and estuarine floral composition, but also resulted in stunted growth of some mangrove species. Neo-tectonic movements in the Bengal Basin between the twelfth and fifteenth century AD resulted in an easterly tilt of the deltaic complex (Chaudhuri and Choudhury 1994). During the sixteenth century, the River Ganga changed its course to shift eastwards and join the Brahmaputra (Deb 1956; Blasco 1975; Snedaker 1991). Later, in the mid-eighteenth century, the combined Ganga (now called Padma) and Brahmaputra again tilted eastwards to empty into the River Meghna (Snedaker 1991). This continuing tectonic activity greatly influenced the hydrology of the deltaic region because of changes in the sedimentation patterns and the reduction in freshwater inflows. Most rivers (distributaries) other than the Hooghly, that contributed to the formation of the Ganga Delta (from west to east: Muriganga, Saptamukhi, Thakuran, Matla, Gosaba and Bidya), have lost original connections with the Ganga because of siltation, and their estuarine character is now maintained by the monsoonal run-off (Cole and Vidyaraman 1966) and tidal actions (Mitra et al. 2009, 2011). The aquatic salinity of these rivers has increased that resulted in the stunted growth of the mangroves. Aquatic salinity has profound influence on the growth and biomass of mangrove floral species. The degree of influence is, however, extremely species-specific depending on the mode of salt regulation. We monitored the aboveground stem biomass (AGSB) in the same set of species in two stations selected in western and central Indian Sundarbans (marked in red dots in Fig. 7.8). Ten experimental plots (of dimension 10 m × 10 m) were selected at random in each of these stations. The species-wise AGSB of individual trees was estimated using non-destructive method in which the diameter at the breast height (DBH) was measured with a measuring tape and height with laser beam (**BOSCH DLE 70 Professional**). Form factor was determined as per the expression outlined by Koul and Panwar (2008) and volume (V) was estimated using the expression $\Pi R^2 HF$, where F is the form factor, R is the radius of the tree derived from its DBH and H is the height of the target tree. Specific gravity (G) was estimated taking the stem cores of each species considered in the study, which was further converted into stem biomass (B_S) as per the expression $B_S = GV$.

To understand the effect of salinity on coastal vegetation, we compared the stem biomass in natural plantation of the selected species ($\sim$11 years plantation age) in the western and central Indian Sundarbans during January 2010 and observed consistency in the data of *Rhizophora mangle* in both the regions. However, for species such as *Sonneratia apetala*, *A. marina*, *Avicennia officinalis* and *Heritiera fomes*, the difference in biomass was significant. The biomass of these species was much lower in the central sector around the Matla River compared to those around the River Hooghly receiving freshwater from the Himalayan glacier after being regulated by barrages on the way. It appears from the AGSB data set (Table 7.2) that salinity has reduced the yield/ha in *S. apetala*, *A. marina*, *A. officinalis* and *H. fomes* by 24.10, 18.03, 33.94 and 64.45 %, respectively, in the central sector, but in case of *R. mangle*, the AGSB has increased by 1.38 % confirming the ability of the species to resist the hypersaline environment. The significant decrease of AGSB of *H. fomes* in central Indian Sundarbans (by 64.45 % compared to western sector) is a clear indication of the sensitivity of the species to high

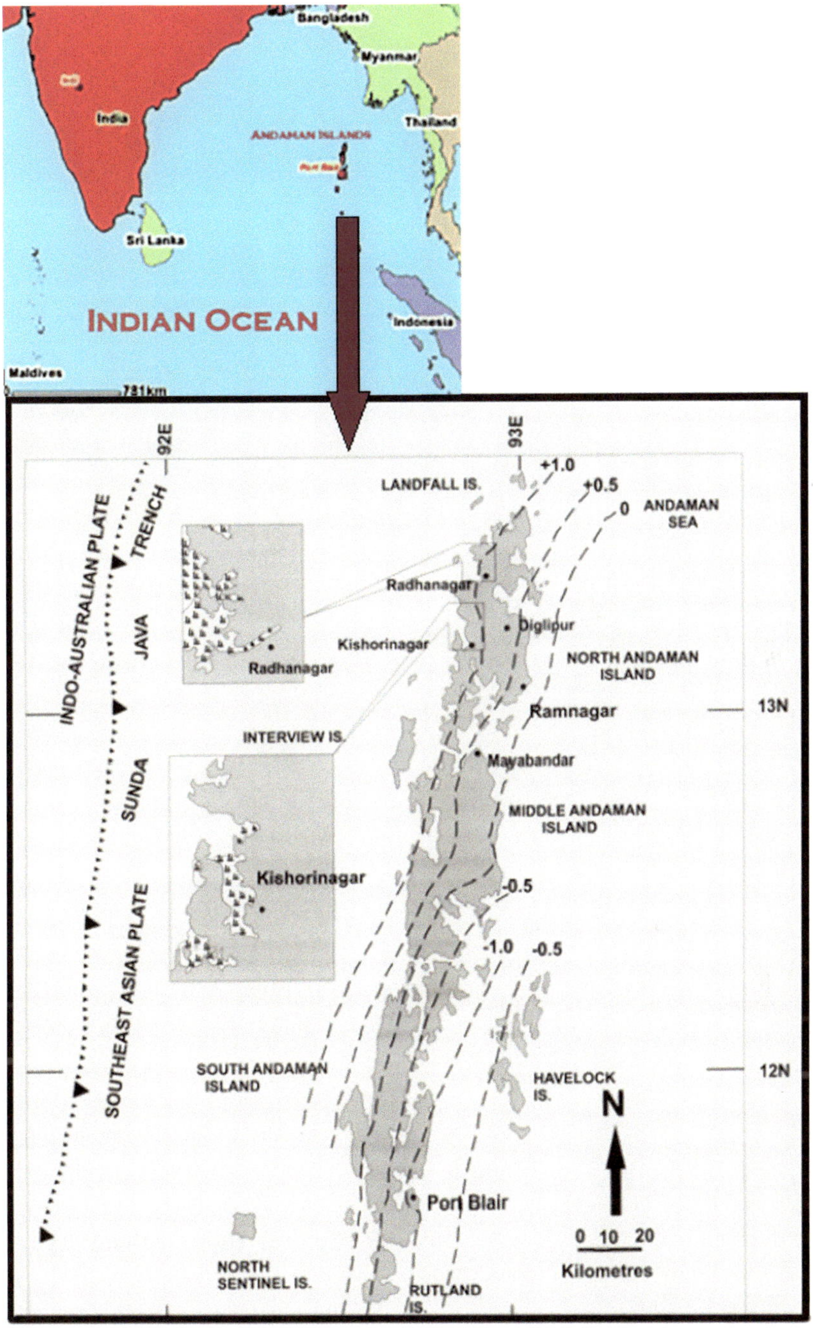

Fig. 7.7 Map showing the location of the Sunda–Java trench and distribution of coseismic vertical ground movement in the Andaman region caused by the 26 December 2004 megathrust earthquake. The *dashed lines* represent the contours of vertical ground movement. Values assigned to the *contour lines* give the estimated vertical offset in metres. The '+' and '–' signs indicate ground uplift and subsidence, respectively. The '0' value contour represents the neutral line

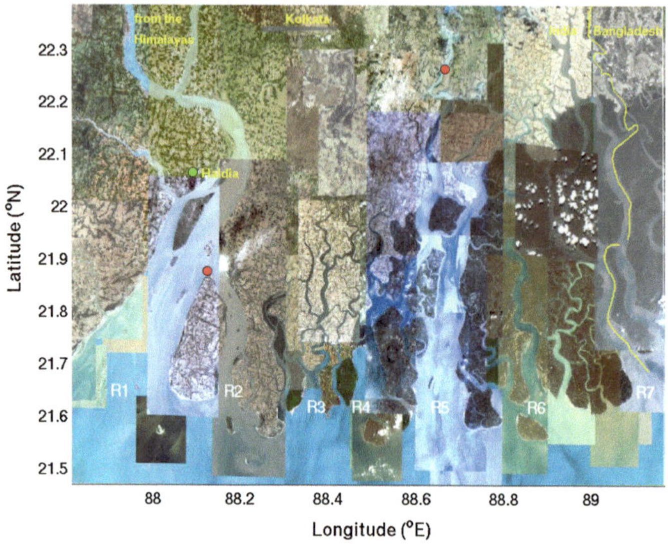

Fig. 7.8 Map of the study region with 2 stations marked in *red*. The western station is located at 21° 52′ 20.78″N, 88° 7′ 29.73″E, at the tip of Sagar Island. The Seven Rivers marked by *R1* through *R7* from west to east are as follows: Hooghly, Muriganga, Saptamukhi, Thakuran, Matla, Gosaba and Harinbhanga. The discharge system of the two metropolises of *Haldia* and *Kolkata* (Calcutta) is connected to the two western rivers, which are also fed by the melt water from the *Himalayas* after being regulated through barrages. The central station, at 22° 15′ 33.97″N, 88° 39′ 34.64″E, is connected to the rivers *R4* and *R5* which do not have any freshwater input

Table 7.2 Aboveground stem biomass (*AGSB*) in t/ha of five selected mangrove species in the western and central Indian Sundarbans

Year	Western (t/ha)	Central (t/ha)
Sonneratia apetala	19.79	15.02
Avicennia marina	18.14	14.87
Avicennia officinalis	7.75	5.12
Rhizophora mangle	13.78	13.97
Heritiera fomes	5.71	2.03

saline conditions and may face extinction from certain pockets of lower Gangetic plain (particularly in and around the Matla River) if the current trend of salinity rise (~2 psu/decade) continues (Mitra et al. 2009).

The carbon sequestration capacity of mangroves (blue carbon) also decreases in hypersaline condition. The carbon content (in %) analysed on a LECO® CHN-600 analyser reflects relatively lower carbon in the mangroves of central sector (Table 7.3) which may be attributed to lower AGSB and relative density of the species.

Table 7.3 Aboveground stem carbon stock (in t/ha) of five selected mangrove species in the western and central Indian Sundarbans

Year	Western (t/ha)	Central (t/ha)
Sonneratia apetala	8.39 (42.4 %)	6.17 (41.1 %)
Avicennia marina	7.22 (39.8 %)	5.84 (39.3 %)
Avicennia officinalis	3.37 (43.5 %)	2.13 (41.7 %)
Rhizophora mangle	6.15 (44.6 %)	6.57 (43.9 %)
Heritiera fomes	2.35 (41.1 %)	0.80 (39.4 %)

The figures in bracket represent the percentage of carbon present

F. Natural Disasters and Extreme Weather Events

Natural disasters and extreme weather events greatly damage the blue carbon community of the planet Earth. People all over the world face the reality of climate variability in different mode and magnitude. More than 530,000 people died as a direct result of almost 15,000 extreme weather events, and loss of more than USD 2.5 trillion (in PPP) occurred from 1993 to 2012 globally. The list of natural disaster is vast and the magnitude of damage caused by such disasters is difficult to assess with extreme accuracy. We present here a picture of few natural disasters that occurred in this decade with special reference to their effect on coastal vegetation.

F.1. Tsunami and Blue Carbon

On 26 December 2004, a colossal shift of tectonic plates triggered a magnitude 9.15 earthquake in the Indian Ocean, 150 km off the west coast of north Sumatra, Indonesia (Fig. 7.9).

The energy released caused the seabed to rise by several metres, displacing an estimated 30 km^3 of water and triggering a series of devastating tsunami waves. More than 230,000 people perished, and millions were left destitute, as ten-metre-high waves struck the coastlines across 13 countries. Among these 13 countries, maximum damages occurred in countries such as Indonesia, Thailand, Sri Lanka and India. The tsunami razed entire towns to the ground, sweeping away buildings, bridges, cars, and, for those that survived its impact, left behind shattered livelihoods. The fisheries sector was particularly badly hit: many thousands of fishing boats and gear were lost, and ports and other landing sites destroyed. For coastal communities reliant on farming, the tsunami ruined crops, drowned livestock and poisoned arable land and water supplies with salt. Mangroves and other coastal habitats, where still in existence, met the tsunami head on. In doing so, they shielded lives and property, but were also heavily impacted. Mangroves and other littoral forests were in places broken and uprooted, and coral reefs spectacularly overturned, but in many cases, initial fears of serious ecological damage were proved untrue. One pattern that has consistently emerged, however, from post-tsunami environmental assessments is that healthy ecosystems fared much better and had a much greater protective function than those denuded by human activities.

A preliminary assessment of the impacts of the tsunami on the ecosystems of Aceh was carried out by the Kuala Lumpur-based Global Environment Centre at the request of the Indonesian Ministry of Environment (Parish and Lee 2005). The GEC found that the original mangrove forests around Banda Aceh, the capital of Aceh province, were cleared during its development and converted to shrimp ponds. When the tsunami hit Banda Aceh, houses built on land that in the past had been sheltered by mangroves were destroyed, and hundreds of hectares of shrimp ponds were swallowed up by the sea. Some pockets of mangroves had survived development, but these were too small to protect the nearby houses (Parish and Lee 2005).

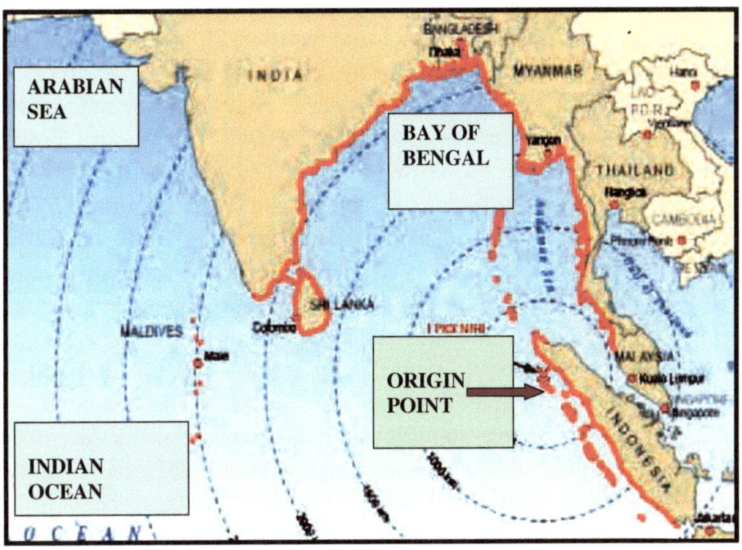

Fig. 7.9 Route of earthquake in the Indian Ocean, 150 km off the west coast of north Sumatra, Indonesia

It is uncertain what would have been the situation in Banda Aceh if the mangroves had not been destroyed. However, five villages 100 km to the southeast of the city were saved by the extensive mangroves in that area (Parish 2005). Also, it has been claimed that Simeulue Island, which is only 41 km from the epicentre of the earthquake, was saved partly by its substantial mangrove cover, coral reefs and seagrass beds and suffered only four deaths in the disaster as a result (IUCN 2005; Parish 2005). Reports from eyewitnesses on Simeulue Island state that no wave penetrated the mangrove forests, and instead the water level increased gently 'like a rising tide' (Parish 2005).

However, in the area of Ulee Lhee, close to Banda Aceh, not even dense, healthy mangroves were capable of withstanding the tsunami, such was the force of the wave (Wetlands International—Indonesia Programme 2005). Prior to the tsunami, there was 10 ha of relatively healthy mangroves at the site but the tsunami did not leave a single tree standing—they were all uprooted and carried inland by the waves and were found in residential areas up to two or three kilometres away (Wetlands International—Indonesia Programme 2005).

Mangrove forests impacted by the 2004 Indian Ocean tsunami were mostly located in sheltered areas (bays, lagoons and estuaries) with very few located on open coast, making it initially difficult to assess whether the areas impacted by the tsunami suffered less because of the intrinsic protective capacity of the forests, or because they were sheltered from direct exposure to the open sea (Chatenoux and Peduzzi 2007). However, several reports based on initial post-impact surveys in south-eastern India, the Andaman Islands and Sri Lanka (Danielsen et al. 2005; Dahdouh-Guebas et al. 2005; Kathiresan and Rajendran 2005) indicated that mangroves offered a significant defence against the full impact of the tsunami. The conclusions of Kathiresan and Rajendran (2005) and Vermaat and Thanpanya (2006) that the presence of mangroves saved lives along the Tamil Nadu coast of south-east India are invalid, however, as inappropriate statistical tests were used (Vermaat and Thampanya 2007). A more proper test of the same data indicated no significant effect of the presence or absence of mangroves on the human death toll (Kerr et al. 2006) and points to the need for caution to avoid overstating the role of mangroves in tsunami protection. Nevertheless, ground surveys and QuickBird pre-tsunami and IKONOS post-tsunami image analysis (Danielsen et al. 2005) and multivariate analysis of mangrove field data (Dahdouh-Guebas et al. 2005) covering the entire Tamil Nadu coast suggest less destruction of man-made structures located directly behind the most extensive mangroves.

F.2. Nargis and Blue Carbon

The blue carbon community has been adversely affected by Nargis. Cyclone Nargis struck Myanmar on 2 and 3 May 2008, making landfall in Ayeyarwady Division, approximately 250 km southwest of Yangon. A category 3 cyclone, Nargis, affected more than 50 townships, mainly in Yangon and Ayeyarwady Divisions, including Yangon, the country's largest city. Strong winds and heavy rain caused the greatest damage in the Ayeyarwady Delta, where a storm surge compounded the impact of the cyclone. Nargis was the worst natural disaster in the history of Myanmar. More than 140,000 people were killed, mainly by the storm surge. The loss and damage of mangrove forests as a result of Nargis is particularly critical (Fig. 7.10), which affected about 16,800 ha (41,514 acres) of natural forest and 21,000 ha (51,892 acres) of forest plantations (PONJA Report 2008). Surveys by Maung (2008) indicated significant destruction of mangroves in the direct path of the cyclone and in adjacent areas. Defoliation and damage to branches ranged from 38.9 to 55.6 % and damage to crowns was between 12.8 and 19.8 % (Fig. 7.10). Uprooting of trees was notably higher in the direct path of the storm (56.7 %) than elsewhere (4.2 %).

F.3. AILA and Blue Carbon

The incidence of **AILA** (a severe tropical cyclonic storm) in the Gangetic plain region of India on 25 May 2009, (Fig. 7.11) posed severe impacts in terms of embankment damage in areas not protected by mangroves.

Fig. 7.10 Damage caused by Nargis on mangroves of the Meinmahla Kyun Island Reserve Forest

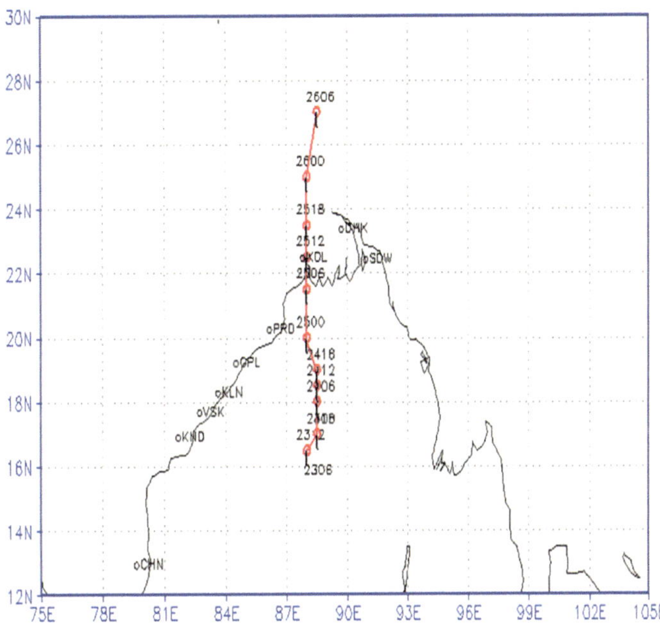

Fig. 7.11 AILA (a severe tropical cyclonic storm) in the Gangetic plain region of India on 25 May 2009

Significant changes in water quality were documented due to AILA (Table 7.4), and in many cases, top-dying disease in mangroves was observed after the exposure of the vegetation to hypersaline condition. Uprooting of mangroves was also observed after AILA in different pockets of Indian Sundarbans.

The blue carbon community in the lower Gangetic region is a notable example of the direct defensive role of mangroves against AILA. The satellite imageries of pre- and post-AILA period (Figs. 7.12, 7.13, 7.14, 7.15, 7.16, 7.17, 7.18, 7.19 and 7.20) clearly exhibit insignificant changes to mangroves although land areas of

Table 7.4 Variations of hydrological parameters in different phases before, during and after AILA

Stations	Surface water salinity (Phase A)	Surface water salinity (Phase B)	Surface water salinity (Phase C)	pH (Phase A)	pH (Phase B)	pH (Phase C)	D.O (Phase A)	D.O (Phase B)	D.O (Phase C)
Stn. 1	3.41	3.99	3.45	7.65	7.65	7.65	5.71	4.98	5.23
Stn. 2	4.88	5.93	4.96	7.68	7.69	7.68	5.43	5.00	5.21
Stn. 3	6.10	7.53	6.18	7.70	7.70	7.70	6.63	6.03	6.50
Stn. 4	13.17	16.32	13.98	8.00	8.00	8.00	6.55	5.89	6.11
Stn. 5	12.05	14.95	12.78	8.10	8.11	8.10	4.80	4.74	5.31
Stn. 6	13.84	17.23	14.01	8.00	8.03	8.01	4.91	4.67	5.12
Stn. 7	15.55	19.40	15.98	8.10	8.11	8.11	4.80	4.65	5.32
Stn. 8	14.43	18.04	14.87	8.10	8.12	8.11	4.75	4.43	5.02
Stn. 9	17.22	21.58	17.96	8.10	8.11	8.11	4.60	4.54	5.99
Stn. 10	19.33	24.30	20.05	8.15	8.15	8.16	4.68	4.60	5.06
Stn. 11	20.58	25.93	21.00	8.20	8.21	8.21	5.05	4.99	5.19
Stn. 12	23.08	29.77	24.67	8.20	8.21	8.21	5.02	4.91	5.27

Phase A pre-AILA period (18 May 2009), *Phase B* AILA phase (27 May 2009), *Phase C* post-AILA phase (4 June 2009)
Units of surface water salinity and DO are psu and ppm, respectively

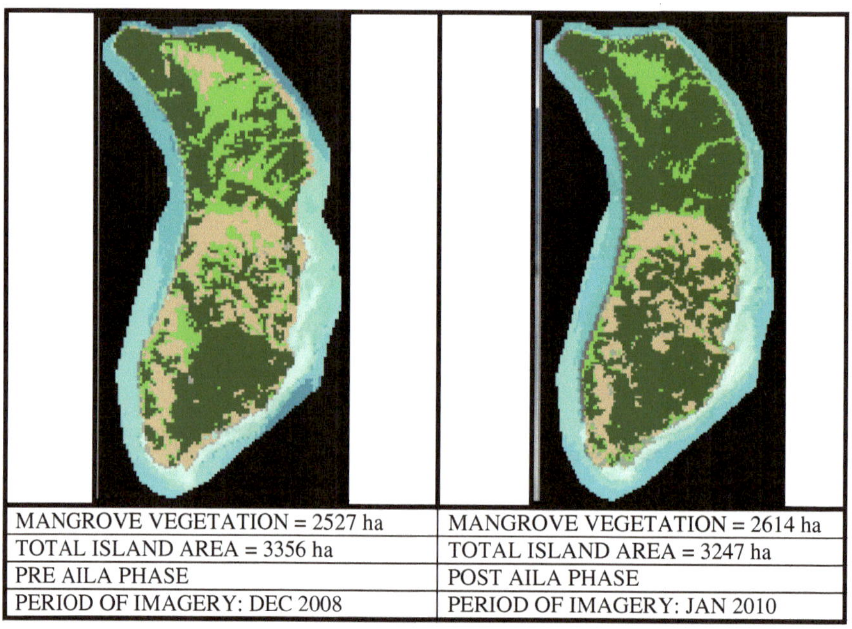

Fig. 7.12 Lothian Island

nine islands in Indian Sundarbans have changed significantly between years (Table 7.5).

The significant change in land area between years ($p < 0.01$) may be the effect of erosion or sea level rise caused by actual increase of sea level and/or subsidence of deltaic region.

F.4.1. Sandy and Blue Carbon

In terms of extreme weather events, the year 2012 will most likely be remembered for the occurrence of Hurricane Sandy in October 2012, which made headlines for several days on end in the media around the world, amounting to

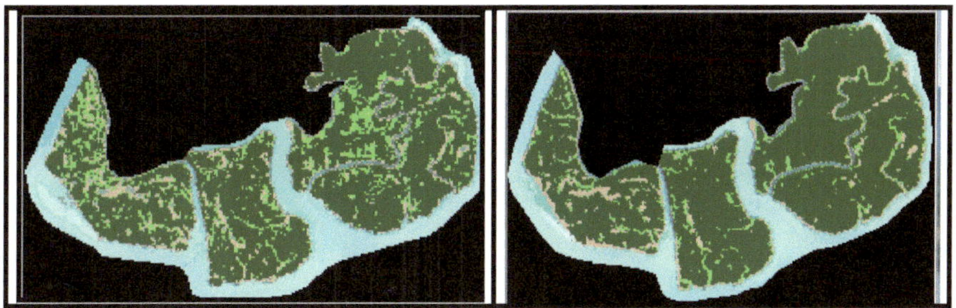

MANGROVE VEGETATION = 6270 ha	MANGROVE VEGETATION = 6359 ha
TOTAL ISLAND AREA = 6899 ha	TOTAL ISLAND AREA = 6931 ha
PRE AILA PHASE	POST AILA PHASE
PERIOD OF IMAGERY: DEC 2008	PERIOD OF IMAGERY: JAN 2010

Fig. 7.13 Baghmara Island

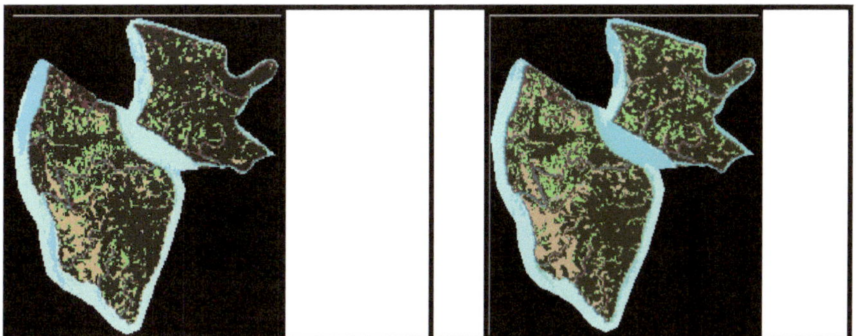

MANGROVE VEGETATION = 6068 ha	MANGROVE VEGETATION = 5725 ha
TOTAL ISLAND AREA = 6825 ha	TOTAL ISLAND AREA = 6645 ha
PRE AILA PHASE	POST AILA PHASE
PERIOD OF IMAGERY: DEC 2008	PERIOD OF IMAGERY: JAN 2010

Fig. 7.14 Gona Island

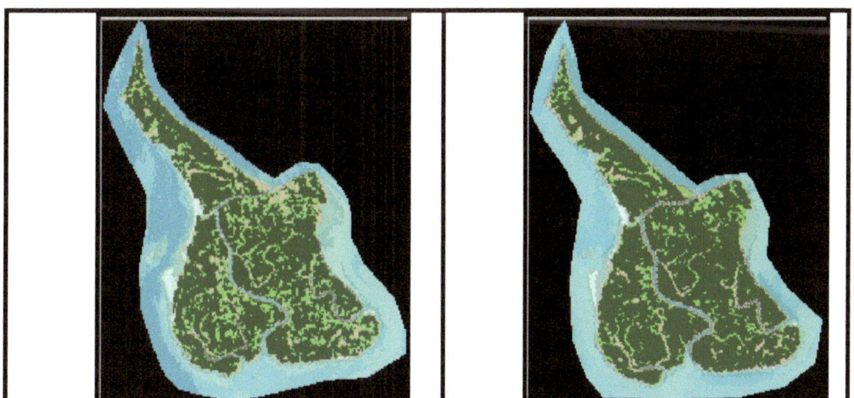

MANGROVE VEGETATION = 5472 ha	MANGROVE VEGETATION = 5458 ha
TOTAL ISLAND AREA = 6002 ha	TOTAL ISLAND AREA = 5939 ha
PRE AILA PHASE	POST AILA PHASE
PERIOD OF IMAGERY: DEC 2008	PERIOD OF IMAGERY: JAN 2010

Fig. 7.15 Mayadwip 1, 2, 3 Island

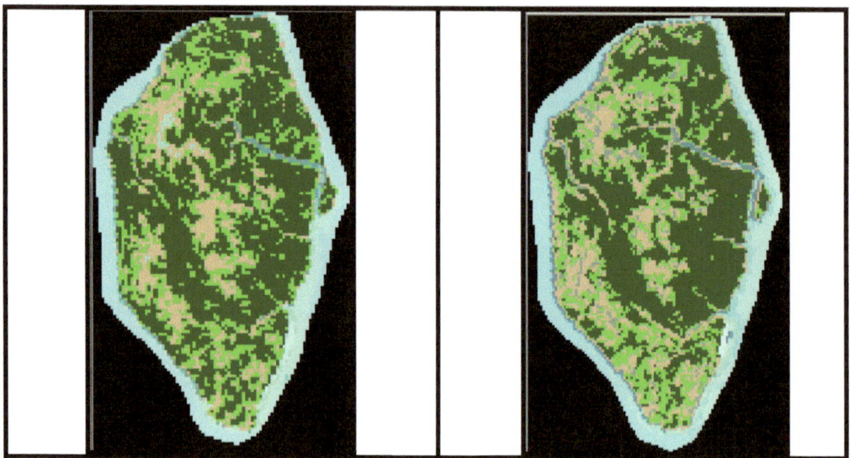

MANGROVE VEGETATION = 2851 ha	MANGROVE VEGETATION = 2544 ha
TOTAL ISLAND AREA = 3332 ha	TOTAL ISLAND AREA = 3164 ha
PRE AILA PHASE	POST AILA PHASE
PERIOD OF IMAGERY: DEC 2008	PERIOD OF IMAGERY: JAN 2010

Fig. 7.16 Thakuran (Dhanchi) Island

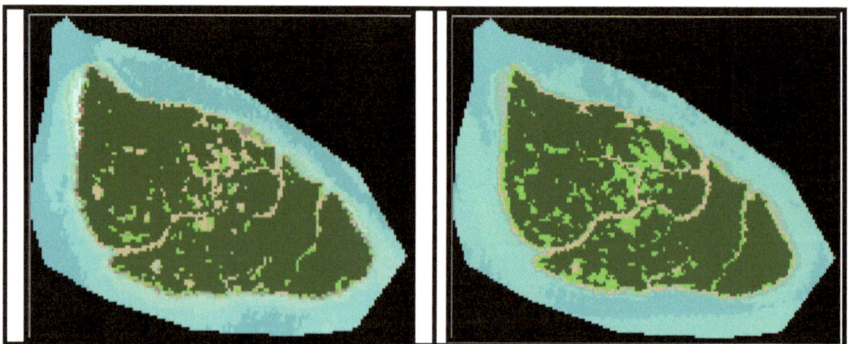

MANGROVE VEGETATION = 2335 ha	MANGROVE VEGETATION = 2333 ha
TOTAL ISLAND AREA = 2599 ha	TOTAL ISLAND AREA = 2550 ha
PRE AILA PHASE	POST AILA PHASE
PERIOD OF IMAGERY: DEC 2008	PERIOD OF IMAGERY: JAN 2010

Fig. 7.17 Mayadwip 4, 5 Island

damages of over USD 68 billions. However, what the media often failed to mention was the impact the 'Frankenstorm' had on countries outside the USA. The hurricane wreaked havoc in the Caribbean with Haiti being hit hardest, thus accounting for the country's rise to the top of this year's Climate Risk Index. In the Caribbean country that is still recovering from the devastating earthquake in 2010, the heavy rainfalls fuelled by Sandy not only left 200,000 people homeless, but also destroyed much of the country's crops, which had already been affected by Hurricane Isaac in late August 2012.

F.4.2. Bhopa and Blue Carbon

In December 2012, the Philippines were hit by Typhoon Bopha, the landfall of which claimed over 1,400 victims, topping the list for most human casualties of the year for the second year in a row.

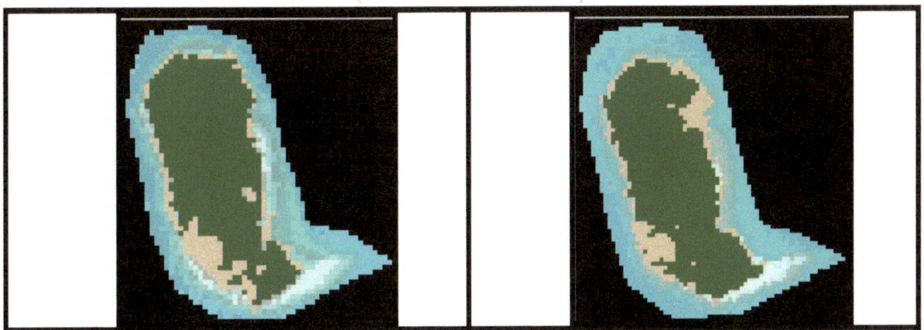

MANGROVE VEGETATION = 368 ha	MANGROVE VEGETATION = 346 ha
TOTAL ISLAND AREA = 449 ha	TOTAL ISLAND AREA = 429 ha
PRE AILA PHASE	POST AILA PHASE
PERIOD OF IMAGERY: DEC 2008	PERIOD OF IMAGERY: JAN 2010

Fig. 7.18 Jambudwip Island

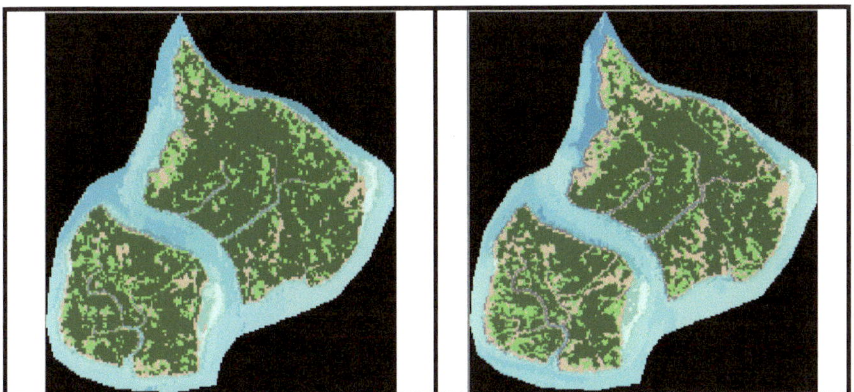

MANGROVE VEGETATION = 5332 ha	MANGROVE VEGETATION = 5197 ha
TOTAL ISLAND AREA = 5894 ha	TOTAL ISLAND AREA = 5845 ha
PRE AILA PHASE	POST AILA PHASE
PERIOD OF IMAGERY: DEC 2008	PERIOD OF IMAGERY: JAN 2010

Fig. 7.19 Chulkati Island

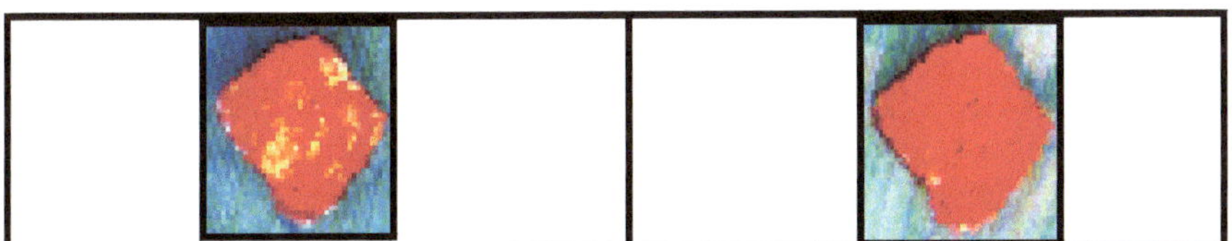

TOTAL ISLAND AREA = 426 ha	TOTAL ISLAND AREA = 436 ha
PRE AILA PHASE	POST AILA PHASE
PERIOD OF IMAGERY: DEC 2008	PERIOD OF IMAGERY: JAN 2010

Fig. 7.20 Ghoramara Island

Table 7.5 ANOVA results showing variation of mangrove vegetation and total land area between years (pre-AILA and post-AILA phases)

Variables	F_{cal}	F_{crit}
Mangrove vegetation (ha) Between years	1.88	5.59
Total land area (ha) Between years	7.29	5.32

F.5. Phailin and Blue Carbon

A Very Severe Cyclonic Storm (VSCS) Phailin originated from a remnant cyclonic circulation from the South China Sea. The cyclonic circulation lays as a low pressure area over the Tenasserim coast on 6 October 2013. It subsequently moved over to the North Andaman Sea as a well-marked low pressure area on October 7. It concentrated into a depression over the same region on October 8 moving west-northwest wards and then intensified into a deep depression on the morning of October 9 and further into a cyclonic storm (CS), 'Phailin' in the evening of the same day. Moving north-westwards, it further intensified into a severe cyclonic storm (SCS) in the morning of October 10 and into a VSCS in the forenoon of the same day over east central Bay of Bengal.

This VSCS crossed Odisha and the adjoining north Andhra Pradesh coast near Gopalpur (Odisha) around 2,230 h IST on 12 October 2013 with a sustained maximum surface wind speed of 200–210 km/h gusting up to 220 km/h (Fig. 7.21). It caused very heavy rainfall over Odisha leading to floods and strong gale winds causing large-scale structural damage and storm surges triggering widespread coastal inundation over Odisha.

A satellite image of Cyclone Phailin is pictured in Fig. 7.22. The impacts of Phailin and ensuing floods affected more than 13.2 million people, left five districts of Odisha under water (Figs. 7.23 and 7.24), and caused hundreds of millions of dollars (GoO 2013) in damage to homes, schools, crops and the fishing industry

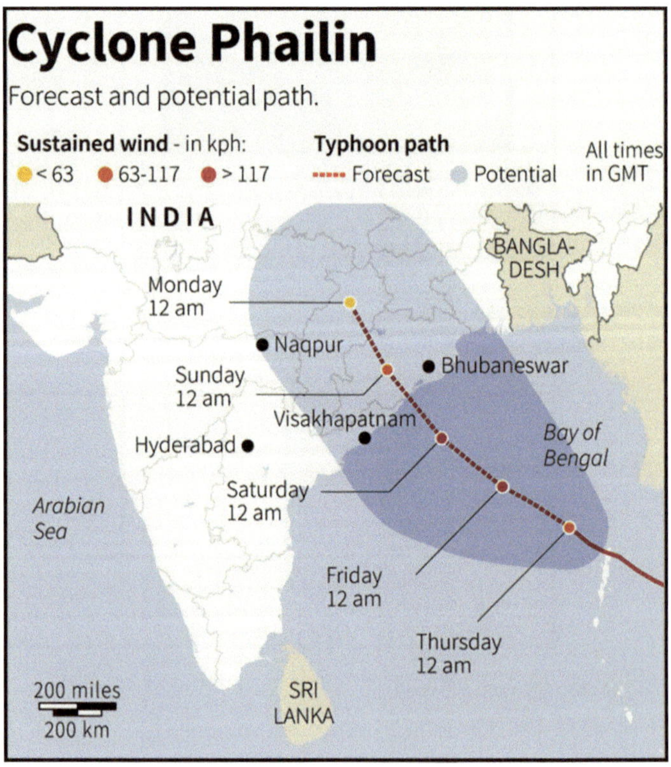

Fig. 7.21 Phailin route

Fig. 7.22 Satellite imagery of Phailin cyclone

Fig. 7.23 False-colour Landsat satellite images of Chilika Lake in 2012, when the coastline was intact. *Image source* USGS/NASA; visualization by UNEP/GRID-Sioux Falls

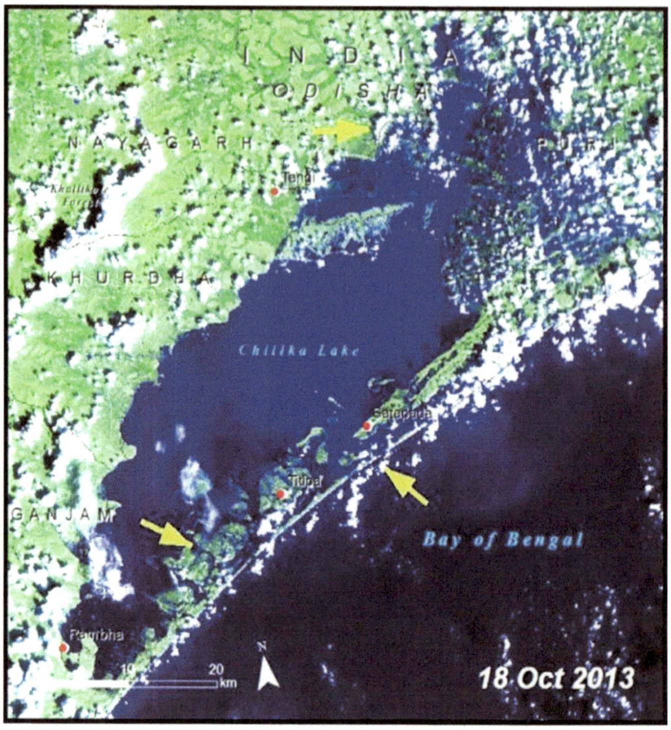

Fig. 7.24 False-colour Landsat satellite images of Chilika Lake after six days of the occurrence of Phailin in 2013, when the coastline had been breached. *Image source* USGS/NASA; visualization by UNEP/GRID-Sioux Falls

(Froberg 2013). However, early warning alerts, disseminated four days before Phailin struck land, allowed for the evacuation of approximately 400,000 people on or by 11 October (Senapati 2013). Ultimately, a total of nearly 1.2 million people were evacuated (GoO 2013), resulting in the largest evacuation operation in India in 23 years (IFRC 2013). Early warning also allowed for the relocation of more than 30,000 animals. A total of 21 lives were lost as a result of the cyclone and an additional 23 lives due to severe flash flooding in the aftermath of the cyclone (GoO 2013).

Chilika Lake in Odisha (a maritime State in the East coast of India) is the largest coastal lagoon and home to a large number of endangered animal and plant species. With a speed of 200–220 km/h, Phailin breached the Chilika coastline (Figs. 7.23 and 7.24) and there was considerable intrusion of saline water. Mangroves were uprooted in large scale.

We conducted a study with the knowledge participation of Prof. K.C. Sahu (Head, Department of Marine Science, Behrampur University, Odisha, India) and Sanjiba Kumar Baliarsingh, Project Scientist B, Ocean Colour Applications Advisory Services and Satellite Oceanography Group, Indian National Centre for Ocean Information Services, Ministry of Earth Sciences, Govt. of India and observed a hike in salinity value in Gopalpur creek (Table 7.6; Fig. 7.25) due to intrusion of saline water from Bay of Bengal (Fig. 7.26). Although the aquatic pH (Table 7.7; Fig. 7.27) did not exhibit any significant alteration, but DO, Chlorophyll *a* and phytoplankton count exhibited significant variation ($p < 0.01$), (Tables 7.8, 7.9, 7.10 and Figs. 7.28, 7.29 and 7.30) which may be attributed due to intrusion of stenohaline phytoplankton from Bay of Bengal along with sea water into the Gopalpur creek. The significant rise in chlorophyll *a*—a consequence of rise in phytoplankton population may have occurred as a result of sea water intrusion.

F.6. Haiyan and Blue Carbon

On 8 November 2013, Haiyan (known locally as Yolanda) with a speed around 250 km h^{-1} became the strongest typhoon to make landfall

Table 7.6 Variation of salinity (in psu) in different phases of Phailan

Stations	Surface water salinity (Phase A)	Surface water salinity (Phase B)	Surface water salinity (Phase C)
Stn. 1	29.91 ± 0.07	39.61	33.56 ± 0.06
Stn. 2	28.48 ± 0.06	37.71	31.98 ± 0.05
Stn. 3	27.63 ± 0.03	36.59	31.00 ± 0.07
Stn. 4	27.11 ± 0.07	35.90	30.98 ± 0.07
Stn. 5	26.88 ± 0.04	35.59	30.00 ± 0.06

Phase A pre-Phailin period (5 October 2013); *Phase B* Phailin phase (16 October 2013); *Phase C* post-Phailin phase (5 November 2013)

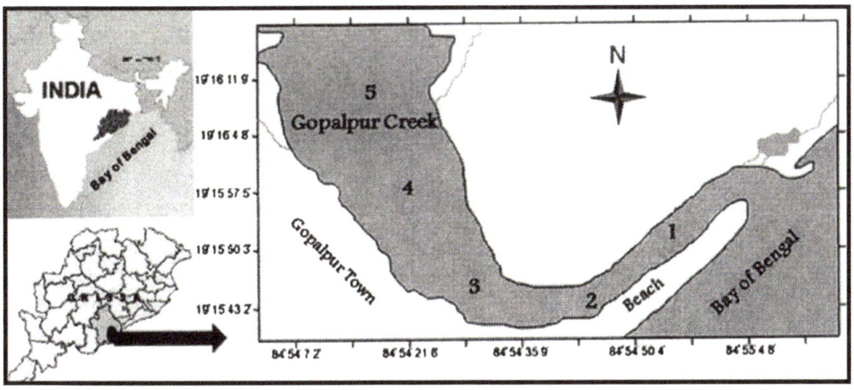

Fig. 7.25 Gopalpur creek; 1, 2, 3, 4 and 5 are the selected sampling stations

Fig. 7.26 Variation of salinity during pre-Phailin (*Phase A*), Phailin (*Phase B*) and post-Phailin (*Phase C*) periods

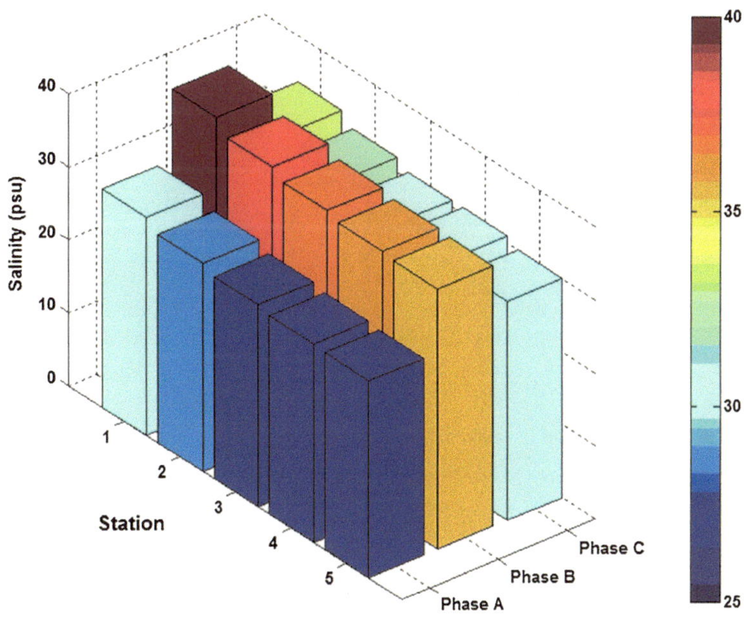

Table 7.7 Variation of pH in different phases of Phailan

Stations	Surface water pH (Phase A)	Surface water pH (Phase B)	Surface water pH (Phase C)
Stn. 1	8.15 ± 0.02	8.17 ± 0.03	8.16 ± 0.03
Stn. 2	8.11 ± 0.03	8.12 ± 0.03	8.11 ± 0.03
Stn. 3	8.10 ± 0.02	8.10 ± 0.03	8.10 ± 0.02
Stn. 4	8.00 ± 0.02	8.00 ± 0.02	8.00 ± 0.02
Stn. 5	7.96 ± 0.03	7.97 ± 0.02	7.96 ± 0.02

Phase A pre-Phailin period (5 October 2013); *Phase B* Phailin phase (16 October 2013); *Phase C* post-Phailin phase (5 November 2013)

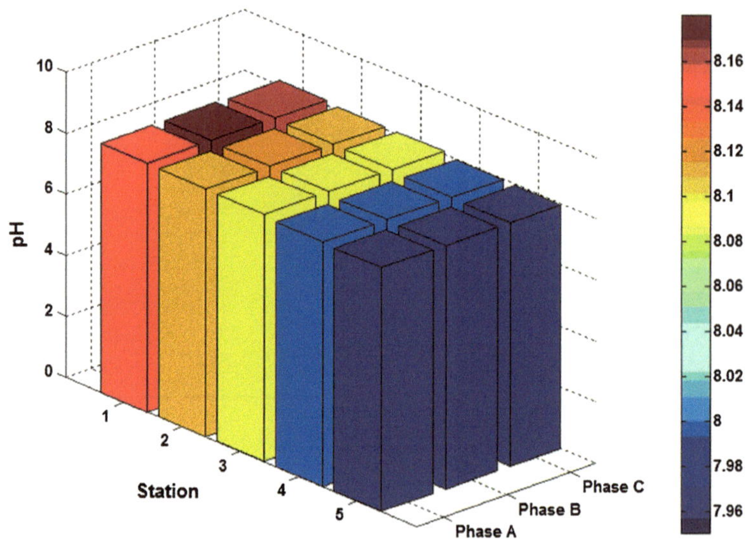

Fig. 7.27 Variation of pH during pre-Phailin (*Phase A*), Phailin (*Phase B*) and post-Phailin (*Phase C*) periods

Table 7.8 Variation of DO (in ppm) in different phases of Phailan

Stations	Surface water DO (Phase A)	Surface water DO (Phase B)	Surface water DO (Phase C)
Stn. 1	5.73 ± 0.02	4.98 ± 0.03	5.33 ± 0.04
Stn. 2	5.43 ± 0.02	5.00 ± 0.02	5.22 ± 0.03
Stn. 3	6.63 ± 0.03	6.03 ± 0.02	5.98 ± 0.02
Stn. 4	6.55 ± 0.03	5.89 ± 0.03	6.10 ± 0.04
Stn. 5	4.80 ± 0.03	4.74 ± 0.03	4.91 ± 0.02

Phase A pre-Phailin period (5 October 2013); *Phase B* Phailin phase (16 October 2013); *Phase C* post-Phailin phase (05 November 2013)

Table 7.9 Variation of chlorophyll *a* and phytoplankton in different phases of Phailan

Stations	Chlorophyll *a* mg/m^3 (Phase A)	Chlorophyll *a* mg/m^3 (Phase B)	Chlorophyll *a* mg/m^3 (Phase C)	Phytoplankton Nos./L (Phase A)	Phytoplankton Nos./L (Phase B)	Phytoplankton Nos./L (Phase C)
Stn. 1	3.98 ± 0.02	6.02 ± 0.05	4.98 ± 0.04	57,004	78,201	69,276
Stn. 2	3.47 ± 0.05	5.14 ± 0.04	4.54 ± 0.03	50,221	75,337	70,114
Stn. 3	3.30 ± 0.05	4.64 ± 0.03	4.33 ± 0.02	42,541	69,881	65,103
Stn. 4	3.29 ± 0.01	4.12 ± 0.06	4.08 ± 0.04	38,697	68,442	62,998
Stn. 5	2.98 ± 0.04	4.09 ± 0.04	3.64 ± 0.03	32,494	65,310	63,812

Phase A pre-Phailin period (5 October 2013); *Phase B* Phailin phase (16 October 2013); *Phase C* post-Phailin phase (05 November 2013)

Table 7.10 ANOVA for selected variables related to Phailin

Variables	F_{obs}	F_{crit}
Salinity		
Between phases	85.53508	3.837853
Between stations	1,531.787	4.45897
pH		
Between phases	1,017	3.837853
Between stations	4.333333	4.45897
DO		
Between phases	38.70268	3.837853
Between stations	10.83638	4.45897
Chlorophyll a		
Between phases	13.80891	3.837853
Between stations	38.45118	4.45897
Phytoplankton		
Between phases	8.521948	3.837853
Between stations	84.37756	4.45897

ever recorded. It was accompanied by a storm surge that smashed through coastal neighbourhoods and farmlands across much of the central Philippines (Fig. 7.31). Preparations and early warnings saved many lives. Despite of this preparedness, thousands died and millions were left in need of urgent assistance. Local officials and emergency response teams were themselves initially shaken, as swamps of sea water and jungles of debris created a logistical nightmare for survivors and those trying to assist them.

7.2 Anthropogenic Threats

Anthropogenic factors are human activities that change the environment and influence climate. The emission of carbon dioxide due to burning of fossil fuels or the increase of greenhouse gas concentrations due to rapid urbanization, industrialization and expansion of unplanned tourism are unquestionable human influences on coastal vegetation through the event of habitat destruction or modification. However, in some cases, the chain of causality is direct and unambiguous (e.g. the effects clearance of the coastal vegetation for setting industries or shrimp farms), while in others, it is less clear (e.g. the effect of acidification on coastal vegetation in terms of bioaccumulation and mortality).

Today, the climate change-related researches unanimously suggest that the biggest factor of alteration of blue carbon reservoir is the increase in carbon dioxide levels due to emissions from fossil fuel combustion, followed by aerosols (particulate matter in the atmosphere), which exert a cooling effect. Other factors, including change of land use, animal grazing, agriculture

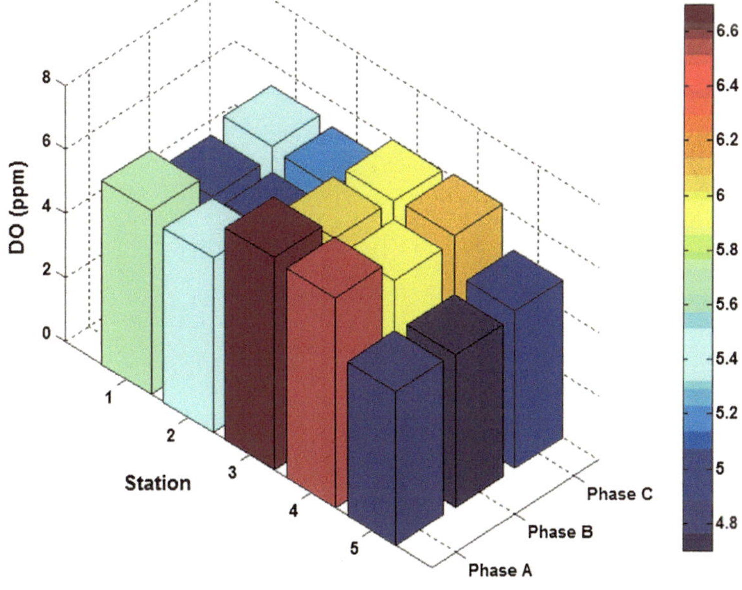

Fig. 7.28 Variation of DO during pre-Phailin (*Phase A*), Phailin (*Phase B*) and post-Phailin (*Phase C*) periods

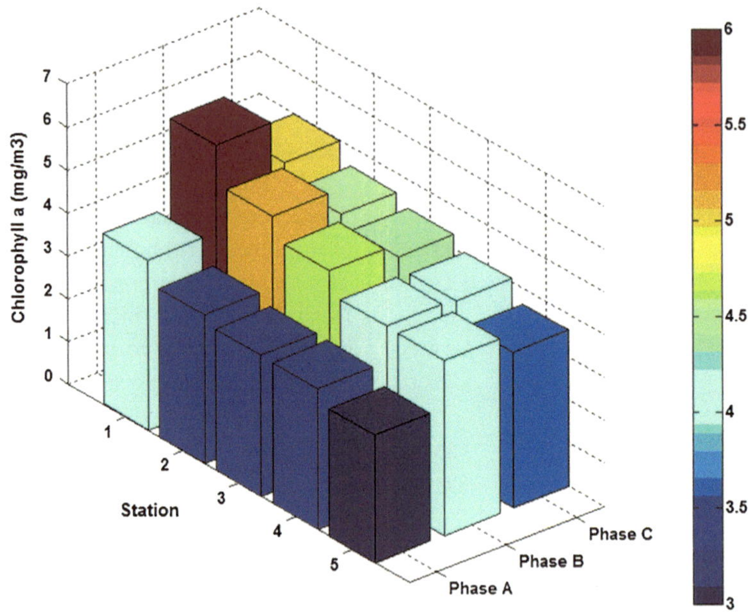

Fig. 7.29 Variation of chlorophyll *a* during pre-Phailin (*Phase A*), Phailin (*Phase B*) and post-Phailin (*Phase C*) periods

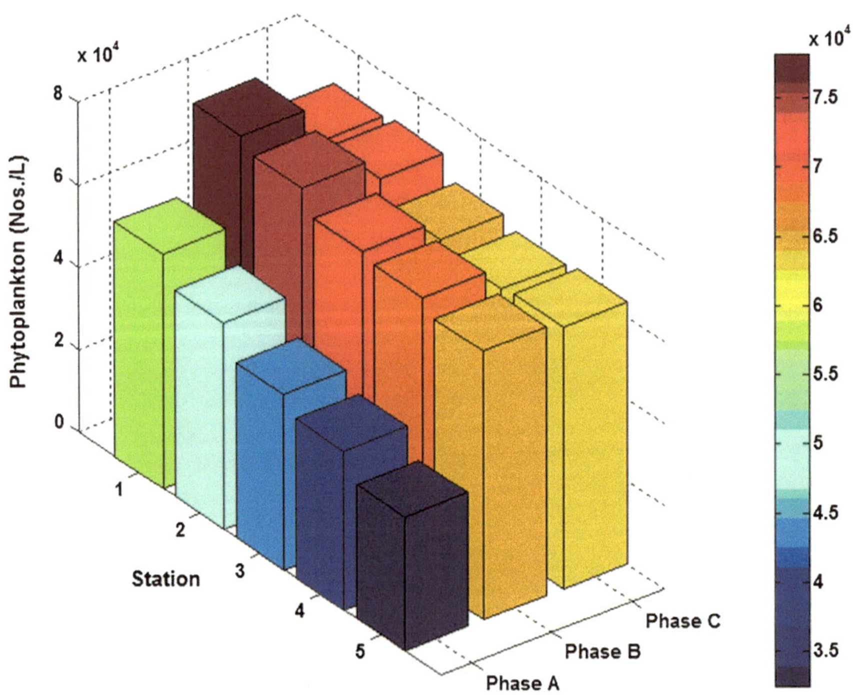

Fig. 7.30 Variation of phytoplankton count during pre-Phailin (*Phase A*), Phailin (*Phase B*) and post-Phailin (*Phase C*) periods

Fig. 7.31 Effect of Haiyan on the coastal region of Philippines

and deforestation, also exert considerable influence on the mangrove forests, seagrass bed, salt marsh ecosystem and seaweeds that are the dominant components of blue carbon. The major anthropogenic factors that alter and influence the coastal vegetation may be broadly classified into six broad headings.

A. Burning of fossil fuels
B. Modification of land use pattern
C. Livestock
D. Development of coastal structures
E. Pollution
F. Construction of dams and barrages.

A. Burning of Fossil Fuels

Coal, oil and natural gas are collectively referred to as fossil fuels because they are formed from the remains of plants and microorganisms that lived millions of years ago. Fossil fuels constitute the backbone of modern civilization. Beginning with the industrial revolution in the 1850s and accelerating ever since, the human consumption of fossil fuels has elevated carbon dioxide levels from a concentration of $\sim$280 ppm to more than 380 ppm today (an increase of 35.7 %). These increases are projected to reach more than 560 ppm (an increase of 100 %) before the end of the twenty-first century. It is known that carbon dioxide levels are substantially higher now than at any time in the last 750,000 years. Along with rising methane levels, these changes are anticipated to cause an increase of atmospheric temperature within 1.4–5.6 °C between 1990 and 2100.

Fossil fuels are burnt to run the wheels of industries. Cement manufacturing is one of the largest causes of human-induced carbon dioxide emissions. The gas is produced when calcium carbonate ($CaCO_3$) is heated to produce the calcium oxide (CaO, also called *quicklime*), which is the main ingredient for cement ($CaSiO_3$) production. The reaction steps in this manufacture are given as follows:

$$CaCO_3 \rightarrow CaO + CO_2$$
$$\Delta$$
$$CaO + SiO_2 \rightarrow CaSiO_3$$

Carbon dioxide produced in the first step has considerable contribution to global warming. While fossil fuel combustion and deforestation each produce significantly more carbon dioxide in the Earth's atmosphere, cement production alone is responsible for approximately 2.5 % of total worldwide emissions from industrial sources. Apart from cement industries, power supply of all the industries is based on the burning of fossil fuels.

The carbon dioxide produced from industrial units and urban sectors has differential effect on coastal vegetation, and many of these reports are contradictory. There are some evidences that elevated carbon dioxide stimulates plant growth at least in agricultural plant species (Kimball 1983; Cure and Acock 1986) where most of the experiments have been carried out with greenhouse grown plants. Eames and Jarvis (1989), in an extensive review, reported some enhancement of growth in juvenile trees and Drake (1992) reported a significant impact of enhanced atmospheric carbon dioxide on a wetland community of sedge and grasses. Though there is evidence that carbon dioxide enrichment will enhance growth in seedling tree species, there is no equivalent evidence that there will be long-term forest growth in response to rising atmospheric carbon dioxide. A growth response to carbon dioxide may be manifest in belowground processes of forest ecosystems which tend to be nutrient and water limited. At the whole plant level, carbon isotope composition data indicate species variation in regulation of water loss with respect to carbon gain. The limited data suggest that not all species will respond similarly in response to elevated atmospheric carbon dioxide levels. In the case of mangroves, Ball and Farquahar (1984a, b) reported that for *Aegiceras corniculatum* and *Avicennia marina,* the rate of photosynthesis was limited by stomatal conductance to carbon dioxide and the internal efficiency of carboxylation involving the enzyme RuBp carboxylase. These results suggest that for these mangroves, photosynthesis would be enhanced if the ambient carbon dioxide levels were increased. Contrary to these results, Cheeseman et al. (1991) after working on *Bruguiera parvifiora, Bruguiera gymnorrhiza* and *Rhizophora apiculata* suggested that the photosynthetic performance was unlikely to be enhanced by increased levels of ambient carbon dioxide. The effect of carbon dioxide enrichment on mangrove forests cannot be interpreted within a simple framework as it will depend on complex interactions between several different physiological and environmental factors. Information is needed from long-term assessments of growth where high carbon dioxide concentration, temperature, water stress and nutrient stress are controlled.

Anthropogenic aerosols, particularly sulphate aerosols from fossil fuel combustion, exert a cooling influence. This, together with natural variability, is believed to account for the relative 'plateau' in the graph of twentieth-century temperatures in the middle of the century. The effect of temperature fall has not been studied critically for coastal vegetation. Researchers observed that the optimum leaf temperatures for photosynthesis in mangroves appear to be 28–32 °C and photosynthetic capacity falls to close to zero at leaf temperatures of 38–40 °C (Clough et al. 1982; Andrews et al. 1984). It is generally accepted that plant development will be accelerated by increased temperature, as long as the temperature reached does not exceed an upper threshold.

Very little is known about the effect of temperature fall on metabolic processes in mangroves and hence the authors are presently not in a position to conclude the cooling effect caused by anthropogenic aerosols on seaweeds, seagrass, salt marsh grass, etc. It has been reported that algae can thrive well in cold environments. Most of the primary productivity of Antarctica involves free-living algae in lakes, and algal symbioses with fungi (lichens, and the green alga *Prasiola* sp.) and, as free-living cyanobacteria, on land. In the surrounding oceans, phytoplankton and sea-ice algae, as well as very large subtidal macroalgae, especially those of the brown algal order Desmarestiales, with kelp (*Laminariales*)—like organisms such as *Himanthothallus granfdifolius* (with blades up to 10 m long and 1 m wide)—are found in plenty (Luning 1990; Fogg 1998; Boyd et al. 2000; Brierley and Thomas 2002; Thomas and Dieckmann 2002a, b; Wiencke and Clayton 2002).

B. Modification of Land Use Pattern

Prior to widespread fossil fuel use, humanity's largest effect on local climate is likely to have resulted from alteration in land use pattern. Irrigation, deforestation and agriculture fundamentally change the environment. These activities alter the amount of water going into and out of a given location. They also may change the local albedo by influencing the ground cover and altering the amount of sunlight that is absorbed. For example, there is evidence to suggest that the climate of Greece and other Mediterranean countries was permanently changed by widespread deforestation between 700 BC and 1 AD (the wood being used for ship building, construction and fuel), with the result that the modern climate in the region is significantly hotter and drier, and the species of trees that were used for ship building in the ancient world are no longer be found in the area.

A controversial hypothesis by William Ruddiman called the early anthropocene hypothesis suggests that the rise of agriculture and the accompanying deforestation led to the increases in carbon dioxide and methane during the period 5,000–8,000 years ago. These increases, which reversed previous declines, may have been responsible for delaying the onset of the next glacial period, according to Ruddimann's overdue-glaciation hypothesis.

In modern times, a 2007 Jet Propulsion Laboratory study found that the average temperature of California has risen about 2° over the past 50 years, with a much higher increase in urban areas. The change was attributed mostly to extensive human development of the landscape. In coastal zone and estuarine villages, modification of landscape is done in large scale to promote aquaculture.

The term aquaculture encompasses the culture of aquatic species ranging from seaweed to fish. Oyster culture, crab culture, mussel culture and clam culture are all the *vital* components under aquaculture. The mangrove ecosystem offers a congenial, natural condition for the growth and survival of aquacultural species. The nutrient provided by mangrove litter and detritus promotes the growth of phytoplankton and subsequently the zooplankton which are the natural feed for the cultured species. This is the main reason why coastal aquaculture farms are constructed on salt marshes and intertidal mudflats of mangroves.

The rate of reclamation of marshes and particularly mangrove swamps has accelerated in recent years in some parts of the tropics due to the rapid expansion of pond farming of shrimps for export. About 50 % of the mangrove forests in the Philippines have been developed into brackish water fish ponds (Saclauso 1989). The area converted in Thailand is estimated to be about 27 % and in Ecuador about 13–14 %. Such large-scale conversions have aroused considerable environmental concern among the public and development agencies.

The floral spectrum in coastal area has been largely altered by aquaculture and more specifically by shrimp culture. Owing to presence of congenial hydrological parameters in terms of salinity and nutrient level, mangrove ecosystem has always attracted the shrimp culturists for short-term gain without caring much about the ambient environment. This attitude has considerably affected the mangrove ecosystem. Tabuchi (2003) estimated that on a global scale, the area under mangroves is shrinking by 100,000 ha annually due to clear cutting of timber and conversion into aquaculture projects (Figs. 7.32 and 7.33). In Thailand, about 65 % of the area of mangroves was lost during 1976–1991 due to an expansion of shrimp farming in the central, east coast and south-east regions of the country (Aksornkoae et al. 1993; MacKinnon 1997) estimated that by the early 1990s, Myanmar had lost almost 75 % of the original extent, Vietnam had lost 37 % and Thailand 84 %. By the mid of 1980s, Brunei and Philippines had lost 20 and 67 % of their original mangrove cover, respectively. Earlier estimates suggested that by early 1980s, Indonesia had lost 55 % of its mangroves. Chan et al. (1993) estimated that Malaysia had lost 12 % of its mangrove forests between 1980 and 1990. According MacKinnon (1997), by 1993, about 74 % of mangrove area in Malaysia was lost from their original extent. In a few regions (i.e. Latin America and the Caribbean), however, the mangrove area is increasing as a result of plantation forestry and natural regeneration. The lack of adequate data on changes in mangrove for some south-east Asian countries prevents efforts to report trends for the region as a whole.

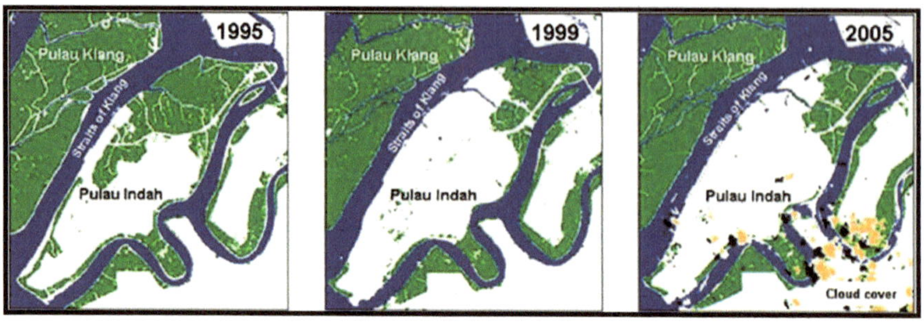

Fig. 7.32 Decline of mangrove forests in Pulau Indah, Malaysia. Area in *dark green* represents mangrove forests (Ahmad and Suratman 2007)

Fig. 7.33 Shrimp culture ponds in mangrove patches: a common scene in tropical mangrove ecosystem; photograph taken on 4 December 2013 by Mr. Tanmay Ray Chaudhuri (Environmentalist and a researcher of Techno India University, Kolkata)

In the most recent study, Ahmad and Suratman (2007) conducted a change detection analysis of mangrove forests utilizing a time series of Landsat TM imagery in Pulau Indah (literally translated means 'Beautiful Island') and its vicinity, Malaysia. This analysis was focused on determining mangrove reduction rates and identifying their spatial patterns within two interval periods, 1995–1999 and 1999–2005 (Fig. 7.32). Results from the analysis suggested there has been a decline in mangrove forests during both intervals at the rates of 14.1 and 1.9 %, respectively. The higher declining rate during the first interval was primarily due to expansion and land development for building seaport infrastructures on this island. The estimated reduction rate of mangrove forests for the 10-year period was 1.6 % per year, which is slightly higher that reported nationally by Ong (1982) over the past 20 years (i.e. 1 % per year).

In many countries, the coastal area is exploited for procuring salt from the sea water. About 30 % of world's supply of salt comes from sea water. In order to keep the cost of production low, natural evaporation is used for extracting salt from the sea water. In the south of France, Puerto Rico, central California, the Bahamas, Hawaii and The Netherland Antilles, sea salt is extracted and reined to produce table salt. The process begins by allowing sea water to enter shallow ponds. Evaporation of the water produces a concentrated salt solution to which more sea water is added. Finally, the water is allowed to totally evaporate, leaving a thick salt layer behind that can be processed commercially. In Indian subcontinent, salt pans (Fig. 7.34) are

(a)

(b)

Fig. 7.34 **a** Coastal waters stocked in shallow ponds for procuring salt through evaporation. **b** Salt extracted after evaporation in coastal region: a major threat to ground water table and adjacent agricultural land through vertical and horizontal migrations of hypersaline water

developed in mangrove patches located along the coastlines of maritime states which causes vertical and horizontal migration of salts in the adjacent areas. This has profound negative impact on the survival and growth of coastal vegetation.

The marshy areas and mangrove patches are also destroyed in few areas of lower Gangetic delta through development brick kilns (Fig. 7.35). The soil excavated for developing shrimp farms is used to manufacture bricks, which a spin-off product of shrimp industry. The construction of brick kilns in swampy areas not only adds carbon dioxide to the atmosphere (a local level effect), but also clears the vegetation and modify the marsh soil, which are rich reservoir of carbon.

C. Livestock

Livestock cause considerable damage to blue carbon reservoir by way of (i) grazing (ii) emission of greenhouse gases (iii) and directly depending on leaves of coastal vegetation for nutrition. According to a 2006 United Nations report, livestock is responsible for 18 % of the world's greenhouse gas emissions as measured in carbon dioxide equivalents. This, however, includes land usage change, meaning deforestation in order to create grazing land. In the Amazon Rainforest, 70 % of deforestation was done solely to make way for grazing land. This is the major factor in the UNFAO report (2006); ftp://ftp.fao.org/docrep/fao/010/A0701E/A0701E00.pdf, which was the first agricultural report to include land usage change into the radiative forcing of livestock. In addition to carbon dioxide emissions, livestock produces 65 % of human-induced nitrous oxide (which has 296 times the global warming potential of carbon dioxide) and 37 % of human-induced methane (which has 23 times the global warming potential of carbon dioxide).

Several case studies can support the issue of damaging the coastal vegetation by livestock. In Indian subcontinent, the Maldharis are a nomadic pastoral community that travels with their herds of cattle or camels from place to place. The over-

Fig. 7.35 Brick kiln in the marshy land near estuary in the lower Gangetic delta region; photograph taken on 4 December 2013 by Mr. Tanmay Ray Chaudhuri (Environmentalist and a researcher of Techno India University, Kolkata)

exploitation of mangroves for fodder and fuel-wood by local communities including the Maldharis is a complex problem with several inter-related causes. Due to increasing government control of natural resources, the people living around mangroves have lost control over the management of common property resources in their areas. Earlier, their traditions and customs ensured the proper stewardship of natural resources but now, these age-old safeguards have unfortunately broken down. The result has been unproductive village commons, and when accompanied by frequent droughts, the pressures on mangroves for fodder and fuel-wood have only been accentuated.

Compared to other common lands, mangrove regions have an advantage, in that cattle are unable to brave the damp, muddy conditions for free-grazing. This advantage is, however, lost in case of camels, as they can freely move about and can even cross the creeks when there are low tides. A mere one instance of free roaming by herds of camel can often devastate large mangrove areas, as their feet trample the pneumatophores of mangroves, blocking respiration (Fig. 7.36).

For feeding cattle, the people visit the mangrove forests to collect the protein-rich mangrove leaves. These leaves are usually preferred over other grasses as fodder due to their high nutritive value. The village folk cut leaves and pack them in large bundles that they haul on their heads (Fig. 7.37).

The grazing pressure on mangroves and salt marsh grass is also witnessed in some islands of lower Gangetic delta. The name of Nayachar Island (Fig. 7.38) is very relevant in this context. The island is located about 56 nautical miles south of Kolkata, directly facing the port-cum-industrial complex of Haldia. The island is bounded by latitudes 21° 54′ 41″N–22° 01′ 28″N and longitudes 88° 03′ 02″E–88° 08′ 43″E. The island has an area of about 47 km^2 and a maximum elevation of 6.5 m above mean sea level (MSL). Length of the island centrally measured from North to South is 15,861 m and from East to West is 4,340 m.

The land topography is generally flat, with elevation ranging from a low of 1.08 m MSL to as high as 3.68 m MSL along the coastal region of the island. Average ground level of the island is 1.366 m MSL. The island is convex upwards with gently sloping eastern and western flanks. The eastern flank is undergoing erosion at present while the western edge is accreting. According to recent estimation, the present accretion rate is about 1 km^2/year on the Haldia port area and the erosion rate is about 0.05 km^2/year on the eastern part of Nayachar. The island is exposed to strong scouring action of the running Hugli River and hence a programme was

Fig. 7.36 Camels in the Gujarat coast going for mangrove grazing

Fig. 7.37 Women folk collecting mangrove leaves for cattle fodder

Fig. 7.38 Location of Nayachar Island in the upper stretch of Hugli River

undertaken by Calcutta Port Trust in early 1990s to stabilize the island and increase the draft of the adjacent navigational channel. Eleven true mangrove species were planted inside and along the border of the island on geo-jute matrix and within a span of 2 years excellent growth of the species (particularly *S. apetala*) stabilized the boulders with considerable accretion of silt particles. However, grazing by buffaloes has adversely affected the vegetation of the island (Fig. 7.39). These buffaloes swim from adjacent villages and graze on the coastal vegetation and thus destroy the blue carbon reservoir of the island.

D. Development of Coastal Structures

Coastal zones face a variety of pressures. Coastal development results in the infilling of lagoons and reclamation of coastline. Historically, the Dutch has fought a long battle with the North Sea and by an extensive system of dykes have extended their landmass and turned the semi-enclosed marine Zuider Zee into the freshwater IJsselmeer. However, most coastal development is piecemeal and insidious and hence difficult to effectively regulate. The cumulative effects of such developments within a particular estuary can be depressing. For instance, about 2,500 ha of mudflats in Scotland, have disappeared over the last 200 years through a series of individually unspectacular schemes, leading to a reduction in the estuary's fish biomass by 50 % (McLusky et al. 1991). This scale of loss of intertidal zone has occurred or is forecast for many UK areas (Prater 1981) and must apply to most of the world's industrialized estuaries. Development, whether by dykes or coastal defences, seaside promenades, residential 'marinas' or dock complexes, tends to shorten the foreshore and reduces the extent of mudflats, upper-shore creek and salt marsh systems.

Rapid pace of industrialization and urbanization in the coastal zone can cause adverse impact on the habitat of blue carbon. Besides increasing nutrient loading, increased development is often accompanied by alteration of coastal habitat that effectively diminishes critical shallow-water habitat. Examples of such alterations include the construction of seawalls, breakwaters, revetments, groins and jetties. Each of these structures, aimed at protecting coastal property, refracts energy away from the shore and can cause erosion and increase currents that can harm seagrasses. Several such structures and hotels have been constructed without giving any importance to the endemic biodiversity in developing countries. A relevant example in this context is the developmental activities in the Goa coast. Goa is India's smallest state by area and the fourth smallest by population. Located in West India in the region known as the Konkan, it is bounded by the state of Maharashtra to the north and by Karnataka to the east and south, while the Arabian Sea forms its western coast. Goa is India's richest state with a GDP per capita two and a half times that of the country as a whole. This maritime state encompasses an area of 3,702 km^2 (1,429 mi^2) and lies between the latitudes 14° 53′ 54″N and 15° 40′ 00″N and

Fig. 7.39 Grazing of mangroves by buffaloes

longitudes 73° 40′ 33″E and 74° 20′ 13″E. Goa has a coastline of 101 km (63 mi). In Goa, several coastal areas have changed from virtual wilderness in 1970s to haphazardly developed stretches, full of concrete buildings and related structures, in less than 20 years. The Baga–Candolim coast is a classic example of frenzied development (Mascarenhas 1997). Several shore fronts have been designed and built in such a manner that they bear little resemblance to the coast that formerly existed. Many coastal communities thus experienced a dramatic growth during this period with constructions of high-rise buildings, resorts, residential dwellings that mushroomed almost all over the coastal zone of Goa. During the last two decades, the advent of tourism, population increase coupled with building activity and modern societal demands, has resulted in large-scale changes in the geological and ecological set-up and has indelibly altered ecosystems, land use patterns and the coastal zone landscape (Mascarenhas 1997). Figures 7.40 and 7.41 show some developmental concrete constructions almost adjacent to the high-tide line (HTL), which can accelerate the rate and magnitude of erosion subsequently leading to destruction of coastal vegetation.

E. Pollution

Oil spills are the most frequent catastrophic type of disaster. The toxicity varies with the type of oil and the degree of breakdown at sea before it arrives on the shore. Physical smothering is often a more important cause of mortality than chemical poisoning, and clean-up operations often kill more marine life than the oil itself. Different habitats have different sensitivities to oil spills: exposed rocky shores are least sensitive, then come sandy beds and finally by sheltered rocky shores, mudflats and salt marshes, which are the most sensitive. The coastal vegetation particularly the mangroves is highly sensitive to mangroves

Mangrove forests are adversely affected by oil pollution and related developments. Oil spills are a serious concern in regard to the health of our planet's remaining mangrove forests. Leaked oil permeates the coastal waters and streams, coating the exposed, air breathing roots of the mangroves. It is difficult, if not impossible, for the plants' breathing lenticels to perform their essential functions, thus in effect slowly suffocating the mangroves. Massive mangrove die-offs are common phenomena plaguing the mangrove regions where coastal oil exploitation occurs.

Fig. 7.40 Hard concrete structures adjacent to the HTL on the Goa coast (15° 29′ 5.0″N and 73° 47′ 40.4″E); photograph taken on 4 April 2013 by Dr. Shubhadra Devi Gadi (Zoologist)

Fig. 7.41 A hotel under construction on the Goa coast (15° 29′ 5.1″N and 73° 47′ 40.9″E); photograph taken on 4 April 2013 by Ms. Ankita Mitra (Environmentalist)

In salt marshes, the light fractions of oil are the most toxic, but the heavier fractions and crude oil can completely smother the vegetation (Baker 1979; reviews: Long ad Mason 1983; Clarke 1992) (Fig. 7.42). Oil readily becomes trapped by both the vegetation and sediments and may persist for a long time. Annual species are most sensitive; either they are killed or their reproduction is severely impaired. Reseeding is required from remote sources, and it can take 2–3 years before the populations recover. Shallow-rooted annuals, such as *Suaeda maritime*, and *Salicornia* sp., have little or no food reserves and are particularly susceptible. Other species (e.g. *Juncus maritimus*) will survive single light exposures, but frequent light oilings will kill them.

Fig. 7.42 Mangrove roots masked with oil

Many oil products are highly viscous. In particular, crude oils and heavy fuel oils can be deposited on shorelines and shoreline resources in thick, sticky layers that may either disrupt or completely prevent normal biological processes of exchange with the environment. Even if a petroleum product is not especially toxic in its own right, when oil physically covers plants and animals, they may die from suffocation, starvation or other physical interference with normal physiological function.

F. Construction of Dams and Barrages

The Bangladesh part of the Sundarbans spreads over the districts of Khulna, Bagerhat and Satkhira. It covers almost 62 % of total land cover, and geographically, it lies between latitudes 21° 31′N and 22° 30′N and between longitudes 89° 18′E and 90° 18′E (Katebi 2001). The rivers of the Bangladesh Sundarbans are more stable than the main stream of the Ganges. The erosion and accretion balance in the Sundarbans have been estimated to be 146 km^2 for the period 1960–1984 (Jabbar 1995; Siddiqi 2001a, b). The deltaic swamp of the Sundarbans is flat with micro-topographical exceptions only. It runs from north to south at a slope of 0.03 m vertically per km of horizontal distance. Micro-topographical variation is due to rivers, streams and tidal flow (Begum 1987; Ansarul 1995).

With commissioning of the Farakka Barrage, the discharge in the downstream was drastically reduced. As a result, all the components of ecosystems development on the availability of water started being affected. The dominant species Sundari (*H. fomes*) and Goran (*Ceriops decandra*) are affected by top-dying disease. Top dying of *H. fomes* in the Sundarbans is considered as the most serious of all the diseases and disorders in Bangladesh mangrove system (Rahman 2001; Islam 2008).

G. Exploitation of Resources

A heavy dependence on foraging for shellfish on the shore was typical of many of the semi-nomadic indigenous peoples encountered by Europeans for the first time in the sixteenth-–eighteenth centuries, for example the bushmen, called ‘strandlopers’ by the Dutch at the Cape, Native Americans and Native Australians (Meehan 1982; Alfredson 1984). Collecting of shellfish was, and often still is, an important element of the activities of peoples with permanent settlements who also fished and farmed, such as the Polynesian cultures and Native Americans. In many cases, seashells or their products (e.g. wampum in north-east America, cowries in Polynesia) were used as currency. In Europe, widespread subsistence collecting of shellfish was common from prehistoric times until the late middle ages.

It has been documented that in Indian coastal zone, about 3,600 tonnes of molluscs are removed annually from Kakinada Bay and Coringa mudflats for lime production (Fig. 7.43). Species of bivalves (*Placuna placenta*, *Anadara granosa*, *Macoma* sp, *Meretrix* sp.) and gastropods (*Cerithidea cingulata*, *Telescopium telescopium*) are regularly handpicked.

Fig. 7.43 Molluscan shell heaped up for lime production

7.3 Case Studies on Blue Carbon Loss

The environment of the planet Earth is undergoing remarkable changes and the rate of such changes is increasing particularly due to unplanned human activities. The root cause lays basically on the unprecedented rise in human population through the twentieth century (7,237,115,621; Indian Standard Time 2:23 PM, dated 01.06.2014) from 1.5 to 6 billion people in 2000. By 2050, there will be another 30 % increase in human population. This will increase the demands for energy, food, water, space and other resources, which will create several social, economic and political complications that will lead to deterioration of the environment. It is interesting to note, in this context, that major cities of the World have developed very near to the coast. The land–sea interface that supports the blue carbon reservoir such as mangroves and salt marshes is under the clutch of intense industrialization and urbanization. In addition to this, factors like sea level rise is also considered as a major threat to blue carbon community (Nicholls et al. 1999). Observed through modelling approaches that there will be a loss of 13–31 % of coastal wetland of which 0–20 % would probably due to sea level rise. More recently, (IPCC 2007) published that a figure around 30 % loss of coastal wetlands worldwide. Data from Valiela et al. (2001) reveal a substantial loss of mangroves in different continents of the world (Table 7.11).

Salt marsh also contributes a very important component of blue carbon community. Several causes have been identified from the loss of this community, which include submerges by sea level rise, erosion, draught, warming, grazing and fungal infection (Flory and Alber 2002; Alber et al. 2008). Another cause of salt marsh loss is due to invasive reed expansion. The invasive growth usually occurs along the upper edges of salt marshes and gradually extends towards the sea. The invasive taxon appears to be more tolerant to salinity (Vazquez et al. 2006) grows better in response to genotype (Packett and Chambers 2006) and seems to be favoured by the urbanization of the adjoining water sheds (King et al. 2007). *Phragmites australis* is a common reed and is highly invasive with wide range of tolerance to salinity.

Nutrient enrichment is also a major driver of salt marsh loss. We conducted a study during 2013 on the impact of nitrate and phosphate concentrations on the aboveground biomass (AGB) of *P. coarctata* in the three sectors of Indian Sundarbans with variable nutrient concentration and observed significant adverse impact of nutrient enrichment on the AGB of the salt marsh grass (Tables 7.12, 7.13 and 7.14). The results are in alignment with works of Deegan et al. (2012) who observed that the recent trend of nutrient enrichment has exceeded the capacity of salt marsh to remove nutrient from the ambient media. The nutrient enrichment in the present time may be attributed to the increased use of fertilizer, urbanization, and

Table 7.11 Current mangrove swamp area, per cent loss, annual loss rate and per cent of original area lost per year, for the mangroves of the continents and the world

Continent	Current mangrove area (km^2)	% loss of mangrove forest area	Annual rate of loss (km^2/year)	% loss of original area loss/year
Asia	77,169	36	628	1.52
Africa	36,529	32	274	1.25
Australia	10,287	14	231	1.99
America	43,161	38	2,251	3.62
World	166,876	35	2,834	2.07

Source Valiela et al. (2001)

Table 7.12 Nutrient concentrations and AGB of *P. coarctata* in the western, central and eastern Indian Sundarbans during April 2013

Variable	Western	Central	Eastern
Nitrate (μgat/l)	29.85	23.40	16.23
Phosphate (μgat/l)	3.14	1.88	0.69
Biomass (gm/m^2)	194.63	201.10	227.53

Table 7.13 Nutrient concentrations and AGB of *P. coarctata* in the western, central and eastern Indian Sundarbans during September 2013

Variable	Western	Central	Eastern
Nitrate (μgat/l)	33.78	27.13	20.38
Phosphate (μgat/l)	3.98	2.02	1.04
Biomass (gm/m^2)	187.55	198.33	205.85

Table 7.14 Nutrient concentrations and AGB of *P. coarctata* in the western, central and eastern Indian Sundarbans during November 2013

Variable	Western	Central	Eastern
Nitrate (μgat/l)	31.66	25.50	18.59
Phosphate (μgat/l)	3.70	1.96	0.93
Biomass (gm/m^2)	190.65	ND	ND

ND means Not Done as the authors could not collect the sample because of tiger straying from the adjacent Reserve Forest

waste released from shrimp farms and these activities are much more pronounced in the western sector compared to the central and eastern sectors. The highest AGB values of *P. coarctata* in the eastern sector coincide with the lowest nutrient concentration in the ambient water, and this is due to minimum intervention of human in this sector owing to the location of this sector in the Reserve Forest Area. Projected increase in nitrogen flux to the coast, related to feed an expanding human population, may rapidly result in a coastal landscape with less marsh, which would reduce the capacity of coastal region to provide important ecological and economic services (Deegan et al. 2012).

We also observed a local level damage to blue carbon (preferably saplings and seedlings of mangroves) by prawn seed collectors of Indian Sundarbans. At present, the major environmental-cum-socioeconomic issue in this mangrove ecosystem is the proliferation of the tiger prawn or shrimp (*Penaeus monodon*) farms (Fig. 7.44).

In the absence of shrimp hatchery in the entire region, the supply of tiger prawn seeds to these farms is done through the wild collection from the

Fig. 7.44 Shrimp farms in Indian Sundarbans: a major threat to mangrove ecosystem

Fig. 7.45 Sorting of tiger prawn seeds from the wild collection of estuarine water

estuarine and coastal waters. Thousands of aged person, women and children collect wild tiger prawn seeds by employing nets of a particular mesh size to haul in the drifting community, irrespective of the tides, and throw away the major portion of the haul (containing the juveniles of finfish and shellfish) after sorting out the post larvae (or seeds) of tiger prawn (Fig. 7.45). This practice results in a great loss of pelagic biodiversity (in terms of stock), which might cause an adverse effect on the delicate ecological balance of the system. This occupation has spread its root

Fig. 7.46 Seedlings of mangroves along the bank of estuary

to such an extent that about 95 % of the coastal population living below poverty line is engaged with this destructive activity. It is very difficult to implement a total ban on this activity, as these huge masses will face extreme economic marginalization.

The dragging of nets along the banks of estuary often uproots the seedlings of mangroves (Fig. 7.46), and thus, the afforestation and eco-restoration programmes of the governments and NGOs often end in smoke.

References

Ahmad S, Suratman MN (2007) Detection of changes in mangrove forests using Landsat TM in Pulau Indah, Malaysia. Paper to be presented at the conference of forestry and forest products (CFFPR 2007), 27–29 Nov 2007, Kuala Lumpur

Alber M, Swenson EM, Adamowicz SC, Mendelssohn IA (2008) Saltmarsh dieback: an overview of recent events in the U.S. Estuar Coast Shelf Sci 80:1–11

Alfredson G (1984) The aboriginal use of St. Helena Island, Moreton Bay-the archaeological evidence. In: Coleman RJ, Ciovacevich J, Davie P (eds) Focus on stadbroke. Boolarong Publications, Brisbane, pp 1–15

Andrews TJ, Clough BF, Muller GJ (1984) Photosynthetic gas exchange properties and carbon isotope ratios of some mangroves in North Queensland. In: Teas HJ (ed) Physiology and management of mangroves, tasks for vegetation science, vol 9. Dr. W. Junk, The Hague, pp 15–23

Ansarul K (1995) FAO/UNDP project BGD/84/056. In: Integrated resource development of the Sundarbans reserved forest. Report on Mangrove Silviculture, vol 1, FAO/UNDP Khulna

Aksornkoae S, Paphavasit N, Wattayakorn G (1993) Mangroves of Thailand: present status of conservation, use and management. In: Clough B (ed) The economic and environmental values of mangrove forests and their present state of conservation in the South-East Asia/Pacific region. Mangrove ecosystems technical reports. International Society for Mangrove Ecosystems, Okinawa, pp 133–183

Ball MC, Cochrane MJ, Rawason HM (1997) Growth and water use of the mangroves *Rhizophora apiculata* and *R. stylosa* in response to salinity and humidity under ambient and elevated concentration of atmospheric CO_2. Plant, Cell Environ 20:1158–1166

Ball MC, Farquhar GD (1984a) Photosynthetic and stomatal responses of two mangrove species, *Aegiceras corniculatum* and *Avicennia marina*, to long term salinity and humidity conditions. Plant Physiol 74:1–6

Ball MC, Farquhar GD (1984b) Photosynthetic and stomatal responses of the grey mangrove, *Avicennia marina*, to transient salinity conditions. Plant Physiol 74:7–11

Baker JM (1979) Responses of salt marsh vegetation to oil spills and refinery effluents. In: Jefferies RL, Davy AJ (eds) Ecological processes in coastal environments. Blackwell Scientific, Oxford, pp 529–542

Banerjee K, Sengupta K, Raha A, Mitra A (2013) Salinity based allometric equations for biomass estimation of Sundarban mangroves. Biomass Bioenergy 56:382–391

Begum K (1987) Tension over the Farakka Barrage: a techno political tangle in South Asia. University Press Limited, Dhaka, pp 1–297

Blasco F (1975) The mangroves of India. Institut Français Pondichery Travaux Section Scientifique et Technologie 14:1–175

Boyd PW, Watson AJ, Law CS, Abraham ER, Trull T, Murdoch R, Bakker R, Bowie AR, Buessler KO, Chang H (2000) A measureable phytoplankton bloom in the polar southern ocean stimulated by iron fertilization. Nature 407:695–702

Brierley AS, Thomas DN (2002) Ecology of Southern Ocean pack ice. Adv in Mar Biol 43:171–276

Chan HT, Ong JE, Gong WK, Sasekumar A (1993) The socio-economic, ecological and environmental values of mangrove ecosystems in Malaysia and their present state of conservation. The economic and environmental values of mangrove forests and their present state of conservation in the South-East Asia/Pacific region, vol 3. Mangrove ecosystems technical reports, pp 41–81

Chaudhuri AB, Choudhury A (1994) Mangroves of the Sundarbans. The World Conservation Union, Dhaka

Chatenoux B, Peduzzi P (2007) Impacts of the 2004 Indian Ocean tsunami: analysing the potential protecting role of environmental features. Nat Hazards 40:289–304

Cheeseman JM, Clough BF, Carter DR, Lovelock CE, Ong JE, Sim RG (1991) The analysis of photosynthetic performance of leaves under field conditions: a case study using Bruguiera mangroves. Photosynth Res 29:11–22

Clarke RB (1992) Marine pollution, 3rd edn. Clarendon Press, Oxford

Clough BF, Andrews TJ, Cowan IR (1982) Physiological processes in mangroves. In: Clough BF (ed) Mangrove ecosystems in Australia: structure, function and management. Australian National University Press, Canberra, pp 193–210

Cole CV, Vaidyaraman PP (1966) Salinity distribution and effect of freshwater flows in the Hooghly River. In: Proceedings tenth conference on coastal engineering, Tokyo. American Society of Civil Engineers, New York, pp 1312–1434

Collen J, Guisle-Marsollier I, Leger JJ, Boyen C (2007) Response of the transcriptome of the intertidal red seaweed *Chondrus crispus* to controlled and natural stresses. New Phytol 176:45–55

Cure JD, Acock B (1986) Crop responses to carbon dioxide doubling: a literature survey. Agric For Meteorol 38:127–145

Dahdouh-Guebas F, Jayatissa LP, Di Nitto D, Bosire JO, Lo Seen D, Koedam N (2005) How effective were

mangroves as a defence against the recent tsunami? Curr Biol 15:R443–R447
Danielsen F, Sørensen MK, Olwig MF, Selvam V, Parish F, Burgess ND, Hiraishi T, Karunagaran VM, Rasmussen MS, Hansen LB, Quarto A, Suryadiputra N (2005) The Asian tsunami: a protective role for coastal vegetation. Science 310:643
Davison IR, Pearson GA (1996) Stress tolerance in intertidal seaweeds. J Phycol 32:197–211
Deb SC (1956) Paleoclimatology and geophysics of the Gangetic delta. Geogr Rev India 18:11–18
Deegan LA, Johnson DS, Warren RS, Fleeger JW, Fagherazzi S, Wolheim WM (2012) Coastal eutrophication as a driver of salt marsh loss. Nature 490:388–392
Drake BG (1992) A field study of the effects of elevated CO_2 on ecosystem processes in a Chesapeake Bay wetland. Aust J Bot 40:579–595
Eamus D, Jarvis PG (1989) The direct effects of increase in the global atmospheric CO_2 concentration on natural and commercial temperate trees and forests. Adv Ecol Res 19:1–55
Eggert A (2012) Seaweed responses to temperature. In: Wiencke C, Bischof K (eds) Seaweed biology. Springer, Berlin, pp 47–66
Ellison J (2000) How South Pacific mangroves may respond to predicted climate change and sea level rise. In: Gillespie A, Burns W (eds) Climate change in the South Pacific: impacts and responses in Australia, New Zealand, and Small Islands states. Kluwer Academic Publishers, Dordrecht, pp 289–301 (Chapter 15)
FAO (2003) Status and trends in mangrove area extent worldwide. Food and agriculture organization of the United Nations. Forest Resources Division, Paris
Farnsworth EJ, Ellison AM, Gong WK (1996) Elevated CO_2 alters anatomy, physiology, growth and reproduction of red mangrove (*Rhizophora mangle* L.). Oecologia 108:599–609
Field C (1995) Impacts of expected climate change on mangroves. Hydrobiologia 295:75–81
Flory J, Alber M (2002) Dead Marsh Information Georgia Coastal Research Council Nov 2002. http://www.fbbva.es/TLFU/dat/ou%20valiela_separatapdf
IFRC (2013) Disaster relief emergency fund (DREF) India: Cyclone Phailin DREF operation no. MDRIN013 14 Oct 2013. International Federation of Red Cross (IFRC) and Red Crescent Societies. http://www.ifrc.org/docs/Appeals/13/MDRIN0013dref.pdf. Accessed on 24 Oct 13
Froberg M (2013) Preparedness saved thousands of lives during cyclone Phailin; now the recovery begins. International Federation of Red Cross and Red Crescent Societies, 17 Oct 2013. http://www.ifrc.org/en/news-and-media/news-stories/asia-pacific/india/recovering-from-cyclone-phailin-survivors-face-massive-hardships–63582/. Accessed on 24 Oct 13
Fog GE (1998) The biology of polar habitats. Oxford University Press, Oxford
Gattuso JP, Frankignoulle M, Bourge I, Romaine S, Buddemeier RW (1998) Effect of calcium carbonate saturation of seawater on coral calcification. Glob Planet Change 18:37–46
Gazeau F, Quiblier C, Jansen JM, Gattuso JP, Middelburg JJ, Heip CHR (2007) Impact of elevated CO_2 on shellfish calcification. Geophys Res Lett L07603. doi:10.1029/2006GL028554
GoO (2013) Memorandum on the very severe cyclone Phailin and the subsequent flood, 12–15 Oct 2013. The Revenue and Disaster Management Department Government of Odisha. http://www.osdma.org/userfiles/file/MEMORANDUMPhailin.pdf. Accessed on 29 Oct 13
IPCC (International Panel on Climate Change) (2007) The physical science basis: contribution of working group I to the fourth assessment report of the intergovernmental panel on climate change. In: Solomon S, Qin D, Manning M, Chen Z, Marquid M, Aueryl KB, Tignor M, Miller HL (eds) Cambridge University Press, Cambridge
Islam SN (2008) Cultural landscapes changing due to anthropogenic influences on surface water and threats to mangrove wetland ecosystems: a case study on the Sundarbans, Bangladesh. Unpublished Ph.D. thesis, Department of Ecosystems and Environmental Informatics, Brandenburg University of Technology in Cottbus, Germany
IUCN (1989) The impact of climatic change and sea level rise on ecosystems. Report for the Commonwealth Secretariat, London
IUCN (2005) Early observations of tsunami effects on mangrove and coastal forests. www.iucn.org
Jabbar MA (1995) Farakka barrage and its impact on the environment of southwest region of Bangladesh. In: Hasna JM (ed) Women for water sharing. Academic Publisher, Dhaka, pp 65–70
Katebi MNA (2001) Sundarbans and forestry. In: Haider R (ed) Cyclone'91: an environmental and perceptional study. BCAS, Dhaka, pp 79–100
Kathiresan K, Rajendran N (2005) Coastal mangrove forests mitigated tsunami. Estuar Coast Shelf Sci 65:601–606
Kerr AM, Baird AH, Campbell SJ (2006) Comments on "coastal mangrove forests mitigated tsunami" by K. Kathiresan and N. Rajendran [Estuarine, Coastal and Shelf Science 65 (2005) 601–606]. Estuar Coast Shelf Sci 67:539–541
Kimball BA (1983) Annual report, U.S. Water Conservation Laboratory, Phoenix, AZ
Kim E, Park HS, Jung Y, Choi DW, Jeong WJ, Hwang MS, Park EJ, Gong YG (2011) Identification of the high-temperature response genes from *Porphyra seriata* (*Rhodophyta*) expression sequence tags and enhancement of heat tolerance of *Chlamydomonas* (*Chlorophyta*) by expression of the Porphyra HTR2 gene. J Phycol 47:821–828
King RS, Deluca WV, Whigham DF, Marra PP (2007) Threshold effects of coastal urbanization on *Phragmites australis* (*common reed*) abundance and foliar nitrogen in Chesapeake Bay. Estuar Coast 30(3):469–481
Kordas RL, Harley CDG, O'Connor MI (2011) Community ecology in a warming world: the influence of

temperature on interspecific interactions in marine systems. J Exp Mar Biol Ecol 400:218–226
Koul DN, Panwar P (2008) Prioritizing land-management options for carbon sequestration potential. Curr Sci 95 (5):658–663
Kleypas JA, Feely RA, Fabry VJ, Langdon C, Sabine CL, Robbins LL (2006) Impacts of ocean acidification on coral reefs and other marine calcifiers: a guide to future research. Report of a workshop held 18–20 Apr 2005, St. Petersburg, sponsored by NSF, NOAA, and the U.S. Geological Survey, p 88
Kurihara K, Shirayama Y (2004) Impacts of increased atmospheric CO_2 on sea urchin early development. Mar Ecol Prog Ser 274:161–169
Komiyama A, Ong JE, Poungparn S (2007) Allometry, biomass and productivity of mangrove forests a review. Aquat Bot. doi:10.1016/j.aquabot.2007.12.006 (this issue)
Langdon C, Takahashi T, Marubini F, Atkinson M, Sweeney C, Aceves H, Barnett H, Chipman D, Goddard J (2000) Effect of calcium carbonate saturation state on the calcification rate of an experimental coral reef. Global Biogeochem Cycles 14:639–654
Lewis S, Donkin ME, Depledge MH (2001) Hsp70 expression in *Enteromorpha intestinalis* (*Chlorophyta*) exposed to environmental stressors. Aquat Toxicol 51(277):91
Long SP, Mason CF (1983) Saltmarsh ecology. Blackie, Glasgow
Luning K (1990) Seaweeds: their environment, biogeography, and ecophysiology. Wiley, New York
MacKinnon J (1997) Protected areas systems review of the Indomalayan Realm. The World Bank, Asian Bureau for Conservation (ABC) and World Conservation Monitoring Centre (WCMC), Canterbury
Mascarenhas B (1997) The order and size of entry into international markets. J Bus Ventur 12(4):287–299
McLusky DS, Bryant DM, Elliot M (1991) The impact of reclamation on benthic production and prey availability in the Forth Estuary, eastern Scotland. In: Pethick J (ed) The changing coastline. Kluwer, Dordrecht
Meehl GA, Tebaldi C, Nychka D (2004) Changes in frost days in simulations of twenty-first century climate. Clim Dyn 23:495–511
Meehan B (1982) Shell bed to shell midden. Australian Institute of Aboriginal Studies, Globe Press, Melbourne
Millennium Ecosystem Assessment (2005) Ecosystems and human well-being: biodiversity synthesis. World Resources Institute, Washington, DC
Mitra A, Banerjee K, Sengupta K, Gangopadhyay A (2009) Pulse of climate change in Indian Sundarbans: a myth or reality? Natl Acad Sci Lett 32:1–2
Mitra A, Mondal K, Banerjee K (2011) Spatial and tidal variations of physico–chemical parameters in the lower Gangetic Delta region, West Bengal, India. J Spat Hydrol, Am Spat Hydrol Union Spring 11 (1):52–69
Maung MT (2008) Devastation of the 2008 cyclonic storms on mangroves and other coastal systems in Myanmar. Country report, Bangkok workshop
Nicholls RJ, Hoozemans FMJ, Marchand M (1999) Increasing flood risks and wetland losses due to global sea level rise: regional and global analysis. Glob Environ Change 9:S69–S87
Ong JE (1982) Mangroves and mariculture. Ambio 11:252–287
Packett CR, Chambers RM (2006) Distribution and nutrient status of halotypes of the marsh grass *Phragmites australis* along the Rappahannock River in Virginia. Estuar Coasts 29:1222–1225
Parish F, Lee D (2005) Preliminary information on impacts of the 24 Dec 2004. Tsunami on selected coastal ecosystems in Aceh Province, Indonesia. Global Environment Centre, Malaysia. www.riverbasin.org
Parish F (2005) Press release: Mangroves might have helped to protect Banda Aceh. Global Environment Centre, Malaysia. http://www.riverbasin.org/ev_en.php?ID=3794_201&ID2=DO_Topic
Prater AJ (1981) Estuary birds of Britain and Ireland, Poyser. T&D Poyser, Calton, Berkhamstead, p 440
Primavera J (1997) Socio-economic impacts of shrimp culture. Aquac Res 28:815–827
PONJA report, July 2008, Annex 12
Portner HO, Langenbunch M, Michaelidis B (2005) Synergistic effects of temperature extremes, hypoxia, and increases in CO_2 on marine animals. From Earth history to global change. J Geophys Res 110:C09S10. doi:10.1029/2004JC002561
Rahman A (2001) Diseases and disorders of tree species in the Sundarbans and their management. In: Siddiqi NA, Baksha MW (eds) Proceeding of the National workshop on mangrove research and development at Bangladesh, 15–16 May. Forest Research Institute Chittagong, Bangladesh, pp 86–97
Riebesell U, Zondervan I, Rost B, Tortell PD, Zeebe RE, Morel FMM (2000) Reduced calcification of marine plankton in response to increased atmosphere CO_2. Nature 407:364–367
Secretariat R (2001) Wetland values and functions: climate change mitigation. Gland, Switzerland
Saclauso CA (1989) Brackish water aquaculture: threat to the environment? Naga ICLARM Q 12(3):6–8
Senapati A (2013) Mass evacuation on in coastal districts. Down to Earth. 11 Oct 2013. http://www.downtoearth.org.in/content/mass-evacuation-coastal-districts. Accessed on 24 Oct 2013
Sengupta K, Roy Chowdhury M, Bhattacharya SB, Raha A, Zaman S, Mitra A (2013) Spatial variation of stored carbon in *Avicennia alba* of Indian Sundarbans. Discov Nat 3(8):19–24
Smith J, Schellinhuber H, Mirza M (2001) Vulnerability to climate change and reasons for concern: a synthesis (Chapter 19), pp 913–967. In: McCarthy J, Canziani O, Leary N, Dokken D, White K (eds) Climate change 2001: impacts, adaptation, and vulnerability. Published for the Intergovernmental Panel on Climate Change. Cambridge University Press, Cambridge and New York

Siddiqi NA (2001a) Mangrove of Bangladesh Sundarbans and accretion areas. In: Lacerda DDL (ed) Mangrove ecosystems functions and management. Springer, Berlin, pp 142–258

Siddiqi NA (2001b) Mangrove forestry in Bangladesh. Institute of Forestry and Environmental Sciences, University of Chittagong, Nibedon Press Limited, Chittagong, p 201

Snedaker C (1991) Notes on the Sundarbans with emphasis on geology, hydrology and forestry. In: Seidensticker J, Kurin R, Townsend AK (eds) The commons in South Asia: societal pressures and environmental integrity in the Sundarbans. The International Center, Smithsonian Institution, Washington, D.C

Snedaker S (1995) Mangroves and climate change in the Florida and Caribbean region: scenarios and hypotheses. Hydrobiologia 295:43–49

Solomon S, Qin D, Manning M, Alley RB, Berntsen T, Bindoff NL, Chen Z, Chidthaisong A, Gregory JM, Hegerl GC, Heimann M, Hewitson B, Hoskins BJ, Joos F, Jouzel J, Kattsov V, Lohmann U, Matsuno T, Molina M, Nicholls N, Overpeck J, Raga G, Ramaswamy V, Ren J, Rusticucci M, Somerville R, Stocker TF, Whetton P, Wood RA, Wratt D (2007) Technical summary. In: Solomon S, Qin D, Manning M, Chen Z, Marquis M, Averyt KB, Tignor M, Miller HL (eds) Climate change 2007: the physical science basis. Contribution of working group I to the fourth assessment report of the intergovernmental panel on climate change. Cambridge University Press, Cambridge and New York

Tabuchi R (2003) The rehabilitation of mangroves in Southeast Asia. Retrieved 1 June 2007 from http://www.unu.edu/env/plec/marginal/proceedings/TabuchiCH12.pdf

Thomas DN, Dieckmann GS (2002a) Antartic sea-ice-a habitat for extremophiles. Science 295:641–644

Thomas DN, Dieckmann GS (2002b) Biogeochemistry of Antarctic sea ice. Oceanogr Mar Biol Anu Rev Lebensraum Meereis: Bremer Forscher untersuchen das Leben im Eis Biologen heute 2:8–9. hdl:10013/epic.16054

UNEP (1994) Assessment and monitoring of climate change impacts on mangrove ecosystems. UNEP regional seas reports and studies no. 154. Regional Seas Programme, Nairobi, Kenya

Valiela I, Bowen J, York J (2001) Mangrove forests: one of the world's threatened major tropical environments. Bioscience 51:807–815

Vermaat JE, Thampanya U (2006) Mangroves mitigate tsunami damage: a further response. Estuar Coast Shelf Sci 69:1–3

Vermaat JE, Thampanya U (2007) Erratum to "mangroves mitigate tsunami damage: a further response" [Estuarine Coastal Shelf Science, 69 (1–2) (2006) 1–3]. Estuar Coast Shelf Sci 75:564

Vasquez EA, Glenn GR, Gutenspergen GR, Brown JJ, Nelson SG (2006) Salt tolerance and osmotic adjustment of *Spartina alterniflora* (*Poaceae*) and the invasive halotype of *Phragmites australis* (*Poaceae*) along a salinity gradient. Am J Bot 93:1784–1790

Vavrus SJ, Walsh JE, Chapman WL, Portis D (2006) The behavior of extreme cold air outbreaks under greenhouse warming. Int J Climatol 26:1133–1147

Vayda ME, Yuan ML (1994) The heat-shock response of an Antarctic alga is evident at 5 °C. Plant Mol Biol 24:229–233

Wells S (2006) Establishing national and regional systems of MPAs: a review of progress with lessons learned. UNEP World Conservation Monitoring Centre, UNEP Regional Seas Programme, ICRAN, IUCN/WCPA—Marine

Weincke C, Clayton MN (2002) Biology of Antarctic seaweeds. Koeltz Scientific Books, Koenigstein

Wetlands International—Indonesia Programme (2005) Tsunami of Aceh and North Sumatra, 26 Dec 2004. www.wetlands.or.id

Woodroffe CD, Grindrod J (1991) Mangrove biogeography: the role of quaternary environmental and sea-level change. J Biogeogr 18:479–492

Internet Reference

ftp://ftp.fao.org/docrep/fao/010/A0701E/A0701E00.pdf

8 Road Map to Expand Blue Carbon Reservoir

...For achieving a successful conservation.
Human beings must have certain reservation...
The Authors

8.1 Conservation Approach

Coastal vegetations are efficient carbon sinks that sequester and store large quantum of carbon in their biomass (above- and below-ground biomass) and underlying sediments for a considerable long period of time. The carbon sequestration capacity of mangroves, seagrass beds, tidal marshes and other marine and coastal vegetated ecosystems has been a focus of considerable attention in recent times, and the carbon stored in their biomass and sediments is coined as 'blue carbon' (Nelleman et al. 2009; Gordon et al. 2011; Pendleton et al. 2012). Blue carbon has been more formally defined as 'the carbon stored, sequestered or released from coastal ecosystems of tidal marshes, mangroves and seagrass meadows' (Herr et al. 2012). The exact amount of carbon stored by coastal vegetated ecosystems is still an active area of research (Pendleton et al. 2012). Many biotic components of coastal ecosystems like phytoplankton, bivalves and gastropods also store carbon in their biomass, but very few literatures reflect their storage potential. The readers may get detail information on these gap areas from Chaps. 5 and 6 of this book. The overall conclusion of several researches, articles and books, however, suggests that coastal ecosystems are among the most intense carbon sinks on the planet Earth (Nellemann et al. 2009). It is therefore extremely essential to conserve and expand the horizon of blue carbon reservoir for the sake of human civilization and sustain an equilibrium climatic condition.

The scientific conservation of coastal vegetation and other biotic reservoir requires a detailed knowledge of their (i) status, (ii) threats and (iii) scientific management. Several literatures available on the latest status do not give a comprehensive picture on global basis; rather, many studies are conducted at extremely local level or regional level. A basic background on blue carbon is depicted in this chapter that can serve as a backbone to formulate a policy to conserve this valuable domain of the planet.

8.1.1 Status

Blue carbon habitats are distributed globally and are concentrated in the coastal regions. Some important figures on the storage capacity of carbon dioxide by coastal vegetation are highlighted below.

1. Mangroves are among the most carbon-rich habitats in the tropics, storing as much as 3,100–4,400 metric tonnes of carbon dioxide equivalent per hectare (CO_2e/ha) in their biomass and soils (Donato et al. 2011). Several articles published on mangrove carbon, however, exhibit different figures in the context of storage. Chmura et al. (2003) found that the average carbon density of mangroves equals 55,000 g/m^3 (2,016.67 metric tonnes CO_2e/ha per metre depth). Ong (2002) estimated that sediments in mangrove forests held 700 tonnes of carbon per metre depth per hectare.
2. The storage capacity of salt marshes and seagrasses is less than that of mangroves. On

A. Mitra and S. Zaman, *Blue Carbon Reservoir of the Blue Planet*,
DOI 10.1007/978-81-322-2107-4_8, © Springer India 2015

average, coastal salt marshes store 362–2,012 tonnes of CO_2e/ha (Chmura et al. 2003).
3. Seagrasses store from 66 to 1,478 tonnes of CO_2e/ha, which is mostly sequestered in the soils of seagrass beds (Mateo et al. 1997; Vichkovitten and Holmer 2005; Gordon et al. 2011).
4. Table 8.1 reflects the standing carbon stock in the plant biomass and soil under the domain of blue carbon, and the figures are compared with other forest types of the planet.

8.1.2 Threats

The present century is the age of human population explosion whose chain reactions are urbanization, industrialization, unplanned change of land use, exploitation of living and non-living resources and several other activities that threaten the existence of human civilization. Expected scenarios of human population growth and consumption levels indicate that cumulative human demands will impose an unsustainable toll on Earth's ecological resources and services accelerating the rate at which biodiversity is being lost (Mora and Sale 2011). The marine, coastal and estuarine ecosystems sustaining a wide spectrum of flora and fauna are no exceptions to this rule. Despite their enormous socio-economic value and ecosystem services, estuarine and coastal ecosystems are some of the most heavily used and threatened natural systems globally (Lotze et al. 2006; Worm et al. 2006; Halpern et al. 2008). Their deterioration due to human activities is intense and increasing and has been the subject of a number of studies during recent years. Researches on this vertical depict that 50 % of salt marshes, 35 % of mangroves and 29 % of seagrasses have been either lost or degraded worldwide over the last 50–100 years (Valiela et al. 2001; MEA 2005; Orth et al. 2006; UNEP 2006; FAO 2007; Waycott et al. 2009). Causes of habitat conversion vary globally, but can be largely traced to human development pressures. They include conversion to aquaculture, agriculture, forest overexploitation, industrial use, upstream dams, dredging, eutrophication of overlying waters, urban development and conversion to open water due to accelerated sea level rise and subsidence

Table 8.1 Comparison of carbon stocks and accumulation of carbon in soils in terrestrial and coastal ecosystems

Ecosystem type	Standing carbon stock (gC/m^2)		Total global area ($*10^{12}$ m^2)	Global stocks ($*10^{15}$ gC)		Soil carbon accumulation rate (gC/m^2/year)
	Plants	Soil		Plants	Soil	
Tropical forests	12,045	12,273	17.6	212	216	2.3–2.5
Temperate forests	5,673	9,615	10.40	59	100	1.4–12.0
Boreal forests	6,423	34,380	13.7	88	471	0.8–2.2
Tropical savanna and grassland	2,933	11,733	22.5	66	264	–
Temperate grassland and shrublands	720	23,600	12.5	09	295	2.2
Deserts	176	4,198	45.5	08	191	0.8
Tundra	632	12,737	9.5	06	121	0.2–5.7
Croplands	188	8,000	16.0	03	128	–
Wetlands	4,286	72,857	3.5	15	225	20
Tidal salt marshes			Unknown (0.22)			
Mangroves	**7,990**		**0.152**		**1.2**	**139**
Seagrasses	184	7,000	0.3	0.06	2.1	83
Kelp forests	120–720	Na	0.02–0.4	0.009–0.02	Na	Na

Source Laffoley and Grimsditch (2009)

(Pendleton et al. 2012). The clearance of blue carbon ecosystems contributes significantly to atmospheric greenhouse gases. Pendleton et al. (2012) estimated that 0.15–1.02 billion tonnes of carbon dioxide are released annually, and if the rate of degradation continues, blue carbon habitat conversion and clearance will release centuries to millennia of accumulated carbon over a few decades (Gordon et al. 2011). Apart from human-induced changes in land use pattern, phenomenon like erosion of river and estuarine banks also dislodge chunk of blue carbon from their survival ground (Fig. 8.1)

Hypersaline condition due to reduced freshwater supply from upstream region also triggers the degradation rate of blue carbon reservoir by drastically reducing the biomass of the mangroves. We made an extensive survey since 2005 in the central part of Indian Sundarbans and observed that even after 8 years, the girth (diameter at breast height) of *Hertiera fomes* have not exceeded an average value of 12.5 cm (Fig. 8.2), whereas in western Indian Sundarbans, trees of the same species exhibited much more growth with an average DBH value of 40 cm (Fig. 8.3). This finding confirms that salinification of water is a major threat to blue carbon community.

Human intervention into the mangrove forest for cutting trees, honey collection, fishing and molluscan shell collection (for lime) also poses threats to blue carbon community, although the incidences are extremely local in nature. In Indian Sundarbans region, poor island dwellers enter into the deep forest after worshipping the Goddess Bonobibi (Fig. 8.4) for collecting mangrove timber and other forest products. They have firm belief that Bonobibi will be their lifesaver even during tiger attack.

In the present *era*, the management and conservation of coastal ecosystems and more particularly the coastal vegetation has been identified as an important missing piece or gap zone in International and National Climate Change mitigation matrix. The United Nations Framework Convention on Climate Change (UNFCCC) primarily aims to reduce green house gases, conservation and enhancement of sinks and reservoirs of coastal ecosystem. This process encompasses the route map to integrate land use

Fig. 8.1 Uprooting of mangroves due to erosion of intertidal mudflats

Fig. 8.2 Low diameter at breast height (*DBH*) of *Heritiera fomes* in the central sector of Indian Sundarbans (where average salinity is ~16.5 psu)

Fig. 8.3 Better growth with considerable DBH of *Heritiera fomes* in the western sector of Indian Sundarbans (where average salinity is ~9.8 psu)

Fig. 8.4 Goddess Bonobibi is worshipped by island dwellers (irrespective of religion and caste) before entering the deep mangrove forest

change and management activities leading to carbon sequestration in or reduced emissions from coastal ecosystems into existing policy and financing processes of the UNFCC.

8.1.3 Scientific Management

The domain of ecosystem services provided by coastal vegetation is broad in magnitude. Apart from providing fuel wood, timber, fishes, honey, wax and several important pharmaceutical compounds, services like bioremediation is also provided by coastal vegetation. We observed that *Porteresia coarctata* (a salt marsh grass found in lower Gangetic delta) can accumulate considerable level of zinc, copper and lead in its vegetative parts (Fig. 8.1) from the ambient media, i.e. water (Fig. 8.5) and sediment (Fig. 8.6). This study was conducted during 2011–2012 by us in three sites of Indian Sundarbans (Fig. 8.7). The study shows the need for conservation of coastal vegetation for getting the benefits of such ecosystem services, and carbon sequestration is an extra bonus to these ecosystem services (Fig. 8.8).

It is to be noted that the conservation policy is not uniform throughout the globe; it varies from place to place depending on the resource base, pattern of exploitation, and nature and magnitude of threat; for example, the pattern of exploitation in Arctic or Antarctic zone is totally different from the tropical seas or mangrove regions. To make it clearer, the example of regional resource base can be cited. Coral is a biological resource

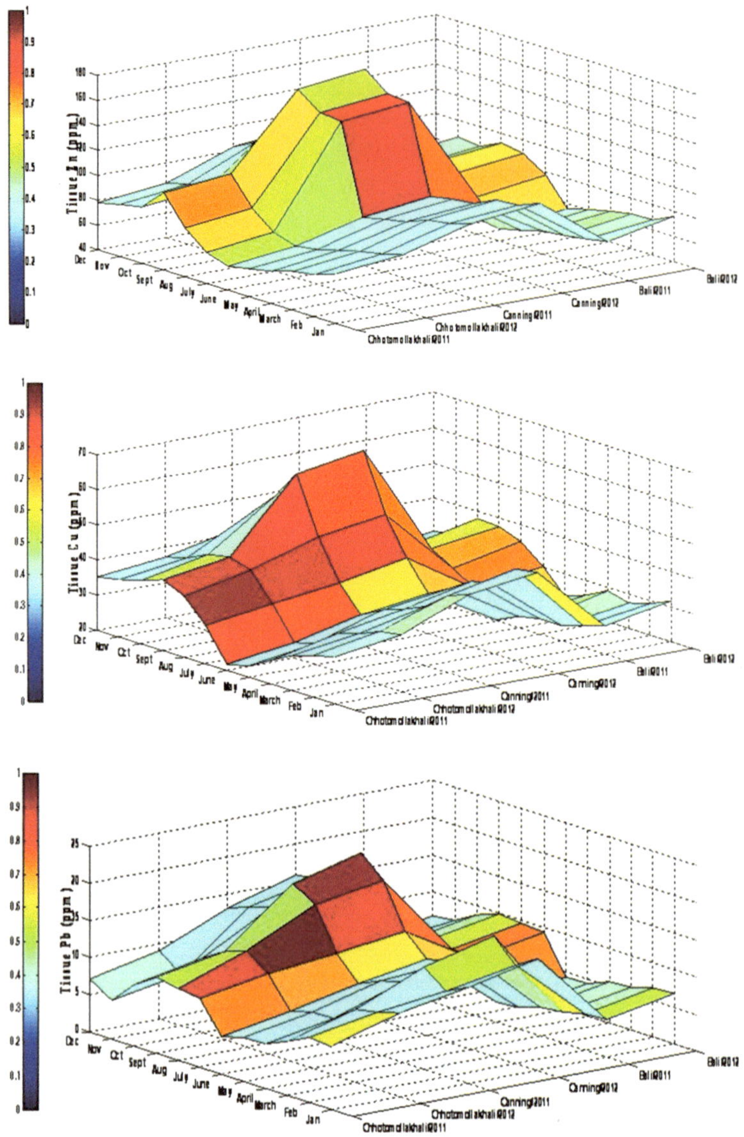

Fig. 8.5 Bioaccumulation of Zn, Cu and Pb in salt marsh grass tissue in three stations of Indian Sundarbans during 2011–2012

scattered around the archipelagos of Andaman and Nicobar Islands of Indian subcontinent, which is gradually eroding due to temperature fluctuation or changes in physico-chemical properties of the seawater. However, in Indian Sundarbans region, there is no existence of coral reefs and in this deltaic complex fish germplasm constitutes an important resource, which is under threat due to wild harvest of prawn seeds, illegal fishing practice, etc. Thus, threats are always site specific in nature and intricately related to the nature of resource, their exploitation pattern and

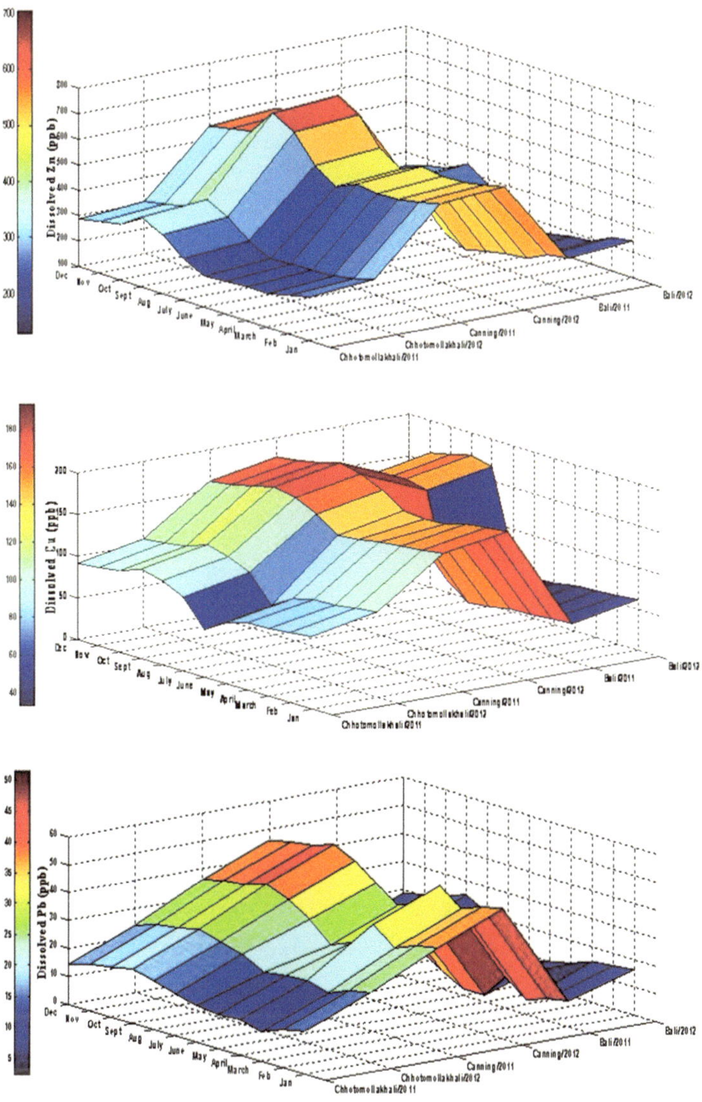

Fig. 8.6 Dissolved Zn, Cu and Pb in three stations of Indian Sundarbans during 2011–2012

socio-economic profile of the area. On this basis, it is important to focus on five main areas of information for systematic management of marine environment including site-based conservation. These are as follows:

1. Data bank on biotic and abiotic resources.
2. Data bank on hydrological parameters.
3. Information on the structure of marine communities and on the key elements in their functioning.
4. Information on natural variability.
5. Information on anthropogenic influences.

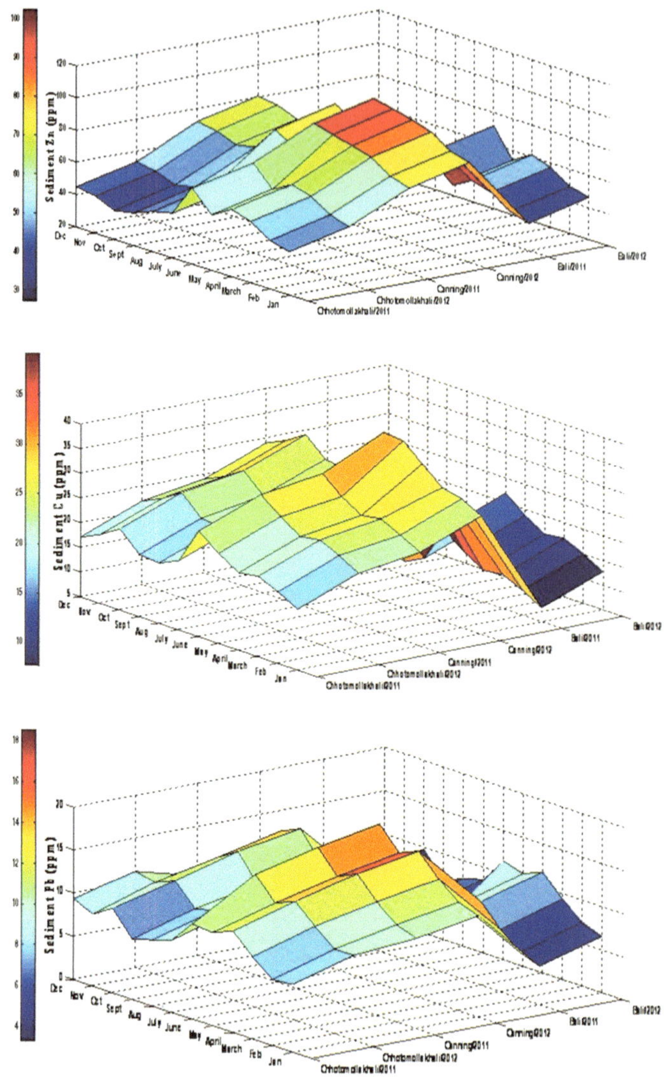

Fig. 8.7 Biologically available Zn, Cu and Pb in surface sediments of three stations of Indian Sundarbans during 2011–2012

All these information will require a structured approach to its application to conservation and management through the following:

1. Classification systems for resource data and
2. Criteria to identify the nature conservation importance of an area including those to identify a comprehensive series of marine protected areas.

The scientific management adopted for conserving mangroves of Indian Sundarbans may be taken as case study in this context. The mangrove ecosystem of Indian Sundarbans is one of the

Fig. 8.8 Map of Indian Sundarbans showing the sampling stations

most biologically productive, taxonomically diverse and aesthetically celebrated gene pools of the country sustaining 34 species of true mangroves along with a variety of associate species (Mitra et al. 1992; Mitra and Pal 2002). This deltaic lobe is the cradle of several species of finfish, nursery of different variety of shellfish and reservoir of various non-living resources of marine origin, which encompasses a wide range of riverine, estuarine, coastal and marine habitats. On the one hand, it exhibits enormous diversity based on its genesis, geographical location, hydrological regimes and substrate factors, and on the other hand, it also exhibits rare endemic genetic material, which demands preservation and proper sustainable utilization for the benefit of mankind. The biodiversity of this unique ecosystem has not yet been comprehensively assessed for several reasons. In general, very little information is available on marine and estuarine microorganisms, which are the key players in releasing carbon dioxide through the process of decomposition. A very important aspect of biodiversity is species-level genetic diversity, which has not received much attention. Both conservation and management of this unique ecosystem therefore demand a holistic approach and are very much dependent on the sciences and skills of geology, pedology, climatology, hydrology, botany, ecology, silviculture, forest technology and economics.

The landscape of Indian Sundarbans has changed remarkably due to large-scale human intervention since the beginning of the last century. As a result of overexploitation, demographic pressure, loss of habitat and change of ecological condition, several of the earlier reported species have become extinct or are in a very threatened or degraded state.

Large-scale clearing of the Sundarbans mangroves in the last 200 years for settlement of human population, aquaculture, etc., has invited a situation of great stress to the pristine mangrove ecosystem in this part of the Indian subcontinent. Hence, effective conservatory measures are essential to protect this gene pool. The integrated management of mangrove resources depends not only on an understanding of the ecological and silvicultural parameters for forest management, but also on the biological role that the primary production from the forest plays in sustaining the food web of the adjacent aquatic subsystem (secondary production). An understanding of the role of key stone species in maintaining the equilibrium of this particular ecosystem is also very essential.

Being a megadiversity country, India accounts for 7–8 % of the recorded species of the world though occupying only 2.4 % of the land area. The convention on biological diversity (CBD), to which India is a contracting party, deals with various aspects, like conservation of biological diversity, their sustainable use and fair and equitable share of benefits. The convention has also identified forests, coastal and marine ecosystems, grasslands, wetlands and deserts as priority ecosystem for biodiversity conservation. Based on this convention, several strategies have been adopted to conserve the vibrating mangrove ecosystem of Indian Sundarbans. The important ones are listed here.

- Encouragement towards alternative livelihood scheme
- Afforestation
- Conservation policies.

Each of the above strategies is discussed in detail.

Encouragement Towards Alternative Livelihood Scheme

Among 4.2 million people in the Sundarbans Biosphere Reserve area of Indian part, 50 % survive below poverty level. Improving the economic conditions of these people through promotion of sustainable utilization of natural resources is an optimistic step towards ensuring the long-term sustainability of the forest ecosystem. According to several research papers and books published on the threats of Indian Sundarbans (Choudhury and Chaudhuri 1994; Mitra and Banerjee 2005; Raha et al. 2013), the blue carbon in this region is under threat due to illegal encroachment into the mangrove forest for collecting timber, honey, wax, fuel wood, etc. Provision of alternative livelihood to the island dwellers of the deltaic complex may reduce the negative pressure on the blue carbon community. These alternative livelihoods should not be mere introduction of piggery, poultry or providing sewing machine to the villagers, but strategies should be formulated considering the salinity profile and environment of Sundarbans. In this context, we conducted a thorough survey after AILA (during June 2009) in 19 islands of Indian Sundarbans and observed intrusion of saline water in all the freshwater ponds of the islands. According to the local villagers, water of freshwater ponds (normally with 0–2 psu salinity) became almost 10–15 psu. This caused a massive damage and mortality to freshwater fishes like *Labeo rohita*, *Catla catla*, etc., on which these villagers used to depend so long for their livelihood. Considering this alteration of salinity due to AILA, we undertook a feasibility study with three species (*Scatophagus argus*, *Etroplus suratensis* and *Macrobrachium rosenbergii*) of commercial importance in an artificial tank at Techno India University, Salt Lake Campus, Kolkata, India, during 2013–2014 and observed high survival rate and growth of these species. We therefore suggest such livelihood scheme befitted with the environment of Sundarbans for further replication so that the earnings of the local villagers may increase. This would also divert the local population from illegal intrusion into the blue carbon belt.

Oyster culture may also be introduced in the area as an alternative livelihood. The oysters are rich in protein and minerals and can be cultured with minimum investment in the creeks and inlets of Sundarbans. However, turbidity and drastic fall of salinity (which is a characteristic feature of the area) need to be scientifically managed to promote oyster culture in the deltaic complex. It is to be noted that a lot of carbon can also be sequestered during the tenure of oyster farming (*Vide* Chap. 6 for details).

Seaweed culture seems to be a befitting livelihood for the island dwellers. The seaweeds have multiple uses, but island dwellers of Indian Sundarbans are not at all aware of the same. Awareness programme on regular basis through government network is a tool to introduce the concept of alternative livelihood to the people of Sundarbans.

Taxonomic position and salient features of some common seaweed found in the mangrove ecosystem of Sundarbans are listed here.

Type I

Systematic Position:
Division—Chlorophyta
Class—Chlorophyceae
Order—Ulvales
Family—Ulvaceae
Genus—*Enteromorpha*
Species—*intestinalis*

Salient features:

1. Plant body is tubular, more or less compressed, constricted and coiled in the form of intestine.
2. Thallus dark green in colour and found attached to the substratum with the help of primary attaching cell.
3. Presence of numerous multinucleated rhizoids growing from lower cell of the thallus.
4. Cells of the thallus are small and elongated.

Type II

Systematic Position:
Division—Chlorophyta
Class—Chlorophyceae
Order—Ulvales
Family—Ulvaceae
Genus—*Enteromorpha*
Species—*compressa*

Salient features:

1. Plant body is tubular, more or less compressed, and constricted.
2. Thallus light green in colour and found attached to the substratum with the help of primary attaching cell.
3. Presence of numerous multinucleated rhizoids growing from lower cell of the thallus.
4. Cells of the thallus are small and round.

Type III

Systematic Position:
Division—Chlorophyta
Class—Chlorophyceae
Order—Ulvales
Family—Ulvaceae
Genus—*Ulva*
Species—*lactuca*

Salient features:

1. Plant body is tubular, more or less compressed, and flattened leaf like.
2. Thallus dark green in colour and found attached to the substratum with the help of primary attaching cell.
3. Presence of numerous multinucleated rhizoids growing from lower cell of the thallus.
4. Cells of the thallus are small and ovoid.

Type IV

Systematic Position:
Division—Chlorophyta
Class—Chlorophyceae
Order—Cladophorales
Family—Cladophoraceae
Genus—*Rhizoclonium*
Species—*hookeri*

Salient features:

1. Filaments rigid and dark green in colour.
2. Thallus simple and intertwined to form a fleecy layer with numerous rhizoids.
3. Branches are formed at right angles to the main axis.
4. Cells measure about 170–82 μ diameter and 72–105 μ long.

Type V

Systematic Position:
Division—Chlorophyta
Class—Chlorophyceae
Order—Cladophorales
Family—Cladophoraceae
Genus—*Rhizoclonium*
Species—*riparium*

Salient features:

1. Filaments pale green and expanded on the substrate.

2. Filaments flexous intertwined into a fleece with thin cell wall.
3. Frequent rhizoid branches.
4. Vegetative cells measure about 23–26 μ, diameter usually 1–2 times as long as broad.

Type VI

Systematic Position:

Division—Chlorophyta
Class—Chlorophyceae
Order—Cladophorales
Family—Cladophoraceae
Genus—*Chaetomorpha*
Species—*aerea*

Salient features:

1. Filaments are unbranched.
2. Filaments 50–55 μ in diameter and 1–1.5 times long.
3. Rhizoids present at the base are unbranched and are disc like in appearance.
4. Cell wall thick, hyaline and lamellate.

Type VII

Systematic Position:

Division—Chlorophyta
Class—Chrysophyceae
Order—Vaucheriaceae
Family—Cladophoraceae
Type—*Vaucheria* sp.

Salient features:

1. Thallus filamentous, aseptate and laterally and irregularly branched.
2. Filaments usually attached by branched rhizoids.
3. Filaments cylindrical with numerous small chloroplasts towards the exterior boundering the central vacuole.
4. Chloroplasts are without pyrenoids.

Type VIII

Systematic Position:

Division—Chlorophyta
Class—Rhodophyceae
Order—Gigartinales
Family—Rhabdoniaceae
Genus—*Catenella*
Species—*repens*

Salient features:

1. Plants with repent and assurgent branches.
2. Branching is ditrichotomous below, but clearly pinnate above.
3. The axis and branches divided into dorsiventrally compressed ellipsoid to ovate segment 3–5 times longer than broad.
4. The haptera terminating into uncorticated flagellar outgrowth chiefly formed at points of forking and not in the plane of branching of the thallus.

Type IX

Systematic Position:

Division—Chlorophyta
Class—Rhodophyceae
Order—Ceramiales
Family—Rhodomelaceae
Type—*Bostrychia* sp.

Salient features:

1. Filiform, black or dull purplish.
2. Thallus with erect branches often distinguishable.
3. Rhizoids polysiphonous, ordinarily regular and bilaterally branched.
4. Branches with several cells of equal length being disposed about the central axis or these pericentral cells regularly transversely divided.

Type X

Systematic Position:

Division—Chlorophyta
Class—Rhodophyceae
Order—Ceramiales
Family—Delesseriaceae
Genus—*Caloglossa*
Species—*leprieurie*

Salient features:

1. Plants dorsiventral, spreading or somewhat erect up to 4 cm across deep reddish violet in colour.
2. Blades 1.5 mm broad, constricted at the forking and elsewhere.
3. Individual segments are lanceolate, 4–6 mm long, sometimes linear–attenuate, rarely ovate.
4. Rhizoids and secondary segments or blades formed at the constrictions; blades formed here are irregularly branched.

Afforestation

The magnitude of blue carbon can be enhanced through afforestation programmes on regular basis. With the growth of the plants, not only the stored carbon in the vegetation will increase, but at the same time, the soil carbon will also enhance due to addition of litter and detritus. We conducted a survey in Indian Sundarbans during 2005–2010 and observed three distinct categories of coastal and estuarine zone that may be taken up for afforestation through relevant government departments or NGOs. These categories are as follows:

1. Degraded areas devoid of any vegetation (Category-1)
2. Degraded areas under scrubby growth of *Ceriops decandra* (Category-2)
3. Degraded areas under scrubby growth of *Excoecaria agallocha* (Category-3).

Category-1 is characterized by saucer-shaped areas with no vegetation. Floodwater remains in the depressions from March to December. Salt accumulates on the top soil, and no plantation can be undertaken. Salinity can be reduced by improved drainage of these areas, and the soil may be improved through the use of gypsum, sulphur and natural grasses.

The area under **category-2** is covered with a scrubby growth of *C. decandra* and requires silvicultural improvement through planting of upper-storey species. Some experiments have been carried out where strips are cleared and upper-storey species have been planted.

Category-3 has hard soil, and apart from a scrubby growth of *E. agallocha*, no seed naturally transported or artificially dibbled germinates in this area.

A unique observation with respect to soil organic carbon is its interrelationship with anthropogenic activities. Researchers found that organic carbon is relatively higher in areas with intense human activities compared to protected areas inside the forest. The reason is that organic matter is introduced more in soil through agricultural run-off, sewage, poultry waste and waste generated from shrimp farms. The litter produced by the coastal vegetation is a vital source of soil organic carbon that greatly increases the fertility of intertidal mudflats (Fig. 8.9).

8.2 Policy Approach

The Kyoto Protocol to the UNFCCC provides Annex I countries to achieve their national emission reduction targets. During the past few decades, as the need for reducing emissions of green house gases grew, the concept and approaches to carbon trading through carbon credits evolved, whereby it became possible for emitters in industrialized and developed countries to offset their abnormally high-carbon emissions targets under the Kyoto Protocol. This innovative market-based mechanism has been provided under the Kyoto Protocol and is being adopted on a commercial basis. It is also helping developing countries in sourcing funds for their sustainable development by adopting green projects.

Carbon Credit Mechanism

The carbon market mechanism has evolved over the past many years after the Kyoto Protocol to the UNFCCC was entered into by member countries. It enables Annex I countries to achieve their national targets of reduction in green house gas emissions by trading carbon. The main features of the mechanism provided for carbon of emission trading under the Kyoto Protocol has been discussed in brief in the text below.

1. **Joint Implementation**

   Under Article 6 of the Kyoto Protocol, there is provision for obtaining carbon credits from other countries. Annex I countries may jointly obtain carbon credits or trade for meeting their national objectives.

2. **Clean Development Mechanism (CDM)**

   The Clean Development Mechanism (CDM) under Article 12 of the Kyoto Protocol provides for promotion of sustainable development in developing countries by allowing Annex I countries either through governments or private firms to invest in CDM projects that reduce or avoid emissions. On the other hand, the Annex I country would gain carbon credits which are taken into account as part of their targets under the Kyoto Protocol.

Fig. 8.9 Mangrove seedlings grow on the carbon-rich soil (contribution of mangrove litter) of intertidal mudflats

Article 17 of the Kyoto Protocol provides transfer of emission credits between Annex I countries. Those Annex I countries which achieve more than their national targets under this protocol may sell their excess carbon credits to other Annex I countries.

Thus CDM is a process under which environment-friendly investments take place in developing countries through the governments and businesses from developed countries which helps both the former and latter. This took shape after the Kyoto Protocol adopted under the UNFCCC came into force in 2005 following ratification by Russia.

Under the CDM mechanism, Annex I parties can achieve their emission reduction targets by taking investments either through their governments or through private businesses in non-Annex I countries. This is a cheaper option and helps both the Annex I country which is directly and indirectly making investments and the non-Annex I country where the investment is being made. The developed country gets credits against their targets for emissions avoided when it makes investments in these projects in non-Annex I countries which are able to achieve sustainable development at a faster pace by following the eco-friendly approach.

CDM Projects: Who Qualify?

Under the Kyoto Protocol, there are a number of projects that qualify as CDM projects. While it is difficult to list all such projects here, the general outline and scope of these projects have been given in the following points:

1. Land use, land use change and forestry sector (LULUCF)—afforestation and reforestation, which function as carbon sinks.
2. Small-scale CDM projects—renewable energy project activities having a maximum output capacity equivalent to up to 15 MW, energy efficiency improvement projects leading to the reduction of energy consumption and other small-scale projects that reduce anthropogenic emissions of green house gases.

3. Normal-scale CDM projects—these include projects which are neither small-scale projects nor afforestation/reforestation projects.
4. Small-scale A/R CDM projects—these include projects that result in net removals of green house gases in the form of sinks by less than 16 kilotonnes of carbon dioxide per year and projects developed and implemented by low-income countries and individuals determined by the host party.
5. Normal-scale A/R CDM projects—these include projects not included in the small-scale A/R CDM projects.
6. Energy sector projects such as renewable and non-renewable sources of energy.
7. CDM projects related to manufacturing industries and chemical industries leading to reduction in emissions of green house gases.
8. Construction sector CDM projects.
9. Transport sector CDM projects.
10. CDM projects related to mining/mineral production, metal production, fugitive emissions, solvent use, waste handling and disposal.
11. CDM projects related to agriculture sector.

Institutions and Bodies Related to Carbon Trading and CDM

CDM is a process under the Kyoto Protocol which involves different institutions and bodies for overseeing its implementation. These include the following:

a. **COP and Executive Board**

The Conference of parties to the UNFCCC is the apex body for CDM and takes the policy decisions and issues guidelines through the Meetings of Parties. The Executive Board or EB oversees CDM mechanisms. It is comprised of ten members, five from UN groups, two each from Annex I and non-Annex I countries and one from small island developing states. The meetings of EB are held after every 2–3 months.

Its functions include the following:

- Lay down and modify existing guidelines for baseline information, monitoring and project process.
- Review definitions and procedures.
- Acceleration of operational entities.
- Defining the operational procedures and standards for CDM.
- Giving recommendations to the EB for various procedures and guidelines about CDM projects.
- Maintaining a register of CDM projects.

b. **Methodologies Panel**

The Methodologies Panel or Meth Panel has been set up for preparing recommendations to the executive board for the methodologies to be adopted for baselines and monitoring plans. It gives recommendations for new baselines and monitoring methods for new proposals. The Methodologies Panel also submits recommendations for expanding the applicability of methodologies and gives options to the project participants for choosing from methodologies of resembling nature.

Among the other functions of this panel is also to maintain a roster of experts the appraisal of proposed new methodologies.

c. **Accreditation Panel**

The accreditation panel is responsible for guiding the EB with the procedure for accrediting operational entities. It assesses the applications of the applicant or the designated operational entity and prepares a report for the EB.

d. **Working Groups**

A number of working groups are also involved in the CDM process. These include the following:

- Afforestation and reforestation working group for A/R CDM projects.
- Small-scale working group for small CDM projects.

e. **Designated Operational Entity**

The designated operational entity or DOE conducts the third-party validation of the project design and also verifies and certifies reductions of emissions. The DOE could be a domestic legal entity or an international body approved by the EB and confirmed by the COP in its meeting of parties. The main functions of the DOE are the following:

- Validation of the CDM projects
- Verification and certification of the CDM projects

- After verification and certification, the DOE requests the EB to issue Certified Emission Reductions.

Project Cycle

There is a well-defined cycle or path through which a CDM project has to pass for qualification under the EB procedure. This cycle is rigorous for obtaining Certified Emission Reductions or Carbon Credits. The steps are highlighted here in brief.

1. **Project Design and Formation**
   The Project Concept Note or PCN is the first step in designing and formulating a CDM project. This leads to the development of a Project Design Document (PDD) which is a legal requirement for CDM project approval. The PDD gives the project baseline leading to the estimation of net reductions in the emissions of carbon.
2. **Approval by DNA**
   The Designated National Authority or DNA in the host country where the project is to be implemented appraises the project against the objectives of sustainable development at the national level. In India, the National CDM Authority under the Ministry of Environment and Forests, Government of India, is the DNA for CDM projects.
3. **Validation and Registration**
   This is the third stage of the CDM project cycle in which it is assessed and registered by the DOE approved by the EM of CDM. The DOE assesses the project and ensures that it complies with the CDM rules and procedures that have been approved by the EB/COP from time to time. After the DOE recommends a particular project, the EB approves its registration.
4. **Financing and Sale**
   This stage involves arranging the finances for the project.

   The sale of carbon may be done in the following ways:

a. Carbon buyers may pay directly for the CDM project, mainly through financing its costs.
b. Carbon buyers may take equity in the project.

After the mode of financing has been worked out, detailed contracts are signed for the sale of carbon from the project.

5. **Implementation and Monitoring**
   After national approval, validation, registration and tying up financing, the CDM project implementation begins. The monitoring of the project is done in accordance with the Project Design Document or PDD that has been approved during the previous stages by the EB under the Kyoto Protocol.

   Recording of the carbon emissions and changes in the baseline are done as a part of the monitoring process. All relevant data for calculating reductions in green house gas emissions or carbon sinks are calculated by the CDM project participants.

The CDM projects are good options to reduce emission, no doubt, but the community-based expansion of blue carbon reservoir can also reduce emission substantially, which requires the participation of local people. Few important recommendations to enhance the blue carbon reservoir in coastal zone have been clearly defined through several programmes taken up in different parts of the world. The primary steps are listed here:

(A) Site selection
(B) Capacity building
(C) Partner selection
(D) Facilitator selection
(E) Linking livelihood with conservation
(F) Marketing facility

(A) **Site Selection**
Expansion of blue carbon reservoir can be done if the sites have congenial environment not only in terms of tidal action and water quality, but also in terms of anthropogenic influences (like sewage discharge, agricultural run-off, industrial waste discharge and grazing pressure). Also lack of clear and legally binding land and carbon ownership rights, along with an absence of strong community support and awareness, have been found to be two major roadblocks leading to stakeholder conflicts prior to

initiation of the payments for ecosystem services (PES).

(B) **Capacity Building**

The expansion of coastal vegetation cannot be done solely by the scientists (preferably the ecologists). Without community desire, support and active participation from the initial stage to final implementation stage, the process of conservation will not succeed. The local community must also be involved during the entire monitoring process. Community-based participation in data collection and monitoring is possible if supplemented with capacity training. Significant community awareness capacity building (preferably for raising nurseries, erosion control, embankment construction, etc.) and assistance flows can be achieved through government and NGOs. Global experience from several large-scale REDD+ projects involving indigenous tribes (as for example in Costa Rica, Brazil) exhibited that multiple years of assistance in clarifying land tenure and carbon ownership rights, inter- and intratribal negotiations, technical training and financial guidance are required. For this purpose, a knowledge hub or incubation centre may be developed (preferably in the sphere of agribusiness, pisciculture and alternative livelihood) where training on regular basis should be imparted by experts to the local community.

C) **Partner Selection**

In order to develop and implement successful carbon bank in the coastal areas, it is extremely essential to procure support from technical, financial and legal experts, i.e. often lacking at both village and government scales. Trustworthy, high-quality NGOs with demonstrable previous experience in carbon sequestration-related projects or REDD+ projects should be identified and vetted to assist in establishing a sound blue carbon conservation programme.

(D) **Facilitator Selection**

It is observed that the movement of conservation often becomes unsuccessful due to deep-rooted corruption in several tiers through which the funds flow. In regions like Solomon Island, disproportionate economic gain of forest resources has benefited well-connected insiders (for example corrupt government and departments). However, successful conservation movement has been observed in Costa Rica, Brazil, where a community driver REDD+ system with the government as facilitator has been observed to be a healthy partnership.

(E) **Linking Livelihood with Conservation**

One of the important mechanisms to reduce the decline of mangrove forests is the introduction of PES. More recently, this has occurred within international and national programmes to reduce emissions from deforestation and degradation (REDD+), whereby developing countries are compensated for maintaining carbon sequestration functions of their forests. Recent assessments indicate that mangrove forest is a unique sink of carbon, and therefore, PES and carbon credit systems may offer the opportunity to achieve the dual goals of poverty reduction and protection of blue carbon reservoir. Development of REDD+ programmes will therefore ensure livelihood and security and sustainable use of the mangrove resources.

(F) **Marketing Facility**

The alternative livelihood programmes designed to reduce the stress on the blue carbon community of the coastal zone often face failure due to absence of concrete market channel and sometimes due to lack of proper storage facilities. The example of jelly prepared from mangrove fruit *Sonneratia apetala* can be cited in this context. This jelly is rich in Vitamin C and other mineral contents and is comparable to several other branded products, but the ground-level manufacturer of this jelly could not market it due to lack of coordination, transportation and bulk storage facilities. The manufacturers of this mangrove fruit product therefore suffered a loss of Rs. 16300/- in one month even the quality of the jelly was certified by experts. Similarly, the freshwater prawn produced through a Department of

Science and Technology Project (DST, Government of India) did not get the actual price as transportation, icing and storage facility was lacking. Dr. Sufia Zaman, the Principal Investigator of the project (Sanction no. SSD/SS/028/2010 dated 9 Sept 2011–2013) therefore had to sale the harvested prawns to the middleman at a price of Rs. 300/kg when the actual market price was Rs. 500/kg. All these instances clearly indicate the need for establishing a proper marketing link to sustain the alternative livelihood generation schemes.

A very important prospect of linking livelihood with the economic profile of Sundarban region can open up if the endemic salt marsh grass (*Porteresia coarctata*) can be used as ingredients of biofertilizer. Study conducted by us reveals that the salt marsh grass species that appear in bulk (Figs. 8.10 and 8.11) in the intertidal mudflats of Sundarbans is a very rich source of macro- and micronutrients needed for the growth of plants.

Considering the biochemical composition, widespread abundances, tolerance to a wide range of salinity and socio-economic profile of the local population (who are mostly below the poverty line), a Kolkata-based NGO known as Progressive Organisation of Rural Service for Health, Education and Environment (PORSHEE) has taken up a pilot project to manufacture biofertilizer from this salt marsh grass species and make aware the farmers and agriculturists regarding the benefits of using marine-based biofertilizer. Insecticides (preferably mosquito larvicide) have been reported to be prepared from sand binders (*Ipomoea pes-caprae*) (Fig. 8.12).

All these mangrove associate floral species are still untapped from the commercial angle and alternative livelihood generation although documented in the taxonomic list of Sundarban biodiversity by several researchers.

8.3 Case Studies

8.3.1 Case Study of Gujarat

Gujarat is located on the west coast of India between 20° 2′–24° 41′N and 68° 8′–74° 23′E and has the Tropic of Cancer passing through the districts of Kachchh and Surendranagar. The state has a long coastline of 1,650 km, constituting over 21 % of the Indian coastline and the longest among all states. There are two major indentations in Gujarat's coast—the Gulf of Kachchh and the Gulf of Khambat. Of the total

Fig. 8.10 *P. coarctata*: an endemic and untapped floral resource of Indian Sundarbans

Fig. 8.11 Huge biomass of *P. coarctata* can be used as important ingredient of biofertilizer

Fig. 8.12 *Ipomoea pescaprae*: a source of natural mosquito larvicide

wetland area 27,175 km^2 in the state, coastal wetlands comprise as much as 92.3 %. These coastal stretches are rich in biodiversity and have the Marine National Park and four sanctuaries located along it. Broadly, Gujarat's coast has been divided into five major regions that include the Gulf of Kachchh, the Gulf of Khambat, the Saurashtra Coast, the South Gujarat Coast and the Rann of Kachchh.

The coastal zone of Gujarat supports mangrove vegetation mainly distributed in 14 districts (Table 8.2).

During the handling of the wide variety of dry and liquid cargo at the ports, various pollutants in low dosage get released in the surrounding marine ecosystems. These pollutants have a chronic impact in differing degrees at the different marine locations. They affect the biotic and abiotic components, changing water quality and the physico-chemical attributes of the specific sites. The impact of pollution on the physical and biological functions and water quality are closely interlinked and influence each other. Dredging and the alteration of physical processes around the ports have been major factors that have had a deep impact on the marine ecosystems through the increase in turbidity and decrease in productivity of the region (Fig. 8.13).

Apart from these, other activities like collection of fuel woods from local communities, grazing of mangroves by cattle or camels, and collection of mangrove leaves for feeding the cattle are some local threats on Gujarat mangroves. Considering the ecological services offered by mangroves, about twenty-three industries participated in ecorestoration programme [for example the restoration site at Kantiyajal has the tallest and most mature trees of all the project sites (Fig. 8.14)] through public–private partnership model (PPP) due to which about 4,675 ha of land has been taken up for mangrove plantation. The area covered by different industries is highlighted in Table 8.3.

8.3.2 Case Study of Indian Sundarbans

The process of ecorestoration or plantation stands on the foundation of ecological condition and

Table 8.2 District-wise mangrove cover (area in km^2)

District	Very dense mangrove	Moderately dense mangrove	Open mangrove	Total	Change w.r.t. 2009 assessment
Ahmedabad	0	1	29	30	3
Aimreli	0	0	1	1	1
Anand	0	0	0	0	−3
Bharuch	0	21	22	43	1
Bhavnagar	0	6	13	19	6
Jamnagar	0	28	131	159	2
Junagadh	0	0	1	1	1
Kachchh	0	118	660	778	3
Navsari	0	0	1	1	0
Porbandar	0	0	0	0	0
Rajkot	0	1	1	2	0
Surat	0	7	13	20	3
Vadodara	0	0	2	2	−2
Valsad	0	0	2	2	−3

Source FSI (2011)

Fig. 8.13 Industrial activity along the Gujarat coast

Fig. 8.14 Satellite imagery shows change in mangrove cover between 2000 and 2007 at Kantiyajal village of Bharuch district, Gujarat

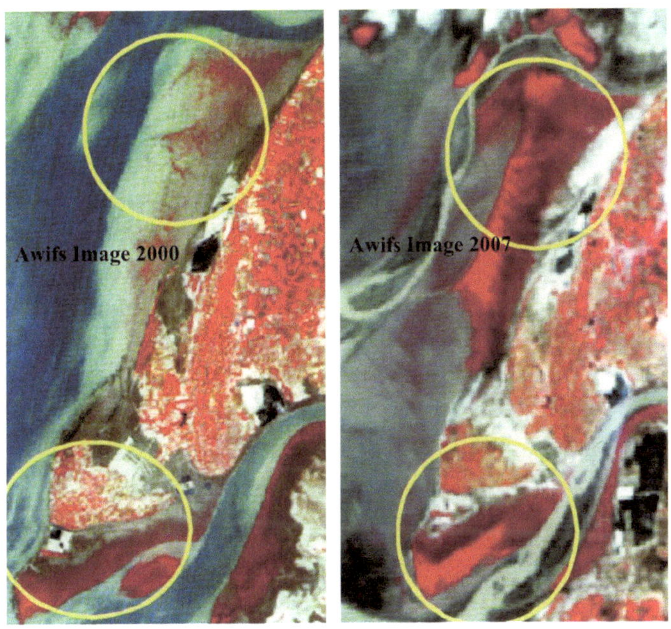

scientific management. In this context, the case study of Sundarbans is a classic example. The region experiences three distinct categories of salinity profile. The western and central sectors of Indian Sundarbans are gradually freshening, while the central sector is experiencing a hike in salinity due to total blockage of freshwater owing to siltation of the Bidyadhari River (since the late fifteenth century) in the upstream zone. Ecorestoration and plantation programmes are taken up in these three sectors on the basis of adaptation of mangrove floral species to salinity which is highly species specific; for example, *Heritiera fomes*, *Sonneratia apetala* and *Bruguiera* sp. are planted after demarcating the hyposaline regions, whereas *Avicennia* spp. are better suited in hypersaline central sector. Nurseries are raised accordingly by State Forest

Table 8.3 Area covered by mangrove plantation by different industries of Gujarat through PPP model during 2006–2012

Sr. No.	Name of company	Name of site	Total area covered (in ha)
1	NIKO Resources Ltd., Vadodara	Dandi, District Surat	250
2	Gujarat Maritime Board, Gandhinagar	Dholai, District Navsari; Magdalla, District Surat; Ghogha, District Bhavnagar Jakhau, District Kachchh	120
3	Gujarat Mineral Development Corporation, Ahmedabad	Nanicher, District Kachchh	30
4	Mundra Port & Special Economic Zone Ltd., Adani, Ahmedabad	Dandi, District Surat	700
5	Adani Petronet Pvt. Ltd.	Dandi, District Surat	100
6	Gujarat Pipavav Port Ltd., Mumbai	Dandi, District Surat Kantiyajal, District Bharuch	500
7	Bayer Crop Science, Bharuch	Kantiyajal, District Bharuch	10
8	Essar Bulk Terminal Ltd., Hazira, Surat	Dandi, District Surat; Ankalva, District Bharuch	300
9	Essar Steel Pvt. Ltd.	Dandi, District Surat	100
10	Hazira LNG Pvt Ltd., Hazira Surat	Karanj, District Surat	300
11	GSPC Pipavav Power Co. Ltd., Gandhinagar	Kantiyajal, District Bharuch; Karanj, District Surat	110
12	AMBUJA Cement Ltd., Ahmedabad	Karanj, District Surat	150
13	Gujarat Heavy Chemical Ltd., Sutrapeda	Rohino Island, Bhavnagar; Tarsara, Bhavnagar	100
14	Pipavav Shipyard	Kantiyajal, District Bharuch	5
15	Petronet, LNG, Dahej	Nada, District Jambusar; Ankalva, District Bharuch	350
16	ABG Shipyard, Dahej	Nada, District Jambusar; Ankalva, District Bharuch	100
17	Ultatech Cement Ltd.	Rohino Island, Bhavnagar; Tarsara, Bhavnagar	100
18	Anjan Cement (Jaypee Group)	Muhadi Kachchh	100
19	LARSEN and TUBRO	Karanj, District Surat	100
20	Coastal Gujarat Power Ltd.	Kantiyajal, District Bharuch	800
21	KRIBHCO	Kantiyajal, District Bharuch	100
22	India Rayon	Madhavad Kotda, District Junagar	50
23	Kandla Port	Nakti Creek, District Kachchh; Satsaida Bed, District Kachchh	200
			4,675

Department (Government of West Bengal, India) and several NGOs involving primarily the women folk of the local villages (Figs. 8.15 and 8.16).

Seedlings from these nurseries are finally planted in the barren intertidal mudflats after testing the soil texture and other environmental factors associated with the growth and survival of the species (Fig. 8.17).

Prior to initiation of afforestation programme, the slope of the intertidal mudflat is also determined as steep slope can lead to failure of the project by way of massive erosion and subsequent uprooting of the planted seedlings (Fig. 8.18).

Fig. 8.15 Nursery of *Ceriops decandra* for afforestation programme

Fig. 8.16 Nursery of *Bruguiera* sp. for afforestation programme

Fig. 8.17 Researcher of Techno India University, Salt Lake, Kolkata (India), monitoring soil moisture and pH of a newly growing island in Sundarbans prior to mangrove plantation

Fig. 8.18 Steep intertidal mudflats are prone to erosion leading to a failure of the afforestation programme

References

Chaudhuri AB, Choudhury A (1994) Mangroves of the Sundarbans. The world Conservation Union, Dhaka

Chmura GL, Anisfeld SC, Cahoon DR, Lynch JC (2003) Global carbon sequestration in tidal, saline wetland soils. Global Biogeochem Cycles 17(4):1–12

Donato J, Daniel C, Boone K, Daniel M, Sofyan K, Melanie S, Markku K (2011) Mangroves among the most carbon-rich forests in the tropics. Nat Geosci 4:293–297

FAO (Food and Agriculture Organization of the United Nations) (2007) The world's mangroves, 1980–2005. FAO, Rome

FSI (Forest Survey of India) (2011) State of forest report. Published by FSI, Ministry of Environment and Forest, Government of India

Gordon D, Murray BC, Pendleton L, Victor B (2011) Financing options for Blue Carbon: opportunities and lessons from the REDD+ experience. Report NI R 11-11, Nicholas Institute for Environmental Policy Solutions, Duke University, Durham. Online at http://nicholasinstitute.duke.edu/sites/default/files/publications/financingoptions-for-blue-carbon-paper.pdf. Accessed 23 Mar 2012

Halpern BS, Walbridge S, Kimberly A, Carrie S, Kappel V, Fiorenza M, Caterina D'A, John F, Bruno K, Casey S, Colin E, Helen E, Fox RF, Dennis H, Hunter S, Lenihan E, Madin MP, Matthew T, Perry E, Selig R, Steneck R, Watson R (2008) A global map of human impacts on marine ecosystems. Science 319:948–952

Herr D, Pidgeon E, Laffoley D (eds) (2012) Blue carbon policy framework: based on the discussion of the international blue carbon policy working group. International Union for Conservation of Nature and Arlington Conservation International. Conservation International, Gland, Virginia

Laffoley D, Grimsditch GD (eds) (2009) The management of natural coastal carbon sinks. IUCN, Gland, Switzerland, p 53

Lotze HK, Lenihan HS, Bourque BJ, Bradbury RH, Cooke RG, Kay MC, Kidwell SM, Kirby MX, Peterson CH, Jackson JBC (2006) Depletion, degradation, and recovery potential of estuaries and coastal seas. Science 312:1806–1809

Mateo MA, Romero J, Perez MM, Littler DS (1997) Dynamics of millenary organic deposits resulting from the growth of the Mediterranean seagrass *Posidonia oceanica*. Estuar Coast Shelf Sci 44:103–110

MEA (2005) The millennium ecosystem assessment. Online at http://www.unep.org/maweb/en/index.aspx. Accessed 23 Mar 2012

Mitra A, Choudhury A, Zamaddar YA (1992) Seasonal variations in metal content in the gastropod *Cerithedia* (Cerithideopsis) *cingulata*. Proc Zool Soc 45:497–500

Mitra A, Pal S (2002) In: Banerjee S, Tampal F (eds) The oscillating mangrove ecosystem and the Indian Sundarbans. World Wild Life Fund for Nature-India, West Bengal State Office

Mitra A, Banerjee K (2005) In: Banerjee Col SR (ed) Living resources of the sea: focus Indian Sundarbans. WWF India, West Bengal, p 96

Mora C, Sale PF (2011) Ongoing global biodiversity loss and the need to move beyond protected areas: a review of the technical and practical shortcomings of protected areas on land and sea. Mar Ecol Prog Ser 434:251–266

Nellemann C, Corcoran E, Duarte CM, Valdes L, DeYoung C (eds) (2009) Blue carbon. A rapid response assessment. United Nations Environment Programme, GRID-Arendal. Online at http://dev.grida.no/RRAbluecarbon/pdfs/update/BlueCarbon_print12.10.09.pdf. Accessed 23 Mar 2013

Ong JE (2002) The hidden costs of mangrove services: use of mangroves for shrimp aquaculture. International Science Roundtable for the Media, Bali

Orth RJ, Carruthers TJB, Dennison WC, Duarte CM, Fourqurean JW (2006) A global crisis for seagrass ecosystems. Bioscience 56:987–996

Pendleton L, Donato DC, Murray BC, Crooks S, Jenkins WA, Sifleet S, Craft C, Fourqurean JW, Kauffman JB, Marba N, Megonigal P, Pidgeon E, Herr D, Gordon D, Baldera A (2012) Estimating global "blue carbon" emissions from conversion and degradation of vegetated coastal ecosystems. PLoS ONE 7(9):1–7

Raha AK, Zaman S, Sengupta K, Bhattacharyya SB, Raha S, Banerjee K, Mitra A (2013) Climate change and sustainable livelihood programme: a case study from Indian Sundarbans. J Ecol 107:335–348

UNEP (United Nations Environment Programme) (2006) Marine and coastal ecosystems and human wellbeing: a synthesis report based on the findings of the millennium ecosystem assessment. UNEP, Nairobi

Valiela I, Bowen JL, York JK (2001) Mangrove forests: one of the world's threatened major tropical environments. Bioscience 51:807–815

Vichkovitten T, Holmer M (2005) Dissolved and particulate organic matter in contrasting *Zostera marina* (eelgrass) sediments. J Exp Mar Biol Ecol 316:183–201

Waycott M, Duarte CM, Carruthers TJB, Orth RJ, Dennison WC (2009) Accelerating loss of seagrasses across the globe threatens coastal ecosystems. Proc Natl Acad Sci USA 106:12377–12381

Worm B, Barbier EB, Beaumont N, Duffy JE, Folke C, Halpern BS, Jackson JBC, Lotze HK, Micheli F, Palumbi SR, Sala E, Selkoe KA, Stachowicz JJ, Watson R (2006) Impacts of biodiversity loss on ocean ecosystem services. Science 314:787–790. doi:10.1126/science.1132294